STEARN'S
Dictionary of Plant Names for Gardeners

STEARN'S
Dictionary of Plant Names for Gardeners

A Handbook on the
Origin and Meaning of the Botanical Names
of some Cultivated Plants

WILLIAM T. STEARN

CASSELL

Cassell Publishers Limited
Villiers House, 41/47 Strand
London WC2N 5JE

This edition first published 1992

This is a revised edition of
A Gardener's Dictionary of Plant Names
revised and enlarged in 1972 by William T. Stearn
from A. W. Smith's
A Gardener's Book of Plant Names
1963

British Library Cataloguing in Publication Data
A catalogue record for this book is available
from the British Library

ISBN 0-304-34149-5

Distributed in the United States by
Sterling Publishing Co. Inc.
387 Park Avenue South, New York, NY 10016–8810

Distributed in Australia by
Capricorn Link (Australia) Pty Ltd
PO Box 665, Lane Cove, NSW 2066

Typeset in Janson by DSC Corporation Ltd, Cornwall, England
Printed in Great Britain by
Hartnolls Ltd, Bodmin, Cornwall

TO THE MEMORY OF

ARCHIBALD WILLIAM SMITH

(1899–1962)

SOLDIER, FORESTER, GARDENER
AND AUTHOR OF

A Gardener's Book of Plant Names
1963

PRODROME OF
THE PRESENT WORK

Contents

Introduction 1

Preface to the 1972 *Gardener's Dictionary* 5
by Isadore L. L. Smith

An Introduction to Botanical Names 7

Some Sources of Further Information 23

Dictionary of Plant Names 27

An Introduction to Vernacular Names 317

Some Sources of Further Information 330

Dictionary of Vernacular Names 333

Introduction

The origin and scope of this *Dictionary of Plant Names for Gardeners* merits explanation.

Archibald William Smith (1899–1962), author of *A Gardener's Book of Plant Names*, posthumously published in 1963, had a remarkable and varied career, as the preface by his widow Isadore L. L. Smith to the second edition (1972) of that work, entitled *A Gardener's Dictionary of Plant Names*, makes evident. Collecting information about plant names amused him as a hobby over many years. His book aimed to provide fellow gardeners with basic information about their origin and meaning. Unfortunately, there are few botanical scholastic matters on which it is easier to go astray through repetition and accumulation of past errors, particularly manifest in the extensive misinformation with which G. C. Wittstein's apparently authoritative *Etymologisch-botanisches Handwörterbuch* (1852–1853, reissued 1971) provided an uncritical public awed by nineteenth-century German scholarship. Hugo von Mohl pointed out its many defects, but with little effect, in reviews as long ago as 1852 to 1854. Smith accepted, like many other authors, material derived ultimately from this source. He did not consult the Dutch work of unrivalled scholarship on plant names, both generic and specific, *Verklarend Woordenboek der wetenschappelijke Namen* (1936) by Cornelis Andries Backer (1874–1963) and he was probably unaware of its existence. Thus Smith's praiseworthy intent, his enthusiasm and public spirit did not totally compensate for his reliance on too many untrustworthy works.

When I undertook revision of Smith's 1963 book for a second edition, this soon became obvious and the task of correction and extension proved much more extensive than anticipated, bearing in mind Smith's purpose. After completing this and adding the gender of generic names in 1970, I added two chapters on botanical and vernacular names, now revised.

Any book on plant names necessarily has its limitations. The number of accepted generic names of flowering plants and ferns, probably about 1,500, is now so great that no book of reasonable size can include even half of them; however many of these, relating, for example, to tropical families such as *Chrysobalanaceae*, *Sapindaceae* and *Sapotaceae*, are unlikely ever to trouble gardeners. The number of specific epithets is just as unwieldy, with even those of a geographical nature reaching about 11,500 and with descriptive and commemorative epithets probably as many.

Hence the selection of those for inclusion must necessarily be somewhat arbitrary and omissions a cause of disappointment. The second edition (1972) nevertheless included most of the genera listed by Boom, Hylander, Sander and Zander as being cultivated in the Netherlands, Scandinavia, the British Isles and Germany and likely to be available elsewhere; such genera had come from many regions of the world. The late Mrs Isadore L. L. Smith (Ann Leighton) and the late Dr Averil Lysaght, by painstakingly correlating the first edition (1963) with the lists of these authors, provided a number of extra generic names meriting inclusion. Dr Lysaght also helped by inserting accents and provided information relating to Pacific Ocean exploration, of special interest to her as a New Zealander. To those and other names I added derivations etc.

The need to keep this book within bounds has made it impossible to include documentation for single entries but in many instances I have consulted the original publication of a name or an authentic source of information to ascertain the author's intent when coining a name; even so the meaning of a number remains unknown or doubtful. A few examples will indicate what search may reveal. Thus Rumphius himself gave the derivation, a play on words, of his curious name *Quisqualis* (see p. 254) in his *Herbarium Amboinense* 5:71(1747). With this may be contrasted the incorrect explanation given in the Royal Horticultural Society's *Dictionary of Gardening* 4: 1733 (1950): '*quis*, who, *qualis* what kind; from original uncertainty as to its family'. The derivation of *Anemone* is discussed by Toy in *Rhodora* 1:40 (1887); unbeknown to him Paul de Lagarde in 1878 had published in a Göttingen periodical the same conclusion as to its Semitic origin. Thiselton-Dyer provided the most plausible derivation for *Elaeagnus* (*elaiagnos*) in *Journal of Philology* 34: 79 (1915). Schott published many names for new genera of *Araceae* without explanation of their origin but he later provided etymologies for them in his *Genera Aroidearum* (1857). Wittstein stated that the orchid name *Disa* was the 'Name der Pfalnze am Cap'; Schultes and Pease more sensibly say 'origin and meaning obscure'; there can be no doubt however, that its author, Bergius, a Swede, had in mind the mythical Queen Disa of Sweden, as pointed out by Hylander in *Bulletin du Jardin botanique de Bruxelles* 28: 451–453 (1958). A paper in *Names Journal* 1: 210 (1952) by E. Gudde provided information about *Sequoia*. These are cited here merely to indicate the diversity of publications occasionally relevant to the origin of plant names. For biographic details, such as dates and fore-names, Pritzel's *Thesaurus Literaturae botanicae*, 2nd ed. (1872), Saccardo's *La Botanica in Italia* (1895–1901), Barnhart's *Biographical Notes upon Botanists* (1965), Desmond's *Dictionary of British and Irish Botanists and Horticulturists* (1977) and Stafleu & Cowan's *Taxonomic Literature* 2nd ed (1976–1988) have been particularly helpful and to these an enquirer should refer for more personal information than can be given here; other sources have also been used, for example, Viviam's

1922 biography of Cesalpino. The information about Mitchell's dates comes from Dorman and Lewis's paper in the *Virginia Magazine of History* 76: 437–440 (1968). There exist indeed many such relevant articles scattered through learned journals. The publication of the present *Dictionary of Plant Names for Gardeners*, incorporating notes on plant-names I have gathered during the past sixty years, has thus provided an opportunity of making some of these conclusions more readily accessible.

As indicated above, specific epithets are even more difficult to cover adequately than generic names. The present work includes most of the commonly used descriptive epithets as well as some unusual ones but makes no claim whatever about approaching comprehensiveness. Definitions of many of these botanical terms will be found in my *Botanical Latin* and concise descriptions of botanical families in Mabberley's invaluable *The Plant Book*.

The present work differs primarily from its predecessor by the addition of many more entries, the expansion and emendation of others, the omission of materials considered irrelevant and the insertion of families, and will prove, it is hoped, even more serviceable to gardeners and botanists inquisitive about the plant names they use. Since these changes have made it diverge so greatly from Smith's 1963 *Gardener's Book*, Cassell have decided to publish it as a new book with a distinctive new title.

I thank my good friends the librarians at the Royal Botanic Gardens, Kew and the Department of Botany, British Museum (Natural History), London for their ever-willing helpfulness as also my wife and Harriet Stewart-Jones for devotedly reading the proofs.

William T. Stearn
Kew
May 1992

Preface to the 1972 *Gardener's Dictionary*

by Isadore L.L. Smith

My husband began work many years ago collecting the material which forms the basis for this present book to satisfy his own curiosity. Later, he hoped that it might eventually become a source of reference and pleasure for all gardeners who, however keen, find themselves wondering at the botanical names attached to their plants. What useful information lies there? Of what interest are the personal names involved? What, indeed, of the gardener's particular pleasure, noted in the seventeenth century by William Cole, 'to have the plants speaking Greek and Latin to him and putting him in mind of stories which otherwise he would never think of'. Familiar names of plants fade at even a short distance in space or time from their current bloom and gardeners may miss the just deserts of their labours by not communicating exactly with each other.

Archibald William Smith, 1899–1962, was a man whom Justice Oliver Wendell Holmes would have acknowledged to have lived 'at all', since he had so thoroughly 'shared the passion and action of his time'. He was born in Spain, spent his childhood in Russia, went home to school and passed from Shrewsbury into the Royal Military Academy at Sandhurst. Early in the First World War he arrived in the trenches of the Western Front with the Gloucestershire Regiment. Recovering from wounds towards the end of the war he trained cadets at Cambridge and went from there to serve on General Denikin's staff in South Russia. He rode with the Cossacks. Rejoining his regiment in India, he fought on the North-West Frontier. Tired, as he said, after six years of killing, he considered the Indian Forest Service but, instead, joined his uncle's firm and went to extract teak from the Burmese forests, using elephants. His first articles for *The Atlantic Monthly*, *Blackwood's*, and *The National Geographic Magazine* were about this experience. Later he wrote several novels, short stories, and his autobiography to the age of twenty-one, *A Captain Departed*.

From Burma he sent seeds to his uncle, Reginald Heber Macaulay, for his famous rock garden at Kirnan, and when on leave he met there many of the great collectors. It was then that the need for a book such as this for the average gardener was discussed and the filing cards began to accumulate.

Although throughout his life he had gardened wherever he could, his first permanent garden was in Ipswich, Massachusetts, where he 'settled' in 1934 on a little hill overlooking the marshes and the sea. By this time his personal need for such a book as this *Gardener's Dictionary* was evident and he began to arrange the cards in book form. During the 1938 Munich crisis, however, he was notified that he might be recalled to the Army, and in the summer of 1939 he went to London and the General Staff at the War Office. During the Second World War, while employed in M.I.5, he saw much of Africa and the South Pacific. At the end of the war he returned to the United States and helped to start the Conservation Foundation in New York. In Ipswich he worked for the Children's Medical Centre at Boston and began to get his plant name book ready for the publishers. Peter Green, then at the Arnold Arboretum of Harvard University and now at the Royal Botanic Gardens, Kew, encouraged him in his wish to make it useful on both sides of the Atlantic. He died soon after correcting the final proofs of the American edition of the book, which was published in 1963. The book had, however, already proved its worth to him as a gardener's companion.

Isadore Leighton Luce Smith
February 1971

An Introduction to Botanical Names

Many thousands of plants, among them some 250,000 species of flowering plants and 6,000 of ferns, as well as mosses, liverworts, fungi, lichens and algae beyond estimate, have been given botanical names of Latin form since the Swedish naturalist Carl Linnaeus (1707–1778) instituted modern botanical nomenclature in 1753. Of this huge number, most interest only botanists, who are content to use internationally intelligible botanical names for them. The inhabitants of their native lands have locally intelligible vernacular names only for those specially needing them. Hence relatively few possess or need English vernacular names. Indeed, a single English vernacular name may refer to a diversity of plants, its application depending on the region. The name 'bluebell' is applied to *Campanula rotundifolia* in Scotland, *Hyacinthoides non-scripta* in England; on the American continent it may indicate a species of *Campanula*, *Polemonium*, *Mertensia* or *Penstemon*, all on a regional basis; in Australia it is used of *Wahlenbergia*. Local convenience and tradition determine the use of such names, which may be apt and pleasing and also historically and etymologically interesting, but are limited in utility by geographical and linguistic factors; they can never be other than local substitutes or alternatives for botanical names (see p. 316).

Botanical names such as *Campanula rotundifolia* cut across these local preferences. They are intended to be international, the common property of professional and amateur alike in all countries, and because at their first publication they have to be associated with descriptions based on specimens, they have usually a fixed and precise application. Internationally agreed rules and recommendations forming a periodically revised code control their formation, publication and acceptance.

In the eighteenth century, when the foundations were laid for the present system of naming plants, Latin was still the most widely used international language of science and scholarship. This was not, however, classical Latin but an expanded form of Latin derived from the Latin used for many purposes in the Middle Ages and the Renaissance. Botanical names in Latin form are a legacy from the eighteenth century, derived from much earlier usage. The Latin used by botanists today is very different indeed from that of the Romans because it has to deal with so many plants and with so many structures unknown in classical times. Words such as *corolla* and *anthera* now possess special botanical meanings not recorded in dictionaries of classical Latin. A standard Latin dictionary may in fact be quite misleading when consulted for botanical information.

Classification and the subsequent naming of a plant begin with placing it in its plant **family** (formerly called its natural order). The members of a given family have more in common with one another than with those of another family. Indeed, a family is defined by a unique combination of characters, primarily of the flowers and fruits, which set it apart from other families. Thus knowing to which family a plant belongs usefully indicates its basic features and with these others may be associated. Thus the family *Rutaceae*, consisting of woody plants and a few herbs with flower parts in fours or fives, has gland-dotted aromatic leaves and their juice on skin exposed to strong sunlight may cause a dermatitis; rue (*Ruta*) can do this. A family may contain only one genus, for example *Garryaceae*, *Ginkgoaceae* and *Eucryphiaceae*, only a few genera, for example *Berberidaceae*, *Cistaceae* and *Plantaginaceae*, or many genera, for example *Compositae*, *Labiatae*, *Leguminosae*, *Orchidaceae*, *Rosaceae* and *Umbelliferae*. Most family names consist of the name of one of the genera plus the ending *-aceae* which means 'belonging to, having the nature of'. The few exceptions with descriptive names, such as the *Cruciferae*, literally 'cross-bearers', *Gramineae*, literally 'grasses', and *Umbelliferae*, literally 'umbel-bearers', were the first families to be recognized and have been allowed to retain their time-honoured designations. The name of a **species** does not include the family name but consists of a generic name, for example *Berberis*, followed by a specific epithet (formerly called a trivial name), for example *vulgaris*, the two words in combination forming a specific name, *Berberis vulgaris*. This suffices to distinguish the species by name from the many thousands of other known species, although for greater precision or to prevent confusion the name of its author is often indicated (for example L. for Linnaeus), by citing this after the name, thus *Berberis vulgaris* L.

Most **botanical generic names** are of Latin form and end in *-a*, *-um*, or *-us*, but only a limited number are genuine Latin names used by the Romans.

Thus, analysis of 200 alphabetically consecutive generic names in *Orchidaceae* showed 132 to be of Greek origin, 48 personal, 12 Latin, 4 obscure, 3 geographical, 1 vernacular.

According to their origin, several classes may be distinguished:

1. Latin plant names used by classical authors, such as Virgil, Columella, Pliny: examples are *Acer* (maple), *Arbutus* (strawberry tree), *Bellis* (daisy), *Cornus* (dogwood), *Hedera* (ivy), *Juniperus* (juniper), *Lilium* (lily), *Pinus* (pine), *Quercus* (oak), *Reseda* (mignonette), *Ruscus* (butcher's broom), *Ruta* (rue), *Taxus* (yew), *Ulex* (gorse).
2. Greek names taken into Latin by classical authors: examples are *Achillea*,

Acorus, *Anemone*, *Bupleurum*, *Daphne (spurge laurel)*, *Lychnis*, *Petroselinum*, *Phalaris*, *Scilla*. A number of Greek names were derived from Semitic names or from the lost languages of people inhabiting Greece before the coming of the ancestors of the Ancient Greeks, some 4000 years ago, for example *Anemone*, *Hyacinthus*, *Hyssopus*, *Terebinthus*, or even from Assyrian, for example *Mandragora*.

3. Names taken from classical mythology by later authors, but not employed as plant names in classical times: *Acacallis*, *Achlys*, *Andromeda*, *Calypso*, *Cassiope*, *Hebe*, *Nerine* are examples. Linnaeus derived much of his knowledge of classical mythology from the *Fabulae* of Hyginus.
4. Names of medieval origin: examples are *Alchemilla*, *Aquilegia*, *Borago*, *Calendula*, *Carlina*, *Doronicum*, *Filago*, *Filipendula*, *Linaria*, *Tanacetum*.
5. Modern names formed from Latin words: examples are *Arundinaria*, *Cimicifuga*, *Convallaria*, *Digitalis*, *Fissidens*, *Mobilabium*, *Pennisetum*, *Setaria*.
6. Modern names formed from Greek words: such as *Agrostemma*, *Chionodoxa*, *Chrysophyllum*, *Dimorphotheca*, *Echinochloa*, *Echinodorus*, *Enarthrocarpus*, *Eremurus*, *Galanthus*, *Leontodon*, *Meconopsis*, *Mesembryanthemum*, *Odontoglossum*, *Osmanthus*, *Pychnanthemum*, *Tetragonotheca*, *Trichocentrum*.
7. Modern names formed from personal names: examples are *Adansonia*, *Albizia*, (after Albizzi), *Butea*, *Camellia*, *Carludovica* (after King Carlos and Queen Luisa), *Carmichaelia*, *Carnegiea*, *Cattleya*, *Damrongia*, *Dorotheanthus*, *Fuchsia*, *Kitaibela*, *Linnaea*, *Robinia*, *Sequoia*, *Tradescantia*, *Warscewiczella*, *Welwitschia*, *Yoania*, *Zaluzianskya*.
8. Modern names formed from vernacular names: for example *Ailanthus*, *Akebia*, *Amelanchier*, *Angraecum*, *Antiaris*, *Armeria*, *Bambusa*, *Butia*, *Camassia*, *Cantua*, *Caragana*, *Catalpa*, *Catha*, *Corokia*, *Cortaderia*, *Datura*, *Musa*, *Pandanus*, *Puya*, *Rorippa*, *Sasa*, *Sorghum*, *Swida*, *Tacca*, *Wattakaka*.
9. Modern names formed from geographical names: examples are *Arisanorchis*, *Araucaria*, *Austroamericium*, *Bogoria*, *Caribea*, *Cathaya*, *Cotopaxia*, *Domingoa*, *Hammarbya*, *Indopoa*, *Jacaima* (an anagram of Jamaica), *Librevillea*, *Lobivia* (an anagram of Bolivia), *Papuodendron*, *Patagonula*, *Salweenia*, *Tacazzea*, *Taiwania*, *Utahia*.
10. Anagrams made by altering the arrangement of letters in the names of a related genus: *Docynia*, (from Cydonia), *Ivodea* (from Evodia), *Microtoena* (from Craniotome), *Nosema* (from Mesona), *Poacynum* (from Apocynum), *Tellima* (from Mitella), *Tylecodon* (from Cotyledon), and *Gifola*, *Ifloga*, *Lifago*, *Logfia*, *Oglifa* (all from Filago). Such names excited the wrath of the learned Rev. R. T. Lowe who in his *A Manual of the Flora of Madeira* (1868) earnestly invoked all botanists to unite in utterly repudiating and putting down 'such scandalously childish, bald and witless trickery with names as the anagrammatic formations by Cassini ... such unseemly barbarisms ... such base name-coinage'—but all to no effect!
11. Modern names of no evident meaning, their authors having left them unexplained, perhaps wisely: *Aa*, *Asystasia*, *Ate*, *Iliamna*, *Ipsea*, *Liatris*, *Lonas*, *Ratibida*, *Venidium*.

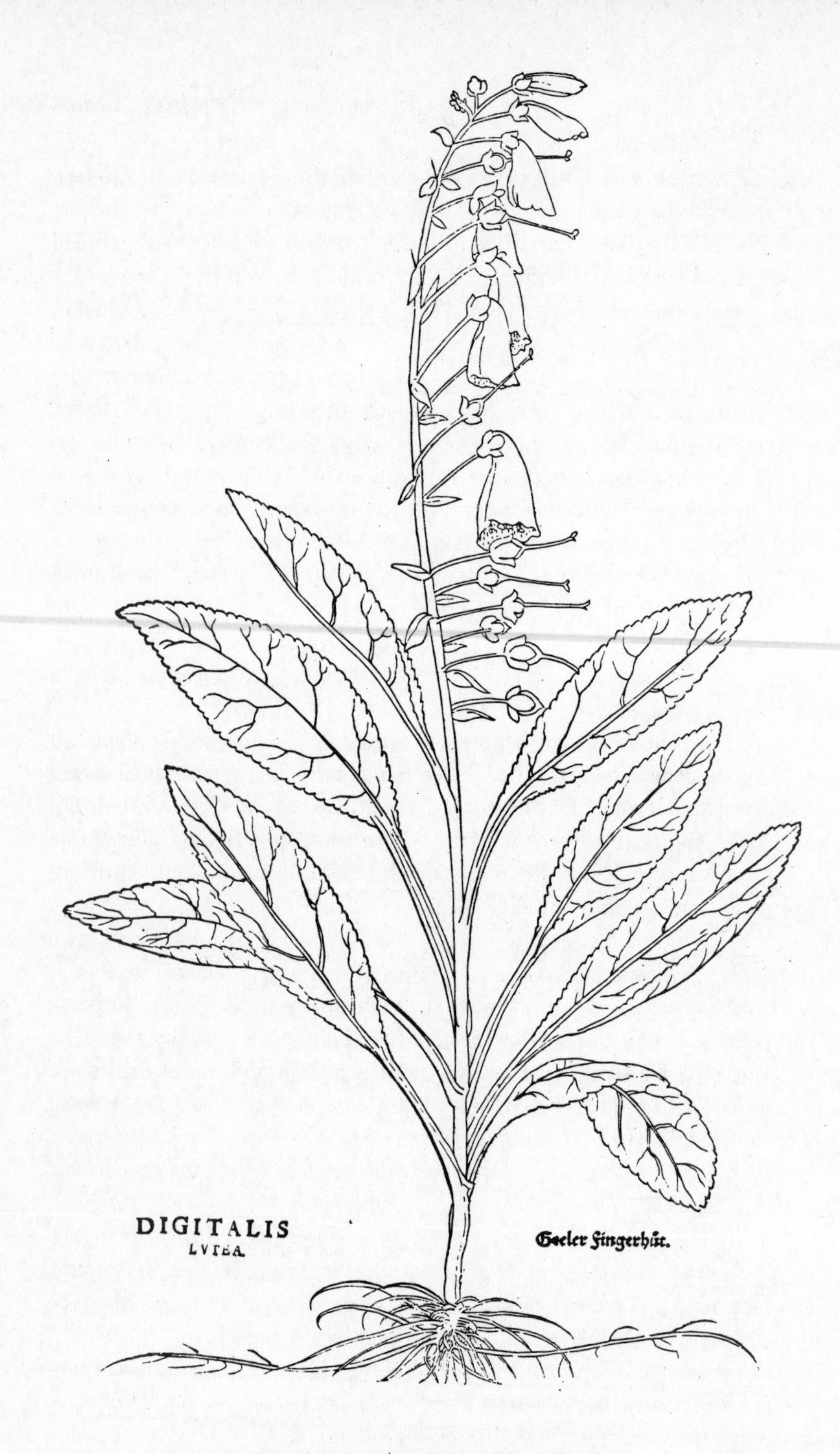

Yellow Foxglove, *Digitalis grandiflora*,
Woodcut from Leonhart Fuchs, *De Historia Stirpium* (1542);
here the first use of the name *Digitalis*.

Ancient Greek and Latin plant names owe their survival into modern botany through their use primarily by three highly esteemed ancient authors Theophrastus, Dioscorides and Pliny, whose works were laboriously copied and copied and copied by hand through the Middle Ages until the sixteenth-century invention of printing made them widely available. They did not invent plant names; they merely recorded names in contemporary use and so preserved them. Theophrastus (*c.* 370–287 B.C.) was a Greek philosopher who taught in Athens and wrote an important enquiry into plants (*Peri phyton historia*) in 9 volumes, wherein he mentioned about 500 plants. Dioscorides (1st century A.D.) was a Greek medical man and herbalist, who apparently served over a wide area with the Roman army and compiled from his own experience and the writings of others an encyclopaedia of medicaments (*Peri hyles iatrikes*) with descriptive accounts in 5 volumes on medicinal plants. His Roman contemporary Pliny the Elder was a military and later naval commander with an intense curiosity about almost everything and stupendous industry in recording it; his natural history (*Naturalis historia*) compiled from over a hundred authors extends to 37 volumes of which 15 deal with plants.

During the Middle Ages other names came into existence to supplement this classical legacy. The first modern maker of Latin-form plant names seems to have been the German physician and herbalist Leonhart Fuchs (1501–1566) who, in 1542, coined the names *Campanula*, *Cruciata*, *Digitalis* and *Ophioglossum* for Central European plants unrecorded by the classical authors. Some Ancient Greek plant names commemorate people. The first to revive this procedure was a French monk, Charles Plumier (1646–1704), a very good botanist, who made three expeditions to the French islands of the West Indies. Here, in this floristically rich area, he distinguished many genera and species hitherto unknown and hence unnamed. To provide them with generic names he commemorated almost all the botanists and herbalists with whose work he was acquainted. The names *Alpina*, *Barleria*, *Bauhinia*, *Bellonia*, *Besleria*, *Bocconia*, *Bontia*, *Breynia*, *Bromelia*, *Caesalpina*, *Cameraria*, *Castorea*, *Clusia*, *Columnea*, *Commelina*, *Cordia*, *Cornutia*, *Cupania*, *Dalechampia*, *Dioscorea*, *Dorstenia*, *Eresia*, *Fuchsia*, *Gerardia*, *Gesnera*, *Hernandia*, *Magnolia*, *Malpighia*, *Maranta*, *Marcgravia*, *Matthiola*, *Mentzelia*, *Muntingia*, *Parkinsonia*, *Pereskia*, *Petiveria*, *Plinia*, *Plukenetia*, *Pisonia*, *Renealmia*, *Rivina*, *Ruellia*, *Sloana*, *Tragia*, *Triumfetta*, *Valdia*, *Vanreedia*, *Ximenia*, almost all of those devised and published by Plumier remain in use today and many are familiar to gardeners.

Carl Linnaeus (1707–1778) followed Plumier in publishing a host of new generic names, many commemorating his benefactors and friends, even more from classical sources and of his coining. He had no scruples about substituting a name of his own for an earlier one, e.g. *Verbesina* for *Eupatoriophalacron* and transferring Ancient Greek names such as *Cactus*, *Ceanothus*, *Ptelea* and *Silphium* to American genera. Classical mythology provided him, as it has done many later botanists, with a rich store of names for generic use.

For the naming of **intergeneric hybrids**, that is those with parents belonging to two or more different genera, special group names are formed by combining parts of the names of the parent genera into a single word not exceeding eight syllables, for example *Mahoberberis* from *Mahonia* and *Berberis*, or by adding the termination *-ara* to the name of a person connected with the group, for example *Burrageara* for orchid hybrids with *Cochlioda*, *Miltonia*, *Odontoglossum* and *Oncidium* in their parentage. Such names are regarded as condensed formulae or as substitutes for formulae. Some examples are:

Amelosorbus (*Amelanchier* × *Sorbus*), *Brassocattleya* (*Brassia* × *Cattleya*), *Chionoscilla* (*Chionodoxa* × *Scilla*), *Cupressocyparis* (*Chamaecyparis* × *Cupressus*), *Cratamespilus* (*Crataegus* × *Mespilus*), *Fatshedera* (*Fatsia* × *Hedera*), *Heucherella* (*Heuchera* × *Tiarella*), *Osmarea* (*Osmanthus* × *Phillyrea*), *Philageria* (*Lapageria* × *Philesia*), *Ruttyruspolia* (*Ruspolia* × *Ruttya*), *Solidaster* (*Aster* × *Solidago*), *Sorbaronia* (*Aronia* × *Sorbus*).

Specific epithets display the same variety. Many descriptive epithets, such as *albus* (white), *bellus* (pretty), *calvus* (bald), *caudatus* (tailed), *centifolius* (hundred-leaved), *foetidus* (stinking), *glomeratus* (crowded), *glutinosus* (sticky), *horridus* (bristly), *laxus* (loose), *multiflorus* (many-flowered), *odoratus* (scented), *roseus* (rose), *sempervirens* (evergreen), and geographical epithets, such as *anglicus* (English), *gallicus* (French), *graecus* (Greek), *hispanicus* (Spanish), *italicus* (Italian) are genuine Latin words; others such as *biltmoreanus*, *canadensis*, *labradoricus*, *araucanus*, *missuricus*, *nantucketensis*, *nepalensis*, *japonicus*, *laurentianus* and *pensylvanicus* have been formed by analogy with them. As, however, the number of species known within certain genera has increased to such an extent that all appropriate Latin epithets have been used, so more and more epithets have been added from Greek, in which it is comparatively easy to combine words and form epithets. Thus in the genus *Rhododendron*, out of 747 alphabetically consecutive epithets, 259 were of Latin origin, 246 of Greek origin, 162 personal, and 80 from other sources, mostly geographical.

Descriptive specific epithets often convey a little information about the plant, its size (usually relative to other species of the genus), habit of growth, colour or habitat. Personal names usually refer to the collector or someone who has studied the group, or they may be purely complimentary; the epithets *armandii*, *davidii*, *kingdonii*, *wardii*, *georgei*, *forrestii*, *ernestii* and *wilsonii* are attached to many plants collected in eastern Asia by Armand David, F. Kingdon-Ward, George Forrest and Ernest H. Wilson, but the epithets *hemsleyi* and *franchetii* commemorate W. B. Hemsley and A. Franchet, who studied the plants of eastern Asia but never had an opportunity of going there.

The two-word system of naming, usually called **binomial nomenclature** is a very old system in vernacular names of plants and animals all the world over. It was first consistently adopted for the scientific names of plants by Linnaeus in his *Species Plantarum* (1753), a two-volume work of 1,200 pages in which he set out to provide a concise guide to all the plants then known. In 1768 in his *Systema Naturae* he named all the known animals likewise. He used in these and later works two sets of names, a descriptive name which stated the main distinguishing features of the species, for example *Ranunculus seminibus aculeatis, foliis superioribus decompositis linearibus*, and a more convenient but less informative two-word name, for example *Ranunculus arvensis*, for everyday use. Many of his contemporaries were reluctant to adopt Linnaeus's binomials but by the end of his life they were accepted everywhere. A primary reason for this was the encyclopaedic nature of his works. They brought together so much information about the kinds of plants in such a very concise, methodical and practical manner that they became indispensable. Through their use people became accustomed to Linnaeus's binomial nomenclature and gradually realized its advantages over the cumbersome phrase-names. Linnaeus's system of nomenclature has persisted, but his temporary and artificial system of classifying the genera into large groups based on the number and relation of their stamens and styles or stigmas, the 'sexual system', has long been superseded by other classifications based on a greater range of structures.

When necessary, **varietal** or **cultivar epithets** are used to distinguish variants within a species or group of interspecific hybrids; a third epithet is added to the binomial. This is of Latin form when referring to a wild subspecies or variety (e.g. *Consolida regalis* subsp. *paniculata*, *Ranunculus marginatus* var. *trachycarpus)*, but should nowadays be a vernacular or fancy name, for example 'Melody', 'Daydream', 'Pixie', 'Rosy Morn', or a personal or geographical one, such as 'Caerhays', 'Laurens Koster', 'Lucy', 'Hester', 'Princess Mary', when referring to a cultivar, that is, an assemblage of cultivated individual plants distinguished by characters significant in agriculture, forestry, or horticulture.

In many groups where there has been much hybridization and selection, it suffices merely to use a generic name such as *Narcissus* or a vernacular equivalent such as daffodil, and to follow this simply by the cultivar epithet, for example 'Fortune'.

Correctness is generally understood to be accordance with an accepted standard of excellence but such a standard may change with time and circumstances. The **correct pronunciation of Latin** accordingly depends upon the kind of Latin taken as a standard. This has long been a matter of debate among scholars. Within the Roman Empire there were differences in speech between the educated and the uneducated classes of Italy and there must have been even greater differences of pronunciation between these and the subject peoples of distant provinces whose native languages with their own speech rhythms had been imperfectly replaced by Latin. Modern languages developed from Latin such as Italian, French, Spanish, Portuguese, Romanian and their dialects indicate the artificiality of linguistic correctness. For modern scholars the aim has been 'primarily to reconstruct the educated pronunciation of Rome in the Golden Age', roughly between 100 B.C. and A.D. 10, the age of Virgil, Cicero, Julius Caesar, Livy and Ovid. There is general agreement on this by classicists but their rules cannot consistently be applied to botanical names, so many of which are of non-Latin origin. Particularly difficult in this respect are names commemorating persons, such as *Dahlia* and *Dalea*, the first of which is often pronounced by English-speaking gardeners as if it were the second. Curiously enough *Clivia* (named for the noble family of Clive) is often pronounced as if 'Clive' rhymed with 'cleave' instead of 'hive'. In general people tend to pronounce botanical names as if these were words in their own language. The pronunciation of Latin as formerly taught in schools in the English-speaking world and as still used by most of its gardeners differs considerably from the reformed academic pronunciation accepted by classical scholars, and it is to the latter that the pronunciation of Continental European botanists more closely approximates. Hence some acquaintance with both systems may help towards international intelligibility. These may be summarized in the following table (taken from Stearn: *Botanical Latin*; 1966):

Reformed academic	Traditional english
ā as in *fāther*	*fāte*
ă as in *ăpart*	*făt*
ae as *ai* in *aisle*	as *ea* in *meat*
au as *ou* in *house*	as *aw* in *bawl*
c always as in *cat*	before *a*, *o*, *u*, as in *cat* before *e*, *i*, *y* as in *centre*
ch (of Greek words) as *k* or *k-h* (if possible)	as *k* or *ch*
ē as in *thēy*	*mē*
ĕ as in *pĕt*	*pĕt*
ei as in *rein*	as in *height*
g always as in *go*	hard before *a*, *o*, *u* as in *gap*, *go* soft before *e*, *i*, *y* as in *gem*, *giro*
ī as in *machīne*	*īce*
ĭ as in *pĭt*	*pĭt*
j (consonant i) as *y* in *yellow*	*j* in *jam*
ng as in *finger*	*finger*
ō as in *nōte*	*nōte*
ŏ as in *nŏt*	*nŏt*

oe as *oi* in *toil*	as *ee* in *bee*.
ph as *p* or *p-h* if possible	like *f*
r always trilled	
s as in *sit*, *gas*	*sit*, *gas*
t as in *table*, *native*	*table* but *ti* within a word as in *nation*
ū as in *brūte*	*brūte*
ŭ as in *fŭll*	*tŭb*
ui as *oui* (French), *we*	*ruin*
v (consonant u) as *w*	as in *van*
ȳ as *u* in French *pur*	as in *cȳpher*
y̆ as in French *du*	as in *cy̆nical*

Vowels such as those in cake, kite, eve, vote are described as 'long'; vowels such as those in cat, kit, egg, pot are 'short'. The stress or emphasis in a Latin word falls on the next to last (penultimate) syllable when this is 'long', for example *for*-mo-*sus*, *cru*-en-*tus*, but on the last but two (antepenultimate) syllable when the last but one is 'short', for example flo-*ri-dus*, *syl*-vat-*ti-cus*.

In the list of plant names which forms the bulk of this book, an accent is used to indicate the stressed syllable, *Ara'lia*, *alternifo'lius*, *rugo'sus*.

To determine accentuation, but opinions on this differ, much use has been made of C. A. Backer: *Verklarend Woordenboek*; N. Hylander: *Våra Prydnadsväxters Namn på Svenska och Latin*; C. Wittstein: *Etymologisch-botanisches Handwörterbuch*; and R. Zander: *Handwörterbuch der Pflanzennamen*.

In transliteration of Greek words *ŏ* represents the Greek *o* (omicron) and *ō* the Greek ω (omega), *ĕ* the Greek ε (epsilon) and *ē* the Greek η (eta), *y* the Greek υ (upsilon).

The scientific names of plants are treated as Latin regardless of their derivation and the **Latin grammatical rules of agreement in gender** are used when applicable. The name of a genus is either a noun or a word treated as such, and has accordingly masculine, feminine or neuter gender. In this book the gender is indicated by the letters 'm' (masculine), 'f' (feminine) or 'n' (neuter) placed after the generic name. There are no completely reliable easy rules for ascertaining the gender but the following notes may help:

1. Generic names ending in *-us* are mostly masculine, for example *Cistus*, *Hibiscus*, *Humulus*, *Lupinus*, *Narcissus*, unless they are names of trees, wherefor *Ailanthus*, *Alnus*, *Cedrus*, *Cornus*, *Fagus*, *Laurus*, *Pinus*, *Sorbus*, for example, are feminine.
   Generic names ending in the Greek *-ōn* are masculine, for example *Cotyledon*, *Dendromecon*, *Lysichiton*, *Platycodon*, *Platystemon*, but those ending in the Greek *-ŏn* (which can also be rendered as *-um*) are neuter, as for example *Acantholimon*, *Clerodendrum*, *Rhododendron*.
2. Most generic names ending in *-a* are feminine, for example *Castanea*,

Diosma, *Fuchsia*, *Hosta*, *Nandina*, *Rosa*, unless they are derived from neuter Greek words ending in *-ma*; such Greek words include *derma* (skin), *loma* (border), *nema* (thread), *paegma*, (sport), *phragma* (partition), *sperma* (seed), *stema* (stamen), *stigma* (stigma); hence *Aglaeonema*, *Atherosperma*, *Ceratostigma*, *Duasperma*, *Pachyphragma*, *Phyteuma*, *Trichostema*, are neuter. Generic names ending in *-ago*, *-ix*, *-odes*, *-oides*, are feminine, for example *Nymphoides*, *Omphalodes*, *Plumbago*, *Solidago*, *Salix*, *Tamarix*, as are most names ending in *-is*, for example *Bellis*, *Cannabis*, *Clematis*, *Nomocharis*, *Orchis*, *Oxalis*, *Pteris*, *Vitis*, or *-es*, for example *Cheilanthes*, *Menyanthes*.

3. Generic names ending in *-dendron* and *-um* are neuter, for example *Allium*, *Halimodendron*, *Epimedium*, *Lilium*, *Spartium*, *Trillium*.

Specific epithets may be either adjectives and participles or nouns. The adjectives agree in gender with the generic names they follow. For example *Philadelphus* is of masculine gender and the epithets *coronarius*, *grandiflorus*, *incanus*, *mexicanus*, *microphyllus* etc. applied to species of *Philadelphus* are likewise masculine. *Rosa* is of feminine gender and the associated epithets *alba*, *floribunda*, *japonica*, *rosea*, etc., are likewise feminine. *Lilium* is of neuter gender and the associated epithets *auratum*, *canadense*, *candidum*, *pumilum*, *sulphureum*, etc., are likewise neuter.

The adjectives and participles form three main groups:

1. Those in which the masculine, feminine and neuter forms each have a different ending:

Masculine–US	Feminine–A	Neuter–UM
albus (white)	*alba*	*album*
florifer (flowery)	*florifera*	*floriferum*
glaber (glabrous)	*glabra*	*glabrum*
graecus (Greek)	*graeca*	*graecum*
hirsutus (hairy)	*hirsuta*	*hirsutum*
triphyllus (three-leaved)	*triphylla*	*triphyllum*

2. Those in which the masculine and feminine forms have the same ending but the neuter a different one:

Masculine	Feminine	Neuter
acaulis (stemless)	*acaulis*	*acaule*
acaulos (stemless)	*acaulos*	*acaulon*
brevis (short)	*brevis*	*breve*
campestris (of plains)	*campestris*	*campestre*
canadensis (Canadian)	*canadensis*	*canadense*
communis (common)	*communis*	*commune*
laevis (smooth)	*laevis*	*laeve*

tenuis (thin)	*tenuis*	*tenue*
triphyllos (three-leaved)	*triphyllos*	*triphyllon*

3. Those in which the masculine, feminine and neuter forms all have the same ending:

Masculine	Feminine	Neuter
bicolor (two-coloured)	*bicolor*	*bicolor*
bryoides (moss-like)	*bryoides*	*bryoides*
macrobotrys (large-clustered)	*macrobotrys*	*macrobotrys*
praecox (early)	*praecox*	*praecox*
pubescens (pubescent)	*pubescens*	*pubescens*
repens (creeping)	*repens*	*repens*
reptans (creeping)	*reptans*	*reptans*
simplex (unbranched)	*simplex*	*simplex*

Nouns used as specific epithets are not affected by the gender of the generic name. They remain the same whatever the gender. Such nouns may be obsolete names as in *Hydrocharis morsus-ranae, Liriodendron tulipifera, Sedum rosea, Teucrium chamaedrys*, or current generic names, as in *Corokia cotoneaster, Diervilla lonicera, Rosa rubus, Rubus rosa*, or vernacular names, as in *Codonopsis tangshen, Gentiana kurroo, Ligustrum ibota, Pinus mugo, Podocarpus totara, Prunus mume, Sabal palmetto, Schinus molle, Silene schafta, Taraxacum kok-saghyz.** They may also be nouns in the singular or genitive plural, for example *deserti* (of the desert), *desertorum* (of deserts), *clivorum* (of the hills), *tectorum* (of roofs), *tinctorum* (of the dyers), or personal names in the singular or genitive plural, for example *baileyi* (of Bailey), *baileyorum* (of the Baileys), *elisabethae* (of Elizabeth), *helenae* (of Helen), *heleni* (of Helenus), *clusii* (of Clusius), *farreri* (of Farrer), *nuttallii* (of Nuttall), *willmottiae* (of Miss Willmott).

The seemingly illogical use of the terms **male (mas, masculus) and female (femineus) in plant names** for plants such as *Cornus mas, Paeonia mascula* and *Anagallis femina*, all known to possess hermaphrodite flowers, is purely metaphorical; it has no real sexual implication. These names are the few survivors into modern botanical nomenclature of many old paired names which came into being long before sex in plants (other than the date palm) comparable to that of animals was recognized. The procedure of naming one plant as 'male' and a rather similar one as 'female' goes back to very ancient times. Thus the Ancient Greeks designated 'male' an Indian bamboo with solid stem and 'female' one with a hollow stem. They distinguished as 'male' a peony with somewhat coarse leaves and 'female' one with more finely cut leaves. The scarlet pimpernel was 'male' the blue pimpernel 'female'. The 'male' dogwood had hard wood from which javelins were made, the 'female' soft wood.

* The cryptic epithet *mitejea* in *Kalanchoe mitejea* of Alice Leblanc and Raymond Hamet is not an obscure African vernacular name, as might be supposed, but a Gallic linguistic prank, being an anagram of 'je t' aime'! The lady involved is commemorated in *Kalanchoe aliciae, Kalanchoe leblanciae* and *Sedum leblanciae*, but is otherwise unknown to botany.

Purple-flowered species of *Cistus* were 'male', white-flowered 'female'. Thus in pre-Linnaean literature occur the names *Cornus mas* and *Cornus femina* (now *C. sanguinea)*, *Paeonia mas* and *Paeonia femina* (now *P. officinalis)*, *Anagallis terrestris mas* (now *A. arvensis)* and *Anagallis terrestris foemina* (now *A. femina)* and *Filix mas* (now *Dryopteris filix-mas)* the male fern and *Filix foemina* (now *Athyrium filix-femina)* the lady fern.

Generic names are always written with a capital letter, specific and varietal epithets with a small letter, cultivar epithets with a capital letter. For example *Viburnum farreri*, *V.* × *bodnantense* 'Dawn', *V. plicatum* var. *tomentosum*, *Bellis perennis* 'Dresden China'.

The **technical legality of botanical names** is, unfortunately, little understood by many gardeners. Botanical names have no standing, that is have no claim to use and acceptance, unless they have been published and are otherwise in accordance with the *International Code of Botanical Nomenclature*. The rules and recommendations in this code have been the subject of debate and consideration at international congresses over a long period. Before 1866 there were no such internationally agreed rules, but long before then botanists had come to accept certain guiding principles and to follow procedures made standard by the example of leading botanists, notably Augustin Pyramus de Candolle (1778–1841) of Geneva. De Candolle, who was the author of numerous monographic works, set out in his *Théorie élémentaire de la Botanique* (Paris; 1813) the main considerations, of which the basic ones were that the names should be (1) either Latin or of Latin form, because Latin was then the common language of civilized people, and (2) formed according to the rules of grammar, avoiding bilingual hybrids, such as a word part Latin, part Greek like *aculeaticarpa*, and that (3) the first to discover or to distinguish a plant had the right to name it, hence names should not be changed without good cause. He accordingly emphasized priority as a means of fixing names.

By 1866 the Candollean regulations and recommendations had been found not to cover all the problems of nomenclature made evident in monographic studies. At an international botanical congress in London a German professor, Karl Koch, proposed that such congresses should deal with matters of nomenclature. Alphonse de Candolle (1806–1893; son of A. P. de Candolle), who had a legal training, accordingly drew up his *Lois de la Nomenclature botanique* (1867), which the Paris botanical congress of 1867 adopted. These *Laws* were replaced in 1905 by a new set of *International Rules of Botanical Nomenclature* adopted by the Vienna botanical congress. They were further modified at the Cambridge botanical congress of 1930 and at the subsequent congresses of 1935 (Amsterdam), 1950 (Stockholm), 1954 (Paris), 1959 (Montreal), 1964 (Edinburgh), 1969 (Seattle), 1975 (Leningrad), 1981 (Sydney) and 1989 (Berlin), the number of significant changes becoming less and less at each successive congress; since

1952 they have formed the *International Code of Botanical Nomenclature*, a detailed and complicated document which must be carefully studied by anyone concerned with establishing the correct name of a plant.

The naming of garden and other cultivated plants is controlled by the *International Code of Nomenclature for Cultivated Plants*, first published in 1953, with revised editions in 1958, 1961 and 1969. It deals with the naming of agricultural, silvicultural and horticultural cultivars which are normally given fancy names such as apple 'Cox's Orange Pippin', barley 'Balder', *Petunia* 'Rosy Morn', etc.

Changes in names which have long been accepted are a source of annoyance and inconvenience to all concerned but are often unavoidable. Contrary to popular belief among gardeners, most botanists detest nomenclatural changes and only make them to conform with the *International Code of Botanical Nomenclature* or revised classification. Many result from **changes in classification**. Thus the Madagascar periwinkle has long been known as *Vinca rosea* but it differs in so many details, including its chemical products, from the true periwinkles (*Vinca*) that it is now put in a separate genus (*Catharanthus*) and its correct name is now *Catharanthus roseus*. For similar reasons the genus *Chaenomeles*, containing the Japanese quince and its Chinese allies, has been separated from *Cydonia*, the true quince. The reverse process takes place when groups kept separate by some botanists are put together by other botanists because the differences once used to separate them appear to be inadequate: thus *Chamaenerium* has been returned to *Epilobium*, and *Siphonosmanthus* to *Osmanthus*. The same applies to species. If, for example, *Portulaca pilosa* is considered to be a widespread and extremely variable species, with the colour and the size of the seeds not important enough to merit breaking up the group into minor species, then the well-known *P. grandiflora* becomes *P. pilosa* subsp. *grandiflora*. Such changes are called taxonomic changes.

Other changes result from **application of the principle of priority**. Thus the name *Hypericum inodorum* published by Miller in 1768 has priority over *Hypericum elatum* published by Aiton in 1789 for the same plant. Similarly the name *Scilla mischtschenkoana* published by Grossheim in 1927 has priority over *Scilla tubergeniana* published by Hoog in 1936 and accepted by Stearn in 1950. Moreover, a specific name once validly published for a given plant cannot later be used for a different species in the same genus. Similar names applied to different plants are called homonyms. The **rejection of later homonyms** has caused a number of unavoidable but regrettable name changes. Thus in 1833 the Russian botanist Bunge named *Viburnum fragrans*, a species cultivated at Peking, which Farrer and Purdom found growing in a wild state in 1914 on hillsides of Kansu (Gansu), Western China, and introduced into European gardens. Bunge was unaware that in 1824 a French botanist, Loiseleur, had

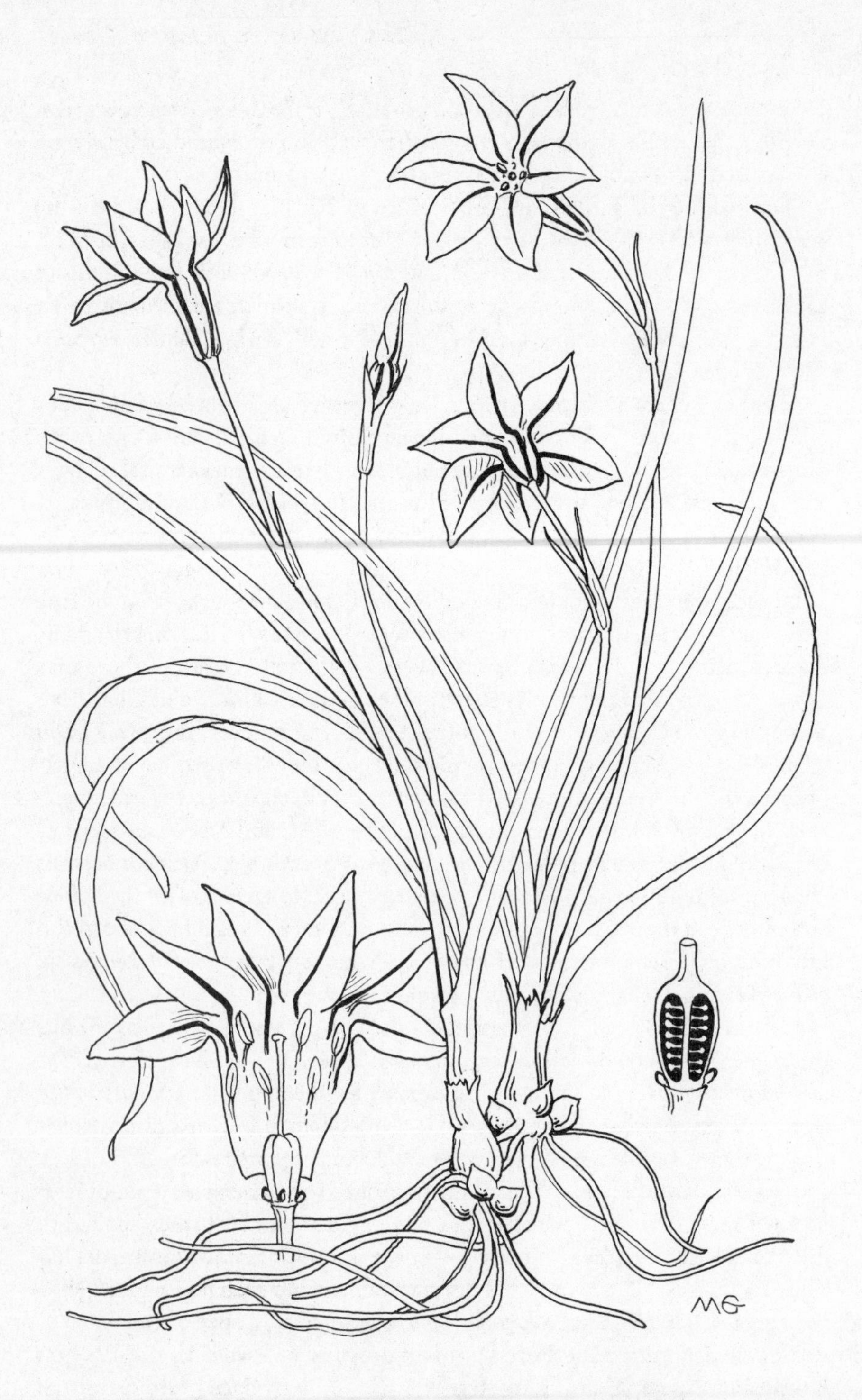

Spring Starflower, *Tristagma uniflorum*, syn. *Ipheion uniflorum*.
(Drawing by Mary Grierson)

used the name *Viburnum fragrans* for a different species (which had, however been already named *V. odoratissum* in 1820). Bunge's *Viburnum fragrans*, the popular garden plant, was accordingly renamed *Viburnum farreri* in 1966.

Some changes are due merely to **correction of past mis-identifications or mis-applications of names**. For example, the genera *Tritonia*, *Montbretia* and *Crocosmia* are considered to be distinct genera of *Iridaceae*. The common montbretia of gardens lacks the special features of *Montbretia* proper and *Tritonia* and is a hybrid between two species of *Crocosmia*, *C. aurea* and *C. pottsii* (also known as *Tritonia* or *Montbretia pottsii*); its correct name is accordingly not *Montbretia* or *Tritonia crocosmiiflora* but *Crocosmia* × *crocosmiiflora* because it must be put in the same genus as that in which both its parents are now placed. Under the name *Nepeta mussinii* two distinct kinds of catmint are known in gardens. The less ornamental one, which sets seed freely, is the Caucasian species to which the name *N. mussinii* was originally given (unless this name, published by Henkel in 1806, is considered a synonym of *N. racemosa*, published by Lamarck in 1785), and for which it must be retained. The other, which is sterile, more ornamental and more widely grown, had accordingly to be distinguished from it by receiving a new name, *N.* × *faassenii*; unknown in a wild state, it is a garden hybrid between *N. mussinii* and *N. nepetella*.

Difficulty over classification and nomenclatural complexity, underlying the names of some well-known plants and producing uncertainty as to which name should be adopted, can be exemplified by the history of the Spring Starflower (*Tristagma uniflorum* or *Ipheion uniflorum*) which has been placed in six different genera under eight different generic names since 1830. This is a delightful low-growing one-flowered South American bulbous plant, with an oniony smell, cultivated in Britain since 1832; its six stamens are placed at two levels within the flower tube. John Lindley, an English botanist, received a dried specimen collected in Argentina by John Gillies and described it briefly in 1830 as *Triteleia uniflora*. In 1832 John Tweedie, like Gillies a Scot resident in Argentina, sent living plants from Buenos Aires to Edinburgh and from these a Scottish botanist, Robert Graham, described it in detail in 1834 as a new species *Milla uniflora*, impressed like Lindley by its one-flowered state but unaware of Lindley's earlier publication. A few years later an American botanist C.S. Rafinesque put it in a genus by itself as *Ipheion uniflorum*, even though he knew it only from Graham's illustration; his *Flora Telluriana* (1837–1838) is so scarce that down to 1946, when it was reprinted, there existed only one copy in Britain; even if it had been widely available, no-one then would have taken seriously a work of such amazing eccentricity. Later botanists transferred this plant to the genera *Brodiaea* and *Leucocoryne*, in both of which it is out of place. Then in 1943 Wilhelm Herter proposed a new genus *Beauverdia* and named it *B. uniflora*, only to have Stearn point out, later in 1943, that his *Beauverdia* was the same as Rafinesque's *Ipheion* and the correct name thus *I. uniflorum*. That, one hoped, was the end of this generic confusion, but no! Among the 17,000 specimens collected in America between 1822 and 1832 by the German traveller Eduard Poeppig was a little bulbous plant from southern Chile named in 1833 *Tristagma nivale*. With its very narrow flower segments and several-flowered umbel this looks very different from our Spring Starflower, but other species related to both are now known and all considered to belong to the same genus; if this view is accepted, the correct name for the Spring Starflower becomes *Tristagma uniflorum*. The sequence of names bestowed is as follows:

Triteleia uniflora Lindley (1830)
Milla uniflora Graham (1834)
Ipheion uniflorum (Graham) Rafinesque (1834)
Brodiaea uniflora (Lindley) Engler (1888)
Leucocoryne uniflora (Lindley) Greene (1890)
Hookera uniflora (Lindley) O. Kuntze (1894)
Beauverdia uniflora (Lindley) Herter (1943)
Tristagma uniflorum (Lindley) Traub (1963)

All of the botanists named above considered they had adequate grounds for classifying this well-distinguished species as they did, with consequent name changes. This case history illustrates several matters which gardeners, understandably irritated by different names being used for the same plant, tend to overlook. Thus botanists in one country have often been unaware of work in another country owing to slowness of communication. Moreover, the naming of a given plant may depend upon the manner in which its relatives are classified, for example, as above, how the genera *Triteleia*, *Milla*, *Brodiaea* (*Hookera*), *Ipheion* (*Beauverdia*), *Leucocoryne* and *Tristagma* are defined, and to which its characters indicate it as belonging. This may be far from simple.

Despite such difficulties, the general effect of applying the accepted rules of nomenclature has been to prevent arbitrary changes and to stabilize the names of plants.

Some Sources of Further Information

Alcock, R. H. 1876. *Botanical Names for English Readers.*

Allen, W. S. 1978. *Vox Latina, a Guide to the Pronunciation of Classical Latin.* 2nd edition. Cambridge (Cambridge University Press).

Allen, W. S. 1987. *Vox Graeca, a Guide to the Pronunciation of Classical Greek.* 3rd edition. Cambridge (Cambridge University Press).

André, J. 1956. *Lexique des Termes de Botanique en Latin.* Paris (Klincksieck).

Backer, C. A. 1936. *Verklarend Woordenboek der wetenschappelijke Namen.* Groningen & Batavia.

Beckmann, J. 1801. *Lexicon botanicum exhibens Etymologiam Orthographiam et Prosodiam Nominum botanicorum.* Göttingen.

Brown, R. W. 1956. *Composition of Scientific Works.* Washington D. C.

Carnoy, A. 1959. *Dictionnaire étymologique des Noms grecs des Plantes.* Louvain (Publications Universitaires).

Drewitt, F. D. 1927. *Latin Names of common Plants.* London.

Du Molin, J. B. 1856. *Flore poétique ancienne ou Etudes sur les Plantes des Poètes anciens, grecs et latins.* Paris.

Else, G. F. 1967. The pronunciation of classical names and words in English. *Classical Journal* 62:210–14.

Featherly, H. I. 1954. *Taxonomic Terminology of the Higher Plants.* Ames, Iowa (Iowa State College Press).

Gilbert-Carter, H. 1964. *Glossary of the British Flora.* 3rd edition. Cambridge (Cambridge University Press).

Gledhill, D. 1985. *The Names of Plants.* Cambridge (Cambridge University Press).

Green, M. L. 1927. History of plant nomenclature. *Kew Bulletin* 1927: 403–415.

Gunawardena, D. C. 1968. *Genera et Species Plantarum Zeylaniae. An etymological and historical Account of the Flowering Plants of Ceylon.* Colombo (Lake House Investments).

Hart, T. S. 1954. Labillardière's plant names. *Victoria Naturalist* 70:173–175.

Hylander, N. 1954. A pans stege och Pyrrhas hår. Några reflexioner över artnamen med twåordsepitet. *Svensk botanisk Tidskrift* 48: 521–549.

Jackson, B. D. 1928. *A Glossary of Botanic Terms.* 4th edition (reprinted 1960). London (Duckworth); New York (Hafner).

Jackson, W. P. U. 1990. *Origins and Meanings of Names of South African Plant Genera.* Rondebosh.

Kobuski, C. E. 1955. A revised glossary of the more common botanical and horticultural terms. *Arnoldia* 15:25–44.

Kunkel, G. 1990. *Geography through Botany. A Dictionary of Plant Names with a geographical Meaning.* The Hague (SPB Academic Publishing).

Lemprière, J. 1949. *Lemprière's Classical Dictionary of Proper Names mentioned in Ancient Authors.* New Edition by F. A. Wright. London (Routledge & Kegan Paul).

Manara, B. 1991. Some guidelines on the use of gender in generic names and species epithets. *Taxon* 40:301–308.

Marafioti, R. L. 1970. The meaning of generic names of important economic plants. *Economic Botany* 24:189–207.

McClintock, D. 1969. *A Guide to the Naming of Plants, with special reference to Heathers.* Horley, Surrey (Heather Society).

Murr, J. 1890. *Die Pflanzenwelt in der griechische Mythologie.* Innsbruck.

Rose, H. J. 1958. *A Handbook of Greek Mythology including its Extension to Rome.* 6th edition (reprinted 1965). London (Methuen).

Saint-Lager, J. B. 1884. Recherches historiques sur les mots Plantes males et plantes femelles. *Annales de la Société botanique de Lyon* 11 (Mém.): 1–48.

Schultes, R. E. & Pease, A. S. 1963. *Generic Names of Orchids, their Origin and Meaning.* New York and London (Academic Press).

Stearn, W. T. 1992 *Botanical Latin.* 4th edition. Newton Abbot etc. (David & Charles).

Stearn, W. T. 1985. Hookerianus or hookeranus? Notes on the ending –*erianus* in plant names. *The Garden* 110:463–465.

Stearn, W. T. 1991. A Chinese puzzle ... names of plants introduced from China. *The Garden* 116:85–89.

Thiselton-Dyer, W. T. 1914–15. On some ancient plant names. *Journal of Philology* 33: 195–207 (1914), 34: 78–96 (1915), 290–312 (1915).

Thompson, R. Campbell. 1924. The migration of Assyrian plant-names into the West. *Classical Review* 38:148–149.

Wittstein, G. C. 1852. *Etymologisch-botanisches Handwörterbuch.* Ansbach. (Reprinted 1971, Niederwalluf bei Wiesbaden).

Zander, R. 1979. *Handwörterbuch der Pflanzennamen*, 11th edition by F. Enke, G. Buchheim & S. Seybold. Stuttgart (Eugen Ulmer).

Zimdahl, R. L. 1985. *Etymology of the scientific Names of Weeds and Crops.* Ames (Iowa State University Press).

Note

Gr.	=	Ancient Greek
L.	=	Latin

In this book the original Greek letters are romanized as follows:

γ gamma	=	g
γγ	=	ng
ε epilson	=	ĕ
η eta	=	ē
θ theta	=	th
κ kappa	=	k
ο omicron	=	ŏ
υ upsilon	=	y
ψ psi	=	ps
ω omega	=	ō

Dictionary of Plant Names

a- Prefix in compound words of Greek origin signifying a negative, lacking, or contrary to. Thus *apetalus*, lacking petals; *Alyssum*, against madness.

abbrevia'tus, -a, -um Shortened; abbreviated.

Abe'lia f. Ornamental shrubs named for Dr. Clarke Abel (1780–1826), who, at the suggestion of Sir Joseph Banks, accompanied Lord Amherst on his embassy to Peking (1816–1817) as botanist. Much of his collection was lost by shipwreck on the way home to Kew. Except for Alexander von Bunge who accompanied a Russian ecclesiastical mission in 1830, no European naturalist was to visit China for nearly thirty years thereafter, Robert Fortune (see *Fortunella*) being among the first to follow. Abel died in India while serving as personal physician to Lord Amherst, who had become Governor-General. CAPRIFOLIACEAE.

Abeliophyl'lum n. *Abelia*; Gr. *phyllon*, a leaf; from its resemblance in foliage to an Abelia. OLEACEAE.

Abelmo'schus m. Apparently from Arabic *abu-l-mosk*, father of musk; in allusion to the smell of the seeds; often included in *Hibiscus*, but distinguished by the split calyx. MALVACEAE.

aberconwa'yi In honour of Henry Duncan McLaren (1879–1953), 2nd Baron Aberconway, owner of a fine garden (now National Trust Property) at Bodnant, Gwynedd, North Wales, President of the Royal Horticultural Society 1931–1953.

aber'rans Deviating from the normal.

A'bies f. Fir. The classical Latin name. PINACEAE.

abieti'nus, -a, -um Resembling the fir tree.

Abob'ra f. Latinized form of the Brazilian name for this climber of the cucumber family. CUCURBITACEAE.

aborti'vus, -a -um With parts missing; imperfect; producing abortion.

Abro'ma n. Gr. *a*, not; *broma*, food. These evergreen trees are mildly poisonous, in contrast to *Theobroma*. STERCULIACEAE.

Abro'nia f. Sand verbena, wild lantana. Gr. *abros*, delicate; in allusion to the appearance of the bracts beneath the flower of these herbs. NYCTAGINACEAE.

Abrophyl'lum n. Gr. *abros*, delicate; *phyllon*, a leaf. The leaves of these Australian shrubs have a delicate appearance. ESCALLONIACEAE.

abrotanifo'lius, -a,-um Having leaves resembling southernwood.
abrotanoi'des Resembling southernwood.
abro'tanum Ancient Latin name and now the specific epithet of southernwood (*Artemisia abrotanum*, originally spelled *abrotonum*).
abrup'tus, -a, -um Ending suddenly; abrupt.
Ab'rus m. Wild licorice. Possibly from Gr. *abros*, delicate; more probably an Arabic name. The poisonous seeds of *A. precatorius* (coral pea), bright scarlet with black spots, are used in India as weights and are strung as beads. LEGUMINOSAE.
abscha'sicus, -a, -um Referring to Abkhasia, a region of the Caucasus.
abscis'sus, -a, -um Ending abruptly; cut off.
absin'thium Latin and pre-Linnaean name for wormwood, the botanical name for which is now *Artemisia absinthium*. It is used to flavour absinthe. In biblical days it was a symbol of calamity and sorrow.
abundiflo'rus, -a, -um Profusely flowered.
Abu'tilon n. From the Arabic name for a mallowlike plant. MALVACEAE.
abyssin'icus, -a, -um Abyssinian; native to Ethiopia (Abyssinia).
Acacal'lis f. Gr. *Akakallis*, name of a nymph in Gr. mythology. ORCHIDACEAE.
Aca'cia f. The Greek name for the tree *Acacia arabica*. Derived from Gr. *akis*, a sharp point. LEGUMINOSAE.
acaciifo'lius, -a, -um Leaved like an Acacia.
Acae'na f. New Zealand bur, biddy-bid. Gr. *akaina*, a thorn; in allusion to the spines on the calyx. ROSACEAE.
Acaly'pha f. Copper leaf. Variant of *akalēphē*, ancient Greek name for nettle but applied by Linnaeus to this genus because of the nettlelike appearance of the leaves. EUPHORBIACEAE.
Acam'pe f. Gr. *akampes*, inflexible or brittle, presumably in allusion to the brittle flowers. ORCHIDACEAE.
acanth- In compound words of Greek origin signifying spiny, spiky, or thorny, from Gr. *akantha*, thorn, prickle.
acanthifo'lius, -a, -um With leaves like *Acanthus*.
acanthocar'pus With spiny fruit.
Acanthoce'reus m. Trailing or climbing cactus. Gr. *akantha*, a thorn; *Cereus*, cactus. CACTACEAE.
acanthoco'mus, -a, -um Having spiny hairs.
Acantholi'mon n. Prickly thrift. Gr. *akantha*, a thorn; *limonium*, sea-lavender, to which it is related. PLUMBAGINACEAE.
Acanthopa'nax m. Gr. *akantha*, a thorn; *Panax*, ginseng, which this genus of spiny trees and shrubs resembles. ARALIACEAE.
Acanthophoe'nix f. Spine-areca. Gr. *akantha*, a thorn; *Phoenix*, both the Greek and the modern generic name for the date palm, which *Acanthophoenix* resembles. PALMAE.

Acanthorrhi'za f. Gr. *akantha*, a thorn; *rhiza*, a root. The rootlets of these palms are spiny. PALMAE.

Acanthosta'chys f. Gr. *akantha*, a thorn; *stachys*, a spike. In allusion to the spiny bracts of the flowerhead. BROMELIACEAE.

Acan'thus m. Latin form of Gr. *akanthos*, from *akantha*, thorn, prickle. In Europe it is sometimes called 'bear's breech' from the size and appearance of the leaf which in some species is big, broad, and hairy. The acanthus leaf was a favourite decoration in classical sculpture, as in the capital of the Corinthian column. ACANTHACEAE.

acau'lis, -e Stemless or with only very short stems.

Ac'ca f. Peruvian name. MYRTACEAE.

-aceae A plural adjectival ending used in the names of plant families denoting 'belonging to'. Thus *Rosaceae* means belonging to the rose family. The eight exceptions such as *Compositae* (daisies), *Gramineae* (grasses), *Labiatae*, *Leguminosae* (legumes), *Umbelliferae*, refer to families which were first recognized as natural groups and given descriptive names.

ace'phalus, -a, -um Without a head.

A'cer n. Latin name for the maple tree. The word also means sharp and refers to the hardness of the wood, which the Romans used for spear hafts. ACERACEAE.

a'cer, a'cris, a'cre Sharp; pungent. Used both in relation to taste and touch.

A'ceras n. Gr. *a*, without, lacking; *kĕras*, horn, spur; referring to the spurless flowers. ORCHIDACEAE.

acer'bus, -a, -um Bitter; sour. Also rough to touch.

acerifo'lius, -a, -um With leaves resembling maple.

acerinus, -a, -um Relating to maple.

aceroi'des Resembling maple.

acero'sus, -a, -um Needlelike.

aceto'sa, acetosel'la Pre-Linnaean names for common sorrel and other plants with acid leaves. From L. *acētum*, vinegar.

Achille'a f. Yarrow or sneezewort. The Greek name honours Achilles, heroic warrior of the Trojan wars. As a youth, he was taught the properties of this plant in healing wounds by his tutor Chiron the Centaur, who was half horse and half man. This useful piece of knowledge was highly regarded and was regularly applied in medicine. Achilles himself had the good fortune to be almost entirely invulnerable—his mother, Thetis, having dipped him as a baby into the River Styx. As a result he was vulnerable only in the heel where his mother's forefinger and thumb had held him during the process of immersion. In the end, the god Apollo directed the arrow of Paris to this one spot and it proved fatal.

'Sneezewort' derives from the fact that since ancient times one common species, *Achillea ptarmica*, has been used as a kind of snuff. COMPOSITAE.

achilleifo'lius, -a, -um With leaves resembling *Achillea millefolium* (milfoil).

Achime'nes f. Name obscure. Tropical American herbaceous plants. GESNERIACEAE.

Ach'lys f. Deer-foot. Vanilla leaf. Named for a minor Greek goddess of hidden places—an allusion to the woodland habitat of these petalless herbs. BERBERIDACEAE.

Ach'ras f. Name of Greek derivation for a kind of wild pear which Linnaeus transferred to the sapodilla or naseberry (*Manilkara zapota, Achras zapota*) and the sapote, mammee sapote or marmalade plum (*Pouteria sapota, Calocarpum sapota*). SAPOTACEAE.

acicula'ris, -is, -e Shaped like a needle; needlelike.

acicu'lifer, aciculi'fera, -um Needle-bearing.

Acidanthe'ra f. Gr. *akis*, a point; *anthera*, anther. Now included in *Gladiolus*. IRIDACEAE.

acidis'simus, -a, -um Very sour indeed.

acido'sus, -a, -um Acid; full of sourness.

a'cidus, -a, -um Acid; sour.

acina'ceus, -a, -um Shaped like a curved sword or scimitar.

acinacifo'lius, -a, -um With leaves shaped like a curved sword or scimitar.

acinacifor'mis, -is, -e Like a curved sword or scimitar.

Acine'ta f. Gr. *akinētos*, without movement; in allusion to the immobile lip of the flower of these orchids. ORCHIDACEAE.

Aci'nos f. Gr. *akinos*, name of an aromatic herb mentioned by Pliny. LABIATAE.

Acka'ma f. From the Maori name *maka-maka* for these New Zealand trees. CUNONIACEAE.

Acokanthe'ra f. Gr. *akis*, a sharp point; *anthera*, anther. The anthers are pointed. Certain species of these shrubs and small trees are the source of deadly arrow poisons and drugs used by some African tribes. APOCYNACEAE.

aconitifo'lius, -a, -um With leaves like aconite or monkshood.

Aconi'tum n. The ancient Latin name, from Gr. *akŏnitŏn*, for these poisonous herbs. The English name monkshood derives from the shape of the flower and wolf's bane from the use of the plant to poison wolves; the root was used for this purpose. The leaves and roots of all species contain a strong alkaloid poison. Care must be taken on all occasions when handling the roots, which have sometimes been mistaken for horse-radish with disastrous results. RANUNCULACEAE.

A'corus m. Sweet flag. The Latin name, from Gr. *akŏrŏn*, applied both to the sweet flag (*Acorus calamus*) and the yellow flag (*Iris pseudacorus*). The root stocks yield the sweet-scented calamus root once much used in making cosmetics. ARACEAE.

Acroco'mia f. Gr. *akros*, highest, topmost, terminal; *kome*, a tuft of hair. The leaves of this tropical American feather palm are at the top of the stem. PALMAE.

Acrony'chia f. Gr. *akros*, terminal; *onyx*, a claw. The points of the petals of these trees and shrubs are curved and look like claws. RUTACEAE.

Acrophyl'lum n. Gr. *akros*, terminal; *phyllon*, leaf; in allusion to the terminal clusters of leaves. CUNONIACEAE.

Acro'stichum n. Gr. *akrŏs*, terminal; *stichos*, a row; reason for use obscure. ADIANTACEAE.

Actae'a f. Baneberry, Cohosh. Latin name adopted by Linnaeus from Pliny, apparently from Gr. *aktĕa* or *aktē*, elder. The fruit is poisonous; hence the English name. RANUNCULACEAE.

actinacan'thus, -a, -um With radiating spines.

Actini'dia f. From Gr. *aktis*, a ray, in allusion to the styles of these climbing shrubs which radiate like the spokes of a wheel. The gooseberry-like fruits of the Chinese species *A. deliciosa* are marketed as Kiwi fruit. ACTINIDIACEAE.

Actiniopteris f. Gr. *aktis*, a ray; *pteris*, fern; in allusion to the radiating leaf segments. ADIANTACEAE.

Actino'meris f. From Gr. *aktis*, ray; *mēris*, a part. COMPOSITAE.

Actinostro'bus m. From Gr. *aktis*, a ray; *strōbos*, a cone; with reference to the cone scales. CUPRESSACEAE.

Actino'tus m. Flannel flower. Gr. *aktinōtus*, furnished with rays; referring to the showy spreading bracts which make these plants of *Umbelliferae* look like *Compositae*. UMBELLIFERAE.

acu- In compound words signifying sharply pointed. Thus:
aculeatiss'imus (very prickly); **aculea'tus** (prickly); **aculeola'tus** (with small prickles); **acumina'tus** (tapering into a long narrow point); **acuminatifo'lius** (with leaves tapering quickly into long narrow points); **acutan'gulus** (with sharp angles); **acutifo'lius** (with sharply pointed leaves); **acuti'lobus** (with sharply pointed lobes); **acutipe'talus** (with sharply pointed petals); **acu'tus** (with a sharp but not a tapering point); **acuti'ssimus** (very acutely pointed).

Adanso'nia f. Baobab or monkey-bread tree of Africa. Named in honour of a celebrated French botanist, Michel Adanson (1727–1806), who resided in Senegal, west tropical Africa, from 1748 to 1754 and later published *Familles des Plantes* (1763).

As a tree, the baobab is frequently a remarkable sight. Although seldom as much as forty feet in height it may be thirty feet in diameter at the bottom. With this enormous girth, so disproportionate to the height, it ranks as one of the largest and most grotesque trees in the world. Its appearance is often that of a huge bottle with few branches and sparse leaves stuck into the top. BOMBACACEAE.

aden- Prefix which in compound Greek words indicates the presence of glands, the minute bodies common on several parts of many plants, sometimes making them very sticky indeed.

Adenan'dra f. From Gr. *adēn* gland; *anēr*, *andros*, man, hence male organ, stamen. RUTACEAE.

Adenanthe'ra f. Red sandalwood tree of India. From Gr. *adēn*, gland; *anthera*, anther. The anthers are tipped with a gland. The red, lens-shaped seeds are used in necklaces. LEGUMINOSAE.

Adenium n. Latinized from Arabic name *aden* for *Adenium obesum*. APOCYNACEAE.

Adenocar'pus m. From Gr. *adēn*, gland; *karpos*, fruit. The pods of these shrubs are sticky. LEGUMINOSAE.

adenogy'nus, -a, -um With glandular or sticky ovary.

Adeno'phora f. Gland bellflower. From Gr. *adēn*, gland; *phoreo*, to bear. A sticky nectary surrounds the base of the style in the flower of these perennial herbs. CAMPANULACEAE.

adeno'phorus, -a, -um Gland-bearing, generally in reference to a nectary.

adenophyl'lus, -a, -um Having sticky leaves.

adeno'podus, -a, -um Having sticky pedicels.

Adenosto'ma n. Gr. *adēn*, gland; *stoma*, mouth; in allusion to the glands at the mouth of the calyx of these evergreen shrubs. ROSACEAE.

Adhato'da f. Latinized from the native Tamil and Sinhalese name: *ada*, goat; *thodai*, not touch; with reference to the bitter leaves. This genus is now included in *Justicia*. ACANTHACEAE.

adiantifo'lius, -a, -um Having leaves like *Adiantum* or maidenhair fern.

adianto'ides Resembling *Adiantum* or maidenhair fern.

Adian'tum n. Maidenhair fern. Gr. *adiantos*, dry, unwetted; so named because the leaflets repel water in a remarkable way—if plunged into water the fronds remain dry. ADIANTACEAE.

ad'lamii Species named for Richard Wills Adlam (1853–1903) of Johannesburg who sent South African plants to the Royal Botanic Gardens, Kew.

Adlu'mia f. Climbing fumitory, Allegheny vine. Named for John Adlum (1759–1836), a native of Pennsylvania and a well-known grape breeder. In 1819 he put Catawba, the first great American grape, on the market from his Georgetown D.C. nursery. FUMARIACEAE.

admira'bilis, -is, -e Noteworthy.

adna'tus, -a, -um Adnate; joined together. Usually used in connection with two unlike structures growing together naturally but in a way which appears to be abnormal.

adolpi Of Adolphus, usually in honour of Adolf Engler; see *engleri*.

adonidifo'lius, -a, -um With leaves like *Adonis* or pheasant's-eye.

Ado'nis f. Pheasant's-eye. The Greek name. The flower is supposed to have sprung from the blood of Adonis (Semitic Adon, 'lord', Naaman or Tamnuz) who was gored to death by a wild boar. He was beloved by Aphrodite and by some accounts was unsuccessfully wooed by her. Adonis was regarded by the Greeks as the god of plants. It was believed that he disappeared into the earth in autumn and winter only to reappear in spring and summer. To celebrate his return, the Greeks adopted the Semitic custom of making Adonis gardens, consisting of clay pots of quickly growing seeds. See below under *Anemone*. RANUNCULACEAE.

Ado'xa f. Gr. *a*, without; *dŏxa*, repute, glory; referring to the inconspicuous greenish flowers. ADOXACEAE.

adpres'sus, -a, -um Adpressed or appressed; signifying the manner in which scales are pressed against a cone or certain leaves against a stem.

Adromis'chus m. From Gr. *hadrŏs*, thick, stout, thickset; *mischos*, stalk; in allusion to the short pedicels of these South African succulents. CRASSULACEAE.

adscen'dens Ascending, mounting; generally implies a somewhat gradual rise or upward curve from a nearly prostrate base.

adscitus, -a, -um Approved, adopted.

adsur'gens Rising erect; pushing straight upwards.

adun'cus, -a, -um Hooked. The uncus was the hook used by Roman executioners to drag away the bodies of the executed.

adus'tus, -a, -um Sunburnt; swarthy.

ad'venus, -a, -um Adventive; newly arrived; not native.

Aech'mea f. Gr. *aichme*, a point; in reference to the stiff points of the sepals of these epiphytic herbs. BROMELIACEAE.

aegeus, -a, -um Of the Aegean Sea (Aegaeum Mare).

Ae'gilops f. Gr. *aigilōps*, a kind of bearded grass. GRAMINEAE.

Ae'gle f. Bael fruit of India. Named for one of the naiads, those female divinities of Greek mythology who presided over springs, rivers, and lakes. The bael tree, sacred to the god Siva, is widely cultivated in India for its fruit which, besides being edible, is regarded as a specific against dysentery. RUTACEAE.

Aegopo'dium n. Goutweed, bishopsweed. Gr. *aix*, a goat; *podion*, a little foot. It was once thought to cure gout. UMBELLIFERAE.

aegypti'acus, -a,-um; aegyp'ticus, -a, -um Egyptian.

ae'mulus, -a, -um Rivalling; imitating.
aemydia'nus, -a, -um In honour of Luiz Emydio, Brazilian botanist and plant collector, fl. 1957–1974.
ae'neus, -a, -um Of a bronze colour.
Aeo'nium, n. Latin name for one of this genus of succulents sometimes included in *Sempervivum*. CRASSULACEAE.
aequa'lis, -is, -e Equal.
aequi'lobus, -a, -um With equal lobes.
aequinoctia'lis, -is, -e Belonging to the equinoctial zone; from the equatorial regions.
aequitrilo'bus, -a, -um With three equal lobes.
Aeran'gis f. Gr. *aer*, air; *aggos* (*angos*) or *aggeion* (*angeion*), a vessel. These orchids are epiphytes, which are sometimes called air plants or tree perchers. They are parasites only to the extent that they steal a growth site from their hosts. ORCHIDACEAE.
Aeran'thes m. From Gr. *aer*, air; *anthos*, flower. ORCHIDACEAE.
Aer'ides n. From Gr. *aer*, air. These orchids are epiphytes. ORCHIDACEAE.
aer'ius, -a, -um Airy, aerial, high.
aeruginosus, -a, -um Rust-coloured.
Aer'va f. From the Arabic name of the plant. AMARANTHACEAE.
Aeschynan'thus m. From Gr. *aischune*, shame; *anthos*, flower; referring to the red flowers. GESNERIACEAE.
aesculifo'lius, -a, -um With leaves like horse-chestnut.
Aes'culus f. Horse-chestnut, buckeye. The Latin name for a kind of oak bearing edible acorns but applied by Linnaeus to this genus. HIPPOCASTANACEAE.
aestiva'lis, -is, -e Pertaining to summer.
aesti'vus, -a, -um Flowering, ripening or developing in summer. This depends on locality; thus *Leucojum aestivum* flowers in southern England in spring (March–May) but in central Sweden (where Linnaeus its namer lived) in early summer (June).
Aethione'ma n. Stone cress; candy mustard. Origin uncertain, possibly from Gr. *aëthes*, unusual, or *aitho*, to scorch; *nema* , a thread. CRUCIFERAE.
aethio'picus, -a, -um African, usually South African.
aetnen'sis, -is, -e Referring to Mount Etna, the volcanic mountain of Sicily.
aeto'licus, -a, -um Of Aetolia, large district of Greece in southwest of Sterea Ellas.
a'fer, af'ra, af'rum African, often referring to the North African coast (Algeria, Tunis, etc.).
affi'nis, -is, -e Related or similar to.
Aframo'mum n. From *Africa* and *Amōmum*, a related genus. ZINGIBERACEAE.
africa'nus, -a, -um African.

Agalmy'la f. From Gr. *agalma*, a glory, pleasing gift for the gods; in allusion to the showy flowers, plus probably *hylē*, woodland; in allusion to its habitat. GESNERIACEAE.

Agani'sia f. From Gr. *aganos*, gentle; referring to the neat appearance of these orchids. ORCHIDACEAE.

agan'niphus, -a, -um Much snowed on.

Aganos'ma f. From Gr. *aganŏs*, mild, gentle; *ŏsmē*, fragrance. RUTACEAE.

Agapan'thus m. African lily. From Gr. *agape*, love; *anthos*, a flower. ALLIACEAE.

Agape'tes f. Gr. *agapētos*, beloved, desirable, lovable. ERICACEAE.

Aga'ricus m. Old generic name for field and other mushrooms, many of which are now reclassified in other genera.

Aga'stache f. Gr. *agan*, very much; *stachys*, an ear of wheat; in reference to the many flower spikes of these perennial herbs. LABIATAE.

agas'tus, -a, -um Charming.

Ag'athis f. Dammar or kauri pine. Gr. *agathis*, a ball of thread; from the appearance of the catkin on the female trees. The dammar pine is the source of gum dammar or copal used in the manufacture of varnish. ARAUCARIACEAE.

Agathos'ma f. From Gr. *agathos*, good; *ŏsmē*, fragrance. Some species of these evergreen shrubs are aromatic. RUTACEAE.

Aga've f. Century plant, sisal. Gr. *agauos*, admirable, in allusion to the splendid appearance of the plants in flower. These plants are used for rope and fibre. They are also the source of both tequila and pulque, the distilled and fermented liquors much appreciated in Mexico. AGAVACEAE.

agavoi'des Resembling *Agave*.

ageratifo'lius, -a, -um With leaves like *Ageratum*.

ageratoi'des Like *Ageratum*.

Age'ratum n. Greek name *agēratos*, presumably from the Gr. *a*, not; *geras*, old age; presumably because the flowers retain their clear colour for a long time. COMPOSITAE.

agglutina'tus, -a, -um Stuck together.

aggrega'tus, -a, -um Aggregate; clustered in a dense mass. The raspberry and strawberry are aggregate fruits, being the massed product of several ovaries.

Aglai'a f. Named for one of the three Graces because of the beauty and sweet scent of the flowers. MELIACEAE.

Aglaomor'pha, f. Gr. *aglaos*, bright, hence pleasing; *morphe*, shape, form. POLYPODIACEAE.

Aglaone'ma n. Gr. *aglaos*, bright, clear, manifest; *nēma*, a thread; referring to the stamens, according to Schott, author of the name. ARACEAE.

Ago'nis f. Gr. *agōn*, a gathering or an assembly; in allusion to the number of the seeds. MYRTACEAE.

agra'rius, -a, -um Growing in the fields.

agres'tis, -is, -e Growing in the fields.

Agrifo'lium n. A widely used medieval Latin name for holly; a variant of *Aquifolium*.

agrifo'lius, -a, -um With rough or scabby leaves.

Agrimo'nia f. Agrimony. Apparently a misrendering of *argemōnia*, from the Greek name *Argemone*. ROSACEAE.

Agropy'rum n. Gr. *agros*, field; *pyros*, wheat. GRAMINEAE.

Agrostem'ma n. Gr. *agros*, field; *stĕmma*, crown, garland. CARYOPHYLLACEAE.

Agros'tis f. Gr. *agrōstis*, a kind of grass. GRAMINEAE.

Aichry'son n. A classical Greek name for *Aeonium arboreum*, which is related to these plants from the Canary Islands resembling house leeks. CRASSULACEAE.

Ailan'thus f. Tree-of-heaven. The Latinized version of the native Moluccan name *ailanto*, meaning 'sky tree', for a species of this genus. SIMAROUBACEAE.

aiolosalpinx Specific epithet meaning the trumpet of Aeolus, the god of the winds.

Ai'phanes f. Gr. *aiphnes*, ragged or jagged. The leaves of these palms have jagged tips. PALMAE.

Aipyan'thus m. Gr. *aipys*, high and steep; *anthos*, flower; in allusion to mountain habitat of this plant often put in *Arnebia* or *Macrotomia*. BORAGINACEAE.

Ai'ra f. Hair grass. From the classical Greek name for another plant. GRAMINEAE.

aitchi'sonii In honour of Dr. James Edward Tierney Aitchison (1836–1898), British physician and botanist who served with the delimitation commission in Afghanistan (1884–1885) and made important collections in India and Afghanistan.

aizoid'es Like *Aizoon*, a genus related to the ice plants.

aja'cis In honour of Ajax, the Greek hero who at the siege of Troy committed suicide in a fit of pique because the armour of Achilles was awarded to Odysseus. This specific epithet was applied to *Delphinium ajacis* because the markings on the flower were thought to resemble the Greek letters *AIAI*.

ajanen'sis, -is, -e From Ajan on the coast of Siberia.

Aju'ga Bugle. Origin obscure. LABIATAE.

ajugifo'lius, -a, -um With leaves like an *Ajuga*.

Ake'bia f. Latinized version of the Japanese name *akebi* for these twining shrubs. LARDIZABALACEAE.

alabamen'sis, -is, -e From the state of Alabama, U.S.A.

alacriporta'nus, -a, -um From Porto Alegre, Brazil.

Alan'gium n. From the Malabar (Tamil) name *alangi*. ALANGIACEAE.

alata'vicus, -a, -um From the Ala Tau mountains of Turkistan, Central Asia.

alaternus Latin plant name of obscure application.

ala'tus, -a, -um Winged, e.g. the seed of the maple.

alb-, albi-, albo- In compound words signifying white. Thus: **albes'cens** (whitish); **al'bicans** (off-white, becoming white); **albicau'lis** (white-stemmed); **al'bidus** (whitish); **albiflo'rus** (white-flowered); **al'bifrons** (with white fronds); **albiple'nus** (with double white flowers); **albispi'nus** (with white thorns); **albocin'ctus** (with a white crown or girdle); **albomacula'tus** (spotted with white); **albopic'tus** (painted with white); **albopilo'sus** (with white hairs; shaggy white); **albople'nus** (with double white flowers); **albovariega'tus** (variegated with white).

albanen'sis, -is, -e From St. Albans in Hertfordshire, England.

albanicus, -a, -um From Albania, Balkan Peninsula.

albanus, -a, -um From Albana (now Daghestan), Caucasus.

alber'ti In honour of Albert von Regel (1845–1908), explorer in Turkistan, Central Asia, eldest son of Eduard von Regel.

Albi'zia f. Pink siris, nemu tree. In honour of F. del Albizzi, a Florentine nobleman who in 1749 introduced *A. julibrissin* into cultivation. LEGUMINOSAE.

albrech'tii In honour of M. Albrecht, 19th-century Russian naval surgeon.

al'bus, -a, -um White.

Al'cea f. L. name, from Gr. *alkaia*, a kind of mallow. MALVACEAE.

Alchemil'la f. Lady's mantle. Origin uncertain, perhaps the Latinized version of an Arabic name. ROSACEAE.

alcicor'nis, -is, -e Palmated like the antlers of the European elk (*Alces alces*), called moose in America.

alcockia'nus, -a, -um In honour of Sir Rutherford Alcock (1809–1897), British consular official in China.

aldenhamen'sis, -is, -e Originated at Aldenham, Elstree, environs of London, where the Hon. Vicary Gibbs (1853–1932) had a celebrated garden.

Aldrovan'da f. In honour of Ulisse Aldrovandi (1522–1605), Italian botanist in Bologna. DROSERACEAE.

alep'picus, -a, -um Of Aleppo (Haleb) in Syria.

A'letris f. Gr. *alĕtris*, the female slave who ground the meal. The name refers to the powdered appearance of these herbs. LILIACEAE.

Aleuri'tes f. Gr. *aleuron*, floury. The young growth of some species appears to be dusted with flour. The genus includes the Tung Oil or China Oil tree, the seed of which produces an oil valuable in making varnish and high-quality paint. A more recent use, at least as far as Western medicine is concerned, is the use of Tung Oil in treating leprosy. EUPHORBIACEAE.

aleu'ticus, -a, -um Of the Aleutian Islands off the coast of Alaska.

alexan'drae Of Queen Alexandra (1844–1925), wife of King Edward VII of England.

alexandri'nus, -a, um Of Alexandria in Egypt.

algarven'sis, -is, -e From the province of Algarve, southern Portugal.
algerien'sis, -is, -e Algerian.
algi'dus, -a, -um Cold; originating in high mountains.
alie'nus, -a, -um Foreign, not related.
Alis'ma f. Water plantain. The classical Greek name for this plant. ALISMATACEAE.
Alkan'na f. From Arabic *al-hinna*, henna. BORAGINACEAE.
alkeken'gi Bladder cherry. From Arabic *al kakendi*, derived from Gr. *halikakabos*. Epithet for a species of *Physalis*.
Allaman'da f. In honour of Frederik Allamand (b. about 1735, d. after 1776), Swiss botanist, who collected in Surinam about 1770, and sent seeds of this climbing plant to Linnaeus. APOCYNACEAE.
alleghanien'sis, -is, -e Of the Allegheny mountains, eastern North America.
allia'ceus, -a, -um Like onion or garlic in flavour or appearance.
alliariifo'lius, -a, -um With leaves like *Alliaria*, a weedy plant named by Leonhart Fuchs for its garlic smell and of no gardening interest.
Allio'nia f. A North American herb named in honour of Carlo Allioni (1705–1804), Italian botanist, professor at Turin. NYCTAGINACEAE.
Al'lium n. Onion, chive, garlic and related plants. From the classical Latin name for garlic. The whole group was prized by the ancients as possessing medicinal and aphrodisiac qualities as well as flavour. ALLIACEAE.
Alloplec'tus m. Gr. *allos*, diverse; *pleco*, to plait. The allusion is to the overlapping of the calyx segments. GESNERIACEAE.
alnifo'lius, -a, -um With leaves like the alder.
Al'nus f. Alder. The classical Latin name. BETULACEAE.
Aloca'sia f. From Gr. *a*, without; *Colocasia*, the name of a closely allied genus, from which it was separated. Both of these foliage plants are very closely related to the genus *Caladium*. ARACEAE.
A'loe f. From the Arabic name of these perennial succulents. ALOACEAE.
aloi'des (sometimes **alooi'des**) Aloe-like.
aloifo'lius, -a, -um With leaves like an aloe.
Aloino'psis f. From *Aloe* and *opsis*, likeness. AIZOACEAE.
Alon'soa f. Mask-flower. Andean plants named in honour of Alonzo Zanoni, Secretary of State of Colombia when it was a Spanish colony in the 18th century. SCROPHULARIACEAE.
Alopecu'rus m. Foxtail grass. Gr. *alōpĕkŏurŏs*, grass like a fox's tail. GRAMINEAE.
Aloy'sia f. In honour of Maria Louisa (d. 1819), princess of Parma, wife of King Carlos IV of Spain. VERBENACEAE.
alpes'tris, -is, -e Of the lower mountains, with the implication of coming from below the timber line, although not invariably.
Alphito'nia f. From Gr. *alphiton*, barley meal, in reference to the dry, mealy quality of the fruit pulp or mesocarp of these tall trees. RHAMNACEAE.

alpi'cola A dweller in high mountains.

alpi'genus, -a, -um Originating in the mountains.

Alpi'nia f. Perennial tropical herbs, named in honour of Prospero Alpino (1553–1616), Italian botanist, professor of botany at Padua, who wrote on the plants of Egypt. ZINGIBERACEAE.

alpi'nus, -a, -um Alpine; from high mountains above the timber line.

Alseuos'mia f. Gr. *alsos*, a grove; *euosmos*, fragrant. These New Zealand shrubs have strongly fragrant flowers. ALSEUOSMIACEAE.

Alsi'ne f. Gr. *alsinē*, a plant-name of obscure application. CARYOPHYLLACEAE.

Also'phila f. Gr. *alsos*, a grove; *philos*, loving. Shade-loving tree ferns often included in *Cyathea*. CYATHEACEAE.

Alsto'nia f. Named in honour of Dr. Charles Alston (1685–1760), professor of botany, Edinburgh University, 1716–1760. APOCYNACEAE.

Alstroeme'ria f. Peruvian lily. In honour of Baron Claus Alstroemer (1736–1794), a friend of Linnaeus. AMARYLLIDACEAE (ALSTROEMERIACEAE).

altacleren'sis, -is, -e Raised at Highclere, Hampshire, England.

alta'icus, -a, -um Of the Altai Mountains in Central Asia.

alter'nans Alternating.

Alternanthe'ra f. Joy-weed. L. *alternans*, alternating; *anthera*, anther. Alternate anthers in this genus of coloured-leaved perennials are barren. AMARANTHACEAE.

alternifo'lius, -a, -um Having alternate leaves; the leaves on each side of a stem not opposite to each other.

alter'nus, -a, -um Alternate; not opposite.

Althae'a f. Hollyhock. Gr. *althaia*, a cure, something that heals; in allusion to the use of some species in medicine. MALVACEAE.

althaeoi'des Resembling hollyhock.

alti'cola A dweller in the heights.

al'tifrons Having tall fronds or foliage.

altis'simus, -a, -um Very tall; tallest.

al'tus, -a, -um Tall.

alula'tus, -a, -um Narrow-winged.

aluta'ceus, -a, -um Like leather in colour or texture.

Alys'sum n. Madwort. From Gr. *a*, not or against; *lyssa*, rage or madness. This herb was regarded as a specific against madness and the bites of mad dogs. The well-known *A. saxatile* is placed by some botanists in another genus as *Aurinia saxatilis*. CRUCIFERAE.

Alyxi'a f. Gr. *halusis*, chain; apparently referring to the fruits. APOCYNACEAE.

ama-bilis, -is, -e Lovely.

amagia'nus Of Mount Amagi, Japan.

ama'nus, -a, -um Of the Amanus mountains of southern Turkey.

Amara'cus m. From Gr. *amarakos*, marjoram. Now included in *Origanum*. LABIATAE.

amaranthoi'des Resembling amaranth.

Amaran'thus m. Love-lies-bleeding. From Gr. *amarantos*, unfading. The flowers of some species of these annuals retain their colour for a long time like everlastings. AMARANTHACEAE.

amaranti'color Of the colour of amaranth or purple.

Amarcri'num n. From a contraction of *Amaryllis* and *Crinum*, this being a hybrid of these two. The earlier and more correct name is *Crinodonna*. AMARYLLIDACEAE.

amarel'lus, -a, -um Somewhat bitter.

amaricau'lis, -is, -e Having a bitter-tasting stem.

ama'rus, -a, -um Bitter.

Amaryl'lis f. Showy, bulbous plants, named after a beautiful shepherdess Amaryllis in classical poetry and equally irresistible to the English pastoral poets of the 16th and 17th centuries. AMARYLLIDACEAE.

Amaso'nia f. Tropical American sub-shrubs, named in honour of Thomas Amason, a traveller in America in colonial days. VERBENACEAE.

amazo'nicus, -a, -um, amazo'num From the region of the Amazon River, South America.

ambi'guus, -a, -um Uncertain; doubtful.

amboinen'sis, -is, -e From the island of Amboina (Ambon), Indonesia.

Ambro'sia f. Gr. *ambrŏsia*. In ancient mythology, ambrosia was the substance which, with nectar, provided the food and drink of the gods, making immortal those who partook of them. Now the generic name of some weedy herbs with copious wind-borne pollen causing hay fever. COMPOSITAE.

ambrosi'acus, -a, -um Ambrosial; sweet.

ambrosioi'des Resembling *Ambrosia*.

Amelan'chier f. From *amelancier*, the French Provençal name of *A. ovalis* (*A. vulgaris*). It is also called 'sarvis' or 'servis berry' from its resemblance to the service tree, an ignored English fruit (*Sorbus domestica*). ROSACEAE.

amelloi'des Resembling *Amellus*, a genus of no gardening interest.

amel'lus Species name of an Italian aster.

america'nus, -a, -um From America, North or South.

amesia'nus, -a, -um In honour of two members of the Ames family of Boston. Frederick Lothrop Ames (1835–1893) was a well-known horticulturist and amateur grower of orchids. Oakes Ames (1874–1950) was Supervisor of the Arnold Arboretum and Professor of Botany at Harvard, prominent as an orchidologist and teacher of economic botany. On his death he left his orchid herbarium of 64,000 specimens to Harvard University. *Rhododendron amesiae* commemorates Mrs. Mary S. Ames.

amethys'tinus, -a, -um Violet-coloured.

Amhers'tia f. A handsome flowering tree, native to Burma but very rare in the wild, named after Lady Sarah Amherst (d. 1838), wife of the Earl who took Dr. Clarke Abel (for whom *Abelia* was named) on his mission to Peking (1816) and who later became Governor-General of India. An amateur botanist, she collected plants on her travels. LEGUMINOSAE.

Ami'cia f. In honour of Giovanni Battista Amici (1786–1863), professor of astronomy and microscopy at Florence. LEGUMINOSAE.

amico'rum Of the Friendly or Tonga Islands in the South Pacific.

Am'mi n. Gr. and L. *ammi*, name of an umbelliferous plant. UMBELLIFERAE.

Ammo'bium n. Winged everlasting. Gr. *ammos*, sand; *bio*, to live; in reference to the native habitat. COMPOSITAE.

Ammo'charis f. Gr. *ammos*, sand; *charis*, grace, beauty; an allusion both to the habitat and to the quality of these bulbous plants. AMARYLLIDACEAE.

Ammo'phila f. Marram or beach grass. Gr. *ammos*, sand; *philos*, loving. GRAMINEAE.

ammo'philus, -a, -um Sand-loving.

amoe'nus, -a, -um Pleasant; delightful.

Amo'mum n. Gr. *amōmŏn*, an Indian spice-plant. *A. cardamon* is not true cardamon but provides a cheap and fairly acceptable substitute. ZINGIBERACEAE.

Amor'pha f. False indigo. Gr. *amorphos*, shapeless or deformed; in allusion to the corolla of these shrubs lacking wings and keel. LEGUMINOSAE.

Amorphophal'lus m. Gr. *amorphos*, shapeless or deformed; *phallus*; penis. ARACEAE.

ampeloprasum n. Gr. *ampelos*, vine; *prason*, leek; a wild leek found in vineyards.

Ampelo'psis f. Pepper vine. Gr. *ampelos*, a vine; *opsis*, likeness. It looks like and is closely related to the grape vine. VITACEAE.

amphi'bius, -a, -um Growing both in water and on land; amphibious.

Amphi'come f. Gr. *amphi*, both; *kŏme*, a head of hair. The seeds of these herbaceous plants have a hairy tuft at each end. They are now included in *Incarvillea*. BIGNONIACEAE.

amplexicau'lis, -is, -e Stem-clasping.

amplexifo'lius, -a, -um Leaf-clasping.

amplia'tus, -a, -um Enlarged.

ampliss'imus, -a, -um Very large.

amp'lus, -a, -um Large.

ampulla'ceus, -a, -um Flask-like.

Amso'nia f. Blue-flowered herbaceous perennials named in honour of Dr. Charles Amson, 18th-century Virginian physician. APOCYNACEAE.

amuren'sis, -is, -e; amuricus, -a, -um From the area of the Amur River (Heilong) in eastern Asia.

amygdalifor'mis, -is, -e Shaped like an almond.

amygda'linus, -a, -um Almond-like.

amygdaloi'des Resembling almond.

Amyg'dalus f. Greek name for almond, but now generally classified under *Prunus*. ROSACEAE.

Anacam'pseros f. Gr. *anakampserōs*, a herb which brought back love when touched, possibly a stonecrop. PORTULACACEAE.

Anacamp'tis f. Pyramid orchid. From Gr. *anakampto*, bend back; referring to the spur of the flower. ORCHIDACEAE.

anacan'thus, -a, -um Without thorns.

anacardioi'des Resembling the cashew or *Anacardium*.

Anacar'dium n. Cashew nut. A name first used by 16th-century pharmacists for the heart-shaped fruit of the marking-nut tree of India, *Semecarpus anacardium*, from the Gr. *kardia*, heart, but adopted by Linnaeus as the generic name for the cashew, called *Ajacoo* by Tournefort. ANACARDIACEAE.

Ana'charis f. Gr. *ana*, without; *charis*, beauty. A genus of North American water plants usually included in *Elodea*, whose principal importance is to act as oxygenators in aquaria and to provide shelter for young fish. HYDROCHARITACEAE.

Anacy'clus m. Shortened form of *Ananthocyclus*; from Gr. *an*, without; *anthos*, flower; *kuklos*, ring. COMPOSITAE.

Anagal'lis f. Greek plant name. The genus of low-growing herbs includes scarlet pimpernel or poor man's weather glass, so called because it opens only in fine weather. PRIMULACEAE.

Anagy'ris f. Gr. name for *A. foetida*. LEGUMINOSAE.

anagyro'ides Resembling *Anagyris*. Small ornamental trees.

A'nanas f. Pineapple. From the South American (Tupi) Indian name. Many languages use this or some related form of this word. BROMELIACEAE.

Ana'phalis f. Pearly everlasting. From the classical Greek name for another of the everlastings. COMPOSITAE.

Anasta'tica f. Rose of Jericho, resurrection plant. From Gr. *anastasis*, resurrection, in allusion to the fact that, no matter how dry this Near East plant (*A. hierochuntica*) may have become, the plant recovers its shape on being placed in water. CRUCIFERAE.

anato'licus, -a, -um Of Anatolia, Turkey in Asia Minor.

an'ceps Two-sided, hence applied to flattened two-edged stems; in classical L. *anceps* sometimes meant wavering, doubtful, uncertain, and *Arundinaria anceps*, with round stems, was so named because of doubts as to its country of origin.

Anchu'sa f. Bugloss, alkanet. From Gr. *ankousa*, alkanet. Some species of these plants can be used to make rouge. BORAGINACEAE.

ancistr- In Greek compounds means hooked or barbed, e.g. *Ancistrochilus* (hooked lip).

ancyren'sis, -is, -e Of Ankara, Turkey.

andegaven'sis, -is, -e Of Angers, France.

andi'cola Dweller in the Andes.

andi'nus, -a, -um Belonging to the Andes.

Andi'ra f. Cabbage tree. The Latinized form of the Brazilian name for these ornamental evergreen trees. LEGUMINOSAE.

Andrach'ne f. Gr. *andrachne*, ancient plant-name applied both to purslane and strawberry-tree. EUPHORBIACEAE.

andrewsii In honour of the English botanical artist Henry C. Andrews (fl. 1799–1830), illustrator of *The Botanist's Repository* (1797–1815)

andro'gynus, -a, -um Having male and female flowers separate but on the same inflorescence; hermaphrodite.

Androm'eda f. Bog rosemary. Named by Linnaeus after the mythological maiden who was chained to a rock as an offering to the sea-monster and rescued by Perseus; see Gourlie, *The Prince of Botanists*, 62 (1953), for Linnaeus's comical sketch of the type-species (*Andromeda polifolia*) as found by him in Lapland and the trivial incident which suggested the application of the name *Andrŏmĕda* to it. The name has been used for a confusing variety of plants. Today among the plants wrongly so-called in some nursery catalogues and by gardeners are *Pieris japonica*, *Pieris floribunda*, *Enkianthus*, *Gaultheria*, and *Leucothoe*. ERICACEAE.

Andropo'gon m. Beard grass. Gr. *aner*, *andros*, a man; *pogon*, a beard; in reference to the hairs on the spikelets of some species of these grasses. GRAMINEAE.

Andro'sace f. Rock jasmine. From Gr. *aner*, a man; *sakos*, a shield; a name used by Dioscorides for some other plant. PRIMULACEAE.

androsaemifo'lius, -a, -um With leaves like *Androsaemum*, an old generic name of *Hypericum*.

Androsae'mum m. Gr. *androsaimon*, plant-name used by Dioscorides for plant with blood-like sap; from *aner*, *andros*, man; *haima*, blood.

androsa'ceus, -a, -um Resembling *Androsace*.

Androste'phium n. Gr. *aner*, *andros* man; *stephanos*, a crown. The stamens form a corona or crown within the flower of these American bulbous plants. AMARYLLIDACEAE.

Anemarrhe'na f. From Gr. *a*, without; *nēma*, thread, filament; *arrēn*, male; in allusion to the sessile anthers. LILIACEAE.

Ane'mia f. Gr. *aneimōn*, naked, unclad; from *heima*, clothing. SCHIZAEACEAE.

Anemo'ne f. Often said to be derived from Gr. *anemos*, wind, with which there is no evident connexion, but more likely a corrupted Greek loan word of

Semitic origin referring to the lament for slain Adonis or Naaman, whose scattered blood produced the blood-red *Anemone coronaria* or *Adonis*; see above under *Adonis*. RANUNCULACEAE.

Anemonel'la f. Diminutive of *Anemone*. RANUNCULACEAE.

anemoniflo'rus, -a, -um *Anemone*-flowered.

anemonoi'des Resembling *Anemone*.

Anemono'psis f. From Gr. *Anemone*; *opsis*, likeness; because of this Japanese perennial's resemblance to *Anemone*. RANUNCULACEAE.

Anemopaeg'ma n. From Gr. *anemos*, wind; *paigma*, sport. BIGNONIACEAE.

Anemo'psis f. From Gr. *anemone*; *opsis*, likeness; from a resemblance of the inflorescence to a flower of *Anemone* in this aquatic herb. SAURURACEAE.

anethifo'lius, -a, -um With leaves like dill.

Ane'thum n. Greek name for dill. Best known in pickles, dill seeds are also used in making gin and in pharmacy. UMBELLIFERAE.

anfractuo'sus, -a, -um Bent; twisted.

Ange'lica f. Formerly *Herba angelica*. So named on account of its supposed medicinal qualities reputedly revealed by an angel. This herb was regarded as an aphrodisiac, a specific against plague, an antidote to poisons, and the root, chewed, protected against witches. Deep green in colour, the candied stalks of *A. archangelica* have a delectable and unusual flavour. UMBELLIFERAE.

Angelo'nia f. Latinized version of the South American vernacular name of one of the species of these perennial herbs and sub-shrubs. SCROPHULARIACEAE.

Angiop'teris f. Gr. *aggeion*, a vessel; *pteros*, a wing; from the form of the receptacle in which the spores of these ferns are produced. MARATTIACEAE.

ang'licus, -a, -um Of England (*Anglia*), English.

Ango'phora f. Gum myrtle. From Gr. *aggeion*, a vessel; *phoreo*, to carry; in allusion to the form of the fruit. MYRTACEAE.

Angrae'cum n. Latinized version of the Malayan name *angurek* for epiphytic plants. Now the name of a large genus. ORCHIDACEAE.

anguici'dus, -a, -um Snake-killing.

angui'nus, -a, -um Serpentine.

angula'ris, -is, -e; angula'tus, -a, -um Angular.

angu'lidens Having hooked teeth.

angu'liger, anguli'gera, -um Having hooks.

Angulo'a f. Terrestrial orchids named in honour of Don Francisco de Angulo, Spanish botanist of the latter part of the 18th century. ORCHIDACEAE.

angulo'sus, -a, -um Full of corners; many angled; with marked angles.

angu'ria A specific epithet derived from the Greek for a cucumber.

angusta'tus, -a, -um Narrowed.

angustifo'lius, -a, -um Having narrow leaves.

A'nia f. Gr. *ania*, trouble; a name possibly suggested by difficulty in placing this genus of orchids. ORCHIDACEAE.

Anigozan'thos m. Kangaroo's paw, Australian sword lily. Gr. *anoigo*, to open; *anthos*, a flower. The flowers of these perennials flare open almost to the base. HAEMODORACEAE.

anisa'tus, -a, -um Like anise.

anisodo'rus, -a, -um Anise-scented.

anisophyl'lus, -a, -um With unequal leaves, generally implying one of a pair of opposite leaves being much larger than the other.

Aniso'tome f. Gr. *anisos*, unequal; *tŏmĕ*, cut; in allusion to the manner in which the edges of the leaves are cut. UMBELLIFERAE.

ani'sum The old name as well as the specific epithet for anise (*Pimpinella anisum*), much used for flavouring in various condiments, sweets, medicines, and liqueurs.

annamen'sis, -is, -e Of Annam, Vietnam, south-eastern Asia.

Anno'na f. Latinized version of the American Indian (Taino) vernacular name for the cherimoya, soursop or custard-apple. ANNONACEAE.

annotinus, -a, -um Of previous year.

annula'ris, -is, -e Ring-shaped.

annula'tus, -a, -um Furnished with rings.

an'nuus, -a, -um Annual.

Ano'da f. Sinhalese (Ceylon) name for a species of *Abutilon*. MALVACEAE.

Anoectochilus m. Gr. *anoiktos*, open; *cheilos*, lip. ORCHIDACEAE.

ano'malus, -a, -um Deviating from the normal or usual.

Anomatheca f. Gr. *anomalos*, abnormal; *theca*, container, hence capsule; with reference to the unusually papillose capsule. IRIDACEAE.

anope'talus, -a, -um With erect or ascending petals.

Ano'pteris f. From Gr. *ana*, upward; *pteron*, a wing; in allusion to the appearance of the flap, or indusium, over the sori or spore-producing part of this fern. ADIANTACEAE.

anos'mus, -a, -um Lacking scent.

Anre'dera f. Derivation unknown. BASELLACEAE.

Ansel'lia f. Orchids named in honour of John Ansell (d. 1847), English gardener who accompanied Vogel on an expedition (1841–1842) to the river Niger in west tropical Africa. ORCHIDACEAE.

anseri'nus, -a, -um Relating to geese; abundant on areas closely grazed by domestic geese.

antarc'ticus, -a, -um Of the South Polar region, used in botany for the region south of 45° S., including Tierra del Fuego.

Antenna'ria f. Everlasting. L. *antenna*, literally, the yard of a sailing ship. The pappus of the flower of these silvery-leaved plants is supposed to look like the antennae of a butterfly. COMPOSITAE.

anthelmicus, -a, -um; anthelmius, -a, -um Acting against and expelling intestinal worms.

An'themis f. Chamomile, also spelled camomile. *Anthĕmis* is the Greek name for this highly scented plant which has a long history as a flavouring herb; in medicine it was valued for its bitter and tonic properties, and is well known as the old household remedy of chamomile tea. COMPOSITAE.

anthemoi'des Resembling chamomile.

Anthe'ricum n. St. Bernard's lily. Gr. *antherikon*, name for asphodel. LILIACEAE.

anthocre'ne Flower fountain. Gr. *anthos*, flower; *krēnē*, spring.

Antholy'za f. Gr. *anthos*, flower; *lyssa*, rage. The opened flower of the original species looks like the mouth of an angry animal. IRIDACEAE.

anthopo'gon Flower beard; with reference to the hairs on the tube of the corolla.

Anthoxan'thum n. Sweet vernal grass. From Gr. *anthos*, a flower; *xanthos*, yellow; in allusion to the yellow of the spikelets when ripe. GRAMINEAE.

Anthris'cus m. Salad chervil. From the Greek and Latin name for another but unidentified plant. UMBELLIFERAE.

Anthu'rium f. Flamingo flower. From Gr. *anthos*, a flower; *oura*, a tail; in allusion to the tail-like spadix of these foliage plants. ARACEAE.

-anthus In Gr compounds means -flowered e.g. *dolichanthus* long-flowered, *macranthus* large-flowered.

anthyllidifo'lius, -a, -um With leaves like woundwort or ladies' fingers.

Anthyl'lis f. Kidney vetch, woundwort, ladies' fingers. The ancient Greek name. This herb once had a place in medicine as a vulnerary. LEGUMINOSAE.

anti- Gr. *anti*, over against, opposite, instead, in place of, like, in return for; used in forming compounds with other Greek words.

Anti'aris f. Upas tree of the East Indies. From the Javanese word *anjar*, the poisonous juice of the tree which is used as an arrow poison. There was a legend that it was deadly merely to sleep in the shade of the upas tree and that it poisoned its surroundings. MORACEAE.

Antides'ma n. From Gr. *anti*, against; *desma*, actually a head-band, but used by Burman to mean 'poison'; the name thus being intended to refer to the plant's supposed efficacy against snake-bite. EUPHORBIACEAE.

Anti'gonon n. Coral vine. From Gr. *anti*, in place of; *polygŏnŏn*, knot-weed; indicating its affinity to *Polygonum*. POLYGONACEAE.

antilla'rum Of the Antilles, the two groups of islands (Greater and Lesser) which comprise the West Indies.

antipo'dus, -a, -um Of the Antipodes.

antiquo'rum Of the ancients.

anti'quus, -a, -um Ancient, antique.

antirrhiniflo'rus, -a, -um With flowers resembling snapdragon.

antirrhinoi'des Resembling snapdragon.

Antirrhi'num n. Snapdragon. From Gr. *anti*, like; *rhis*, a nose or snout, in allusion to the appearance of the flower, which looks like the snout of a dragon. The mouth can be made to open very satisfactorily by gently pinching the sides of the corolla. SCROPHULARIACEAE.

Anu'bias f. From Anubis, the jackal-headed ancient Egyptian god. ARACEAE.

Ao'tus m. From Gr. *a*, without; *ous*, *ŏtŏs*, ear; in allusion to the lack of two appendages on the calyx characteristic of the related genus *Pultenaea*. LEGUMINOSAE.

Apari'ne f. Greek name for cleavers (*Galium aparine*).

aperan'thus, -a, -um Boundless; of unlimited growth.

ape'rtus, -a, -um Exposed; bare; open.

apet'alus, -a, -um Without petals.

Aphano'stephus m. Gr. *aphanes*, inconspicuous; *stĕphos*, a garland; from the small flower heads of these herbs. COMPOSITAE.

Apheland'ra f. Gr. *apheles*, simple; *aner*, *andros*, male; the anthers are one-celled. ACANTHACEAE.

Aphyllan'thes f. From Gr. *a*, without; *phyllon*, leaf; *anthos*, flower; the flowers appear on the leafless rush-like stems. LILIACEAE (APHYLLANTHACEAE).

aphyl'lus, -a, -um Without leaves or apparently so.

apia'tus, -a, -um Dotted, spotted.

apicula'tus, -a, -um Terminating abruptly in a short and often sharp point.

a'pifer, api'fera, -um Literally, bee-bearing. Specific epithet of *Ophrys apifera*, the bee orchid; the labellum of the flower so resembles a bee that drones attempt copulation.

apiifo'lius, -a, -um Having leaves like celery (*Apium*).

A'pios f. Potato-bean, groundnut. Gr. *apios*, a pear; from the shape of the tubers of *Apios americana* which were eaten sometimes by North American Indians. LEGUMINOSAE.

A'pium n. Celery. The classical Latin name for parsley and celery. UMBELLIFERAE.

Aplect'rum n. Adam and Eve, puttyroot. Gr. *a*, without; *plēktron*, a spur. The flower of this small North American terrestrial orchid has no spur. ORCHIDACEAE.

Apo'cynum n. Dogbane. The classical Greek name for this or a related plant. From *apo*, asunder; *kyon*, a dog; supposed to have been poisonous to dogs. The root of one species, *A. cannabinum* (Choctaw root or Indian hemp), furnishes an emetic and cathartic. APOCYNACEAE.

apodan'thus, -a, -um With stalkless flowers.

apodec'tus, -a, -um Gr. *apŏdĕktŏs*, welcome, acceptable.

apo'dus, -a, -um Footless, sessile, without a stalk.

Apo'gon m. From Gr. *a*, not; *pōgōn*, a beard; applied to the group of beardless irises.

Aponoge'ton m. Cape pondweed. From the Latin name of the healing springs at Aquae Aponi (now Bagni d'Abano), Italy; plus *geiton*, a neighbour; originally applied to a water-plant (now *Zannichellia*) found here, later transferred to the tropical genus. APONOGETONACEAE.

Aporocac'tus m. Rat-tail cactus. From Gr. *apŏria*, difficulty, perplexity, embarassment; with reference to troublesome cactus classification. CACTACEAE.

appendicula'tus, -a, -um Having appendages, such as a crown, crest, or hairs.

appeni'nus, -a, -um Of the Apennine Mountains in central Italy.

applana'tus, -a, -um Flattened out.

appres'sus, -a, -um Pressed close to; lying flat against, as some leaves against a stem, or the scales of a fir cone.

ap'ricus, -a, -um Sun-loving; open to the sun.

Apte'nia f. From Gr. *aptēn*, unfledged, unwinged; in allusion to the absence of wings on the valves of the capsule of these plants formerly included in *Mesembryanthemum*. AIZOACEAE.

ap'terus, -a, -um Wingless.

a'pulus, -a, -um From Apulia (Puglia), southern Italy.

a'pus Footless, sessile, without a stalk.

aqua'ticus, -a, -um; aquat'ilis, -is, -e Growing in or near water.

Aquifo'lium n. Classical name for holly (*Ilex*).

Aquile'gia f. Columbine. From L. *aquila*, an eagle, from the form of the petals. The English name also alludes to the form of the flower which has the appearance of doves (L. *columba*) drinking—especially the short-spurred varieties of these perennial herbs. RANUNCULACEAE.

aquilegiifo'lius, -a, -um With leaves like columbine.

aquili'nus, -a, -um Eaglelike; aquiline.

ara'bicus, -a, -um; a'rabus, -a, -um Of Arabia.

Arabido'psis f. From *Arabis* and Gr. *opsis*, appearance. CRUCIFERAE.

A'rabis f. Rock-cress. Origin obscure. CRUCIFERAE.

Ara'chis f. Peanut, groundnut, monkeynut, earthnut. Old Greek plant name transferred to this plant. One of the most valuable oil seeds. After flowering, the stalks bend down and push the fruits into the soil where they develop and ripen. LEGUMINOSAE.

Arachnan'the f. Gr. *arachnis*, a spider; *anthos*, a flower; from the shape of the flower. ORCHIDACEAE.

arachnoi'des, arachnoi'deus, -a, -um Covered with entangled hairs giving the appearance of a spider's web.

aragoa'nus, -a, -um In honour of J.E.V. Arago (1790–1855), French botanist who accompanied Admiral de Freycinet in his explorations, 1817–1820.

aragonen'sis, -is, -e From the province of Aragon in north-eastern Spain.

araiophyl'lus, -a, -um With thin or slender leaves.

Ara'lia f. Latinization of the old French-Canadian name *aralie*. The genus includes wild sarsaparilla. ARALIACEAE.

araliifo'lius, -a, -um With leaves like *Aralia*.

araneo'sus, -a, -um Cobwebby.

arauca'nus, -a, -um From the district Arauco, Chile, where the monkey-puzzle tree was discovered.

Arauca'ria f. Monkey-puzzle tree. A genus of conifers; now confined to Australasia and South America, taking both the generic name and the specific epithet, *araucana*, of the type species from the Araucani Indians of central Chile to whose territory it is native. The Norfolk Island species (*A. heterophylla*) was discovered on Captain Cook's second voyage (1772–1775) and introduced into cultivation in 1793. ARAUCARIACEAE.

Arauj'ia f. From the Brazilian vernacular name of these vines. ASCLEPIADACEAE.

arbores'cens; arbor'eus, -a, -um Tending to be woody; growing in treelike form.

arbus'culus, -a,-um Resembling a small tree.

arbutifo'lius, -a, -um Having leaves like *Arbutus*.

Ar'butus f. Strawberry tree. The Latin name. Mayflower (*Epigaea repens*) is often popularly called trailing arbutus. ERICACEAE.

Archange'lica f. Referring to the Archangel Raphael, who, according to medieval legend, revealed its virtues. UMBELLIFERAE.

Archontophoe'nix f. Gr. *archon*, *archontos*, a chieftain; *phoenix*, the date palm. In allusion to the majestic appearance of these palms. PALMAE.

arc'ticus, -a, -um From the polar regions; arctic.

Arc'tium n. Gr. *arction*, name of a plant, perhaps a mullein, also called *arcturus* by Pliny. COMPOSITAE.

Arctosta'phylos f. Bearberry. Gr. *arktos*, a bear; *staphylē*, bunch of grapes. ERICACEAE.

Arcto'tis f. African daisy. Gr. *arktos*, a bear; *ous*, *otos*, an ear. The scales of the pappus are supposed to look like the ears of a bear. COMPOSITAE.

arcua'tus, -a, -um Bent like a bow; arched.

ar'dens Glowing.

arder'nei Epithet of a *Watsonia* named for a businessman and plantsman H.M. Arderne at the Cape of Good Hope who introduced it about 1890.

Ardis'ia f. Spearflower, marlberry. From Gr. *ardis*, a point; in allusion to the pointed anthers of these flowering trees and shrubs. MYRSINACEAE.

Are'ca f. Betelnut. From the vernacular name of the palm used by the natives of Malabar on the south-west coast of India. The sliced nut, wrapped in a paan leaf smeared with lime, is chewed by people throughout tropical Asia. It gives the blood-red colour to their frequent and copious expectorations. It stains the teeth black but is considered to aid digestion. PALMAE.

Arecas'trum n. Queen palm. From *Areca* plus *-astrum* indicating resemblance. PALMAE.

Arege'lia f. Ornamental epiphytic plants named in honour of Eduard Albert von Regel (1815–1892), Russian botanist of German origin, Director of the Imperial Botanical Gardens at St. Petersburg. The correct name is *Neoregelia*. BROMELIACEAE.

Arena'ria f. Sandwort. From L. *arena*, sand; in allusion to many species of these herbs preferring sandy places. CARYOPHYLLACEAE.

arena'rius, -a, -um Relating to sand, hence growing in sandy places.

arendsii In honour of Georg Arends (1863–1952), German nurseryman and plant-breeder at Ronsdorf, West Germany.

Aren'ga f. From the Malayan name, *areng*, for this palm from which palm-sugar or jaggery is made as well as palm wine. PALMAE.

areni'cola f. A dweller on sand.

areno'sus, -a, -um Full of sand, sandy.

areola'tus, -a, -um Marked out in small areas, areolate.

Arethu'sa f. Bog-orchid. Named in honour of Arethusa, a wood nymph, who, pursued by Alpheus, a river god, was changed by Artemis (Diana) into a stream, which, running under the sea, came up in Sicily as a fountain. ORCHIDACEAE.

Arga'nia f. Latinized version of the local Moroccan name for this tree whose kernels provide a substitute for olive oil. SAPOTACEAE.

Argemo'ne f. Prickly or crested poppy. Greek name of a poppy-like plant. Once reputed to cure cataract of the eye. The word *argĕmŏn* is Greek for cataract; hence *argĕmōnē*. PAPAVERACEAE.

argentatus, -a, -um Silvered.

argen'teo- In compound words signifying silver. Thus:
argenteogutta'tus (spotted with silver); **argenteomargina'tus** (with silver edges); **argenteovariega'tus** (silver-variegated).

argen'teus, -a, -um Silvery.

argenti'nus, -a, -um Of Argentina.

argilla'ceus, -a, -um Whitish. Literally, pertaining to white or potter's clay.

argophyl'lus, -a, -um Whitish-leaved.

argutifolius, -a, -um With sharply toothed leaves.

argu'tus, -a, -um Sharply toothed or notched.

argyrae'us, -a, -um Silvery.

Argyran'themum, n. From Gr. *argyrōs* silvery; *anthemon*, flower.

Argyre'ia f. Silver morning-glory. Gr. *argyreios*, silvery. The undersides of the leaves are silvery. CONVOLVULACEAE.

argyro- In compound words signifying silver. Thus:
argyroco'mus (silver-haired); **argyroneu'rus** (with silver veins); **argyrophyl'lus** (silver-leaved).

Argyroder'ma n. From Gr. *argyrŏs*, silver; *derma*, skin. In allusion to the whitish leaves of these South African succulent plants. AIZOACEAE.

a'ria An old name for *Sorbus aria*, whitebeam.

a'ridus, -a, -um, Dry, growing in dry places.

arieti'nus, -a, -um Like a ram's head; like the horns of a ram.

arifo'lius, -a, -um With leaves like *Arum*.

ariifo'lius, -a, -um With leaves like *Sorbus aria*, whitebeam.

Ariocar'pus m. Gr. *aria*, the whitebeam; *karpos*, fruit. The fruits of these cacti resemble those of whitebeam. CACTACEAE.

Arisae'ma n. Jack-in-the-pulpit, Indian turnip, dragon arum. Gr. *aris* or *aron*, arum; *haima*, blood, used in the sense of 'relationship', i.e. akin to arum. ARACEAE.

Ari'sarum n. Greek name, *arisaron*, for *Arisarum vulgare*. Tuberous herb mentioned by Dioscorides. ARACEAE.

arista'tus, -a, -um Bearded; furnished with an awn, e.g. oats.

Aris'tea f. Apparently from Gr. *aristos*, best, used by Solander, its author, in the sense of 'pleasing'. IRIDACEAE.

Aristolo'chia f. Dutchman's pipe, snakeroot. Gr. *aristos*, best; *lochia*, childbirth, the curved form of the flower with base and top together recalling the human foetus in the right position prior to birth; hence sometimes called birthwort; it was supposed to ease parturition. The form of the root suggested it as a remedy for snake bite. In modern pharmacy it is the source of an astringent bitter. ARISTOLOCHIACEAE.

aristo'sus, -a, -um Bearded; awned.

Aristote'lia f. A genus of trees and shrubs named in honour of Aristotle (384–322 B.C.), Greek philosopher and naturalist in whom is said to have reposed all the knowledge of his time, which he much extended. ELAEOCARPACEAE.

arize'lus, -a, -um Gr. *arizēlŏs*, conspicuous, notable.

arizon'icus, -a, -um Of Arizona, U.S.A.

arman'dii In honour of the French missionary Armand David (1826–1900); see *Davidia*.

arma'tus, -a, -um Armed with thorns, spines, and other such features.

armenia'cus, -a, -um Of Armenia, Western Asia.

arme'nus, -a, -um Armenian.

Arme'ria f. Statice, thrift, sea-pink. Latinized from the old French name *armoires* for a cluster-headed dianthus. PLUMBAGINACEAE.

ar'miger, armi'gera, -um Arm-bearing, hence spiny or prickly.

armilla'ris, -is, -e Encircled, as with a bracelet or collar.

armilla'tus, -a, -um Like a bracelet or collar.

Armora'cia f. Horse-radish. The classical Latin name of a related plant. CRUCIFERAE.

Arne'bia f. Prophet flower. Arabian primrose. Latinized version of the Arabic name *sagaret el arneb*. BORAGINACEAE.

Ar'nica f. The classical name for this plant. Derivation obscure. Tincture of arnica obtained from the roots of *A. montana* is used in the treatment of bruises and sprains. COMPOSITAE.

arnoldia'nus, -a, -um Originated at the Arnold Arboretum, near Boston, Massachusetts, thus commemorating its benefactor by bequest, James Arnold (1781–1863), New England merchant.

arnol'dii Of Joseph Arnold, M.D. (1782–1818), naturalist under Sir Stamford Raffles (see *Rafflesia*).

aroma'ticus, -a, -um Fragrant; aromatic.

Aro'nia f. Chokeberry, from *aria*, the Greek name for whitebeam (a species of *Sorbus*), the fruits of which resemble chokeberry. ROSACEAE.

Arpophyl'lum n. Gr. *arpe*, a sickle; *phyllon*, a leaf; in reference to the shape of the leaves of these epiphytic orchids. ORCHIDACEAE.

Arrabidae'a f. In honour of Antonio da Arrabida, bishop in Brazil in first half of 19th century. BIGNONIACEAE.

Arraca'cia f. From the Spanish name of these South American perennial herbs. UMBELLIFERAE.

arrec'tus, -a, -um Upright; erect.

Arrhena'therum Gr. *arren*, male; *anther*, a bristle. The male or staminate flowers of this genus of perennial grasses are furnished with awns or bristles. GRAMINEAE.

Arta'botrys m. Gr. *artao*, to support; *botrys*, a bunch of grapes. The bunched fruits are suspended by a tendril. In the East Indies, notably Java, the leaves of these woody climbers are regarded as a preventive of cholera. ANNONACEAE.

Artemi'sia f. Wormwood; lad's love; old man. Named in honour of Artĕmis, the Greek goddess of chastity. Of considerable medicinal value, plants of the genus yield vermifuges, stimulants, and a vulnerary, as well as the active principle of absinthe. COMPOSITAE.

artemisioi'des Resembling *Artemisia*.

Arthroce'reus m. Gr. *arthron*, joint, plus *Cereus*. CACTACEAE.

Arthropo'dium n. Gr. *arthron*, joint; *podion*, little foot. With reference to the pedicels being jointed at the middle. LILIACEAE (ANTHERICACEAE).

articula'tus, -a, -um Jointed, e.g. some bamboos and grasses.

Artocar'pus m. Breadfruit. From Gr. *artos*, bread; *karpos*, fruit. The breadfruit when baked and warm has some resemblance to fine white bread. Sir Joseph Banks, who became familiar with it on Tahiti when accompanying Captain James Cook on his first voyage round the world, proposed to introduce it into Jamaica and the West Indies generally where food for the slaves was always short. He organized the ultimately successful expedition of Captain Bligh in the *Bounty* to bring plants from Tahiti to Jamaica. MORACEAE.

A'rum n. Lords-and-ladies, cuckoo-pint. From *aron*, the Greek name for these poisonous plants which are closely related to the American jack-in-the-pulpit (*Arisaema*). ARACEAE.

Arun'cus m. Goat's beard. The classical name for these herbs. ROSACEAE.

arundina'ceus, -a, -um Resembling a reed.

Arundina'ria f. Southern cane of the cane brakes of the south-east United States. From L. *arundo*, a reed. GRAMINEAE.

Arun'do f. Genus of useful grasses. From L. *arundo*, a reed. *A. donax* provides pipes for musical instruments, bean supports, shelters etc.

arva'lis, -is, -e; arva'ticus From near Arvas (Arbas) in the Cantabrian Mountains, northern Spain, between Oviedo and León.

arven'sis, -is, -e Growing in or pertaining to cultivated fields.

asarifo'lius, -a, -um With leaves resembling *Asarum* or wild ginger.

Asari'na f. Spanish vernacular name for an *Antirrhinum*. SCROPHULARIACEAE.

Asa'rum n. Wild ginger. Latin and Greek name. ARISTOLOCHIACEAE.

ascalo'nicus, -a, -um From the neighbourhood of Ashkelon, seaport city in Palestine, from which the English name 'shallot' is derived; the perennial forms of the common onion (*A. cepa*) to which the name 'shallot' is now applied have, however, no connexion with Ashkelon.

ascen'dens Rising upwards in a sloping fashion.

ascelpia'deus, -a, -um Resembling milkweed or *Asclepias*.

Ascle'pias f. Milkweed; butterfly-weed. The Greek name, in honour of Asklēpiŏs, god of medicine, who in Latin was called Aesculāpius. Some of the species were used in medicine. Asklepios was generally shown with Hygieia, goddess of wise living, on one side and on the other, Panakeia, goddess of cure-alls. In hygiene and panaceas (wonder drugs), the cult of both is still widely commemorated. ASCLEPIADACEAE.

Asclepiodo'ra f. From *Asklepios* and *dorea*, gift, i.e. the gift of Asklepios. These plants closely resemble *Asclepias*. ASCLEPIADACEAE.

Ascy'rum n. Gr. *askyrŏn*, the name of a kind of *Hypericum*.

asia'ticus, -a, -um Asian.

Asi'mina f. Paw-paw. The Latinized version of what is said to be the French form of the Indian name for this North American genus. ANNONACEAE.

asparagi'nus, -a, -um; asparagoi'des Resembling asparagus.

Aspa'ragus m. Classical name for the plant, which was well known to the ancients. Old English gardeners called it 'Sperage'. Nowadays it is often referred to vulgarly as 'sparrow-grass'. LILIACEAE.

Aspa'sia f. Probably commemorating Aspasia, the beautiful mistress of the Athenian statesman Pericles (d. 429 B.C.), from *aspasies*, delightful. ORCHIDACEAE.

as'per, -a, -um Rough.

aspera'tus, -a, -um Roughened.
aspericaulis, -is, -e Rough-stemmed.
asperifo'lius, -a, -um Rough-leaved.
asper'rimus, -a, -um Very rough.
Asperu'go f. From *asper*, rough. BORAGINACEAE.
Aspe'rula f. Diminutive of *asper*; an allusion to the rough stems of an original species. RUBIACEAE.
Asphodeli'ne f. Jacob's rod. Yellow asphodel. Named from its relationship to *Asphodelus*. LILIACEAE (ASPHODELACEAE).
asphodeloi'des Resembling asphodel (*asphŏdēlos*).
Aspho'delus m. Greek name for the true asphodel, *Asphodelus ramosus* and allied species. LILIACEAE (ASPHODELACEAE).
Aspidis'tra f. Gr. *aspidion*, a small round shield; in reference to the shape of the stigma. Because of its ability to withstand unlimited ill-treatment it has been called the cast-iron plant and, sometimes, the beerplant for the reason that bartenders often water it with beer dregs as the most readily available liquid. The flowers appear at ground level. LILIACEAE (CONVALLARIACEAE).
Aspi'dium n. From Gr. *aspidion*, a shield; referring to the indusium of these ferns. ASPLENIACEAE.
asplenifo'lius, -a, -um With leaves like spleenwort which are fine, feathery, and fernlike.
Asple'nium n. Spleenwort. Gr. *a*, not; *splēn*, the spleen; a reference to this fern's traditional virtues in afflictions of the spleen and liver. ASPLENIACEAE.
assi'milis, -is, -e Similar; like unto.
assur'gens Ascending, rising upwards.
assurgentiflo'rus, -a, -um With flowers in ascending clusters.
A'ster m. Starwort, Michaelmas daisy. L. *aster*, a star; an allusion to the form of the flower. The termination *-aster* in many plant names such as *Cotoneaster* has, however, nothing to do with *aster* 'star' but indicates incomplete resemblance or inferiority. COMPOSITAE.
Asteranthe'ra f. From L. *aster*, star; *anthera*, anther; with reference to the anthers joined in a star-like manner. GESNERIACEAE.
asteroi'des Resembling aster.
Astil'be f. False goatbeard. Gr. *a*, without; *stilbe*, brightness; in allusion to the dullness of the leaves of the type species. SAXIFRAGACEAE.
astilboi'des Resembling *Astilbe*.
astraca'nicus, -a, -um Of the Astrakhan region, near the Caspian Sea.
Astra'galus m. Milk vetch. Greek word for some leguminous plants but applicable to this genus of about 2000 species, some of which yield gum tragacanth and some of which extract selenium from the soil and poison cattle. LEGUMINOSAE.

Astran'tia f. Masterwort. Derivation obscure, a medieval plant-name possibly corrupted from *Magistrantia* and so derived from L. *magister*, master, or from *aster*, star, plus the ending *-antria*. UMBELLIFERAE.

Astro'loba f. Gr. *astron*, star; *lobos*, lobe; in allusion to the stellately spreading lobes of the perigon. ALOACEAE.

Astro'phytum n. Gr. *astron*, a star; *phyton*, a plant; in allusion to the flattened form of these Mexican cacti. CACTACEAE.

astu'ricus, -a, -um; asturien'sis, -is, -e From the province of Asturia in Spain.

Asyneu'ma n. From Gr. *a-*, lacking, not; and *syn-*, together; with reference to the free spreading corolla lobes which in the related genus *Phyteuma* are joined together at their tips during flowering. CAMPANULACEAE.

Atalan'tia f. Woody plants, useful as stocks for citrus fruits. Named in honour of Atalanta, daughter of King Schoeneus of Scyros. The fruit of the plant is golden-coloured. The allusion is to the story that Atalanta consented to accept in marriage any suitor who could outrun her. Hippomenes did so, but only by a ruse. He threw down three of Aphrodite's golden apples in front of her. In stopping to gather them up she lost the race. RUTACEAE.

a'ter, a'tra, a'trum Dead-black.

Athaman'ta f. Candy carrot. After Mount Athamas in Sicily where some species are found. UMBELLIFERAE.

Atherosper'ma n. Gr. *athēr*, *atheros*, barb, spine; *sperma*, seed; in allusion to the long-pointed or aristate fruits. MONIMIACEAE.

Athrota'xis f. Gr. *athroos*, crowded together; *taxis*, arrangement; referring to the leaves of these Tasmanian evergreens. TAXODIACEAE.

Athy'rium n. Lady fern, swamp spleenwort. Derivation unknown, possibly from Gr. *athuros*, spiritless. ASPLENIACEAE.

atlan'ticus, -a, -um Of the shores of the Atlantic Ocean, or, alternatively, from the Atlas Mountains in North Africa.

atra'tus, -a, -um Blackened.

At'riplex f. The Greek name for orach, a species of this genus of herbs and shrubs which can be used like spinach. The genus also includes salt-bush which is much appreciated by Australian sheep. CHENOPODIACEAE.

atriplicifo'lius, -a, -um With leaves like salt-bush.

atro- In Latin compound words signifying dark. Thus:
atropurpu'reus (dark purple); **atro'rubens** (dark red); **atrosangui'neus** (dark blood-red); **atroviola'ceus** (dark violet); **atro'virens** (dark green).

atrocar'pus, -a, -um With black or very dark fruit.

atrocau'lis, -is, -e With black or very dark stems.

At'ropa f. Deadly nightshade, belladonna. Named for Atrŏpŏs, one of the three Fates whose particular business it was to snip the thread of life. This genus of herbs, while very poisonous, has important uses in medicine. The specific epithet *belladonna*, meaning pretty lady, arises from the fact that it was used in cosmetics to make the eyes appear larger and brighter by dilating the pupils. SOLANACEAE.

Atta'lea f. L. *attalus*, splendid or magnificent; in allusion to the beauty of these palms. PALMAE.

attenua'tus, -a, -um Narrowing to a point; attenuated.

at'ticus, -a, -um Of Attica, the classical name for that part of Greece of which Athens is the chief city.

Aubrie'ta f. (not **Aubretia**) Purple rock-cress. Named in honour of Claude Aubriet (1668–1743), French botanical artist. CRUCIFERAE.

aubrietioi'des Resembling *Aubrieta*.

Aucu'ba f. Latinized version of the Japanese name *aukubi* of these ornamental evergreen shrubs. CORNACEAE.

aucupa'rius, -a, -um Specific epithet for the rowan tree, *Sorbus aucuparia*. From L. *aucupor*, to go bird-catching, from *auceps* or *aviceps*, a bird-catcher; *avis*, bird; *capere*, catch.

Audiber'tia f. Herbs and small shrubs named for Urbain Audibert, nurseryman of Tarascon, France, in the 19th century.

augusti'nii Of Dr. Augustine Henry (1857–1930); see *henry*.

augustis'simus, -a, -um Very notable or majestic.

augus'tus, -a, -um Majestic; notable.

aulacosper'mus, -a, -um From Gr. *aulax*, furrow; *sperma*, seed; with furrowed seeds.

auranti'acus, -a, -um; auran'tius, -a, -um Orange-coloured.

aurantifo'lius, -a, -um With leaves like the orange tree (*Citrus aurantium*).

aura'tus, -a, -um Ornamented with gold.

aubertii Commemorating Georges Eleosippe Aubert (b. 1871), French missionary from 1895 to 1901 in western China (Sichuen) whence he introduced *Polygonum aubertii* into cultivation.

aurelianen'sis, -is, -e Of Orleans (formerly Aurelia and Aurelianum), France.

aureo- In compound words signifying golden. Thus:
aureomac'ulatus (golden-spotted); **aureomar'ginatus** (edged with gold); **aureoregi'na** (golden queen); **aureoreticula'tus** (veined with gold); **aureovariega'tus** (gold-variegated).

au'reus, -a, -um Golden.

auri'comus, -a, -um Having golden hair.

auri'cula Pre-Linnaean name for plants now included in the genus *Primula* but still commonly used in England where auriculas are favourites both for show and for Alpine gardens.

auricula'tus, -a, -um Eared; with an ear-shaped appendage or basal lobe.

Auri'nia f. From *aureus*, golden, in allusion to the flowers of *A. saxatilis* (*Alyssum saxatile*). CRUCIFERAE.

auri'tus, -a, -um Having long or large ears.

au'rum Gold (the metal), represented by the symbol *Au*. It was once believed that gold was of botanic origin, the reason being that the metal chiefly occurs embedded in igneous rocks in a crystalline, wiry, branchlike form which gives it the appearance of a plant in the rocks. This pleasant misconception persisted until the 16th century when Agricola, the Swiss metallurgist, refuted it.

australien'sis, -is, -e Australian.

austra'lis, -is, -e Southern.

austri'acus, -a, -um Austrian.

austri'nus, -a, -um Southern.

autumna'lis, -is, -e Pertaining to autumn.

avella'nus, -a, -um Of Avella Vecchia near Naples, southern Italy.

Ave'na f. Oats. The classical name. GRAMINEAE.

avena'ceus, -a, -um Oat-like.

Averrho'a f. Fruit trees, named after a celebrated Moorish physician and philosopher Ibn-Ruschd known as Averroes (1126–1198) who was born in Cordoba, Spain. *A. carambola* is the carambola or star-fruit. OXALIDACEAE.

avicula'ris, -is, -e Relating to small birds.

a'vium Of the birds. Appropriately, the common sweet cherry is *Prunus avium*.

axilla'ris, -is, -e Borne in the axil; axillary.

Aza'lea f. Gr. *azaleos*, dry; in allusion to the occurrence of the original species, *Azalea procumbens* of Linnaeus (now *Loiseleuria procumbens*), in dry places in Lapland. The name was later transferred to the deciduous rhododendrons, for which it is not appropriate. While these azaleas are technically classified as rhododendrons, they are generally kept apart by gardeners and often by nurserymen.

azaleoi'des Resembling *Azalea*.

Aza'ra f. Evergreen shrubs with fragrant flowers, named in honour of J. N. Azara, a Spanish patron of science of the early 19th century. FLACOURTIACEAE.

Azaro'lus f. Latinized form of Italian vernacular *azzerulo*. ROSACEAE.

aze'darach Contracted form of Persian vernacular name *azaddhirakt*, noble tree, for *Melia azedarach*.

Azol'la f. Fairy moss. Genus of floating minute but quickly spreading aquatics. Possibly from Gr. *azo*, to dry; *olluo*, to kill; the plants die when they become dry. AZOLLACEAE.

azo'ricus, -a, -um From the Azores Islands in the eastern Atlantic.

azu'reus, -a, -um Sky-blue; azure.

B

Babia'na f. Latinized version of an Afrikaans word, *babiaan* (baboon). Baboons are said to eat this bulbous plant which is sometimes called baboon-root. IRIDACEAE.

babingto'nii Commemorating Charles Cardale Babington (1808–1895), Professor of Botany at Cambridge.

babylo'nicus, -a, -um Babylonian. Specific name of the weeping willow (*Salix babylonicus*) which was thought by Linnaeus to be a native of south-west Asia. It belongs, in fact, to the Far East. The biblical 'willows' of the waters of Babylon are generally considered to be *Populus euphratica*.

bac'cans Berrying; becoming berry-like.

bacca'tus, -a, -um Berry-like; having fruits with a pulpy texture.

Bac'charis f. Deciduous and evergreen trees, shrubs, and herbs named in honour of Bacchus, the god of wine. Groundsel tree is *B. halimifolia*. COMPOSITAE.

bac'cifer, bacci'fera, -um Berry-bearing.

bacilla'ris, -is, -e Stick- or staff-like.

Backhou'sia f. In honour of James Backhouse (1794–1869), English nurseryman who travelled as a Quaker missionary in Australia. MYRTACEAE.

Baco'pa f. From a South-American Indian name. SCROPHULARIACEAE.

Bac'tris f. From Gr. *bactron*, a walking staff; suggesting the use to which the young stems of these palms are often put. PALMAE.

Bae'ria Herbs, mostly annuals, named in honour of Karl Ernst von Baer (1792–1876), celebrated Estonian anatomist and embryologist of German origin, from 1819 to 1834 professor at Königsberg, then academician in St. Petersburg, best known for his discovery of the mammalian egg. This is often included in *Lasthenia*. COMPOSITAE.

bahia'nus, -a, -um Of Bahia (Salvador), Brazil.

baicalen'sis, -is, -e From Lake Baikal in eastern Siberia.

bai'leyii, baileyanus, -a, -um Of Bailey, honouring one of the following:

(*a*) Frederick Manson Bailey (1827–1915), Australian botanist, author of the *Queensland Flora*.

(*b*) Lieut.-Colonel Frederick Marshman Bailey (1882–1967), Indian Army, explorer of western China, western and south-eastern Tibet, author of *No Passport to Tibet*. His most celebrated botanical contribution was *Meconopsis betonicifolia*, for a time also known as *M. baileyi*.

(*c*) Major Vernon Bailey, U.S. Army, who collected cacti from 1900 onwards.

(*d*) Liberty Hyde Bailey (1858–1954), from 1888 to 1903 Professor of Horticulture at Cornell University, founder of the Bailey Hortorium and author of many authoritative books (*Standard Cyclopedia of Horticulture, Hortus, Manual of Cultivated Plants*, etc.).

Baillo'nia f. South American shrub named in honour of the French botanist

Henri Baillon (1827–1895), author of *Histoire des Plantes* (1866–1895). VERBENACEAE.

ba'keri Usually in honour of John Gilbert Baker (1834–1920), English botanist, for many years a leading authority on petaloid monocotyledons (lilies, etc.), Keeper of the Kew Herbarium 1890–1899. *Tulipa bakeri* commemorates George Percival Baker (1856–1951), English horticulturist and mountaineer.

Ba'laka f. The vernacular name for these feather palms from the Fiji Islands and Samoa. PALMAE.

balan'sae In honour of Benedict Balansa (1825–1891), French botanical collector in Asia Minor, North Africa, Indo-China and Paraguay.

balca'nicus, -a, -um Of the Balkan Peninsula.

Baldel'lia f. In honour of a 19th-century Italian nobleman Bartolomeo Bartolini-Baldelli. ALISMATACEAE.

balden'sis, -is, -e Of Monte Baldo, northern Italy, east of Lake Garda.

baldschua'nicus Of Baljuan, Turkistan, Central Asia.

balea'ricus, -a, -um Of the Balearic Islands off the Mediterranean coast of Spain.

balfouria'nus, -a, -um; balfou'rii In honour of Sir Isaac Bayley Balfour (1853–1922), professor of botany and director (Regius Keeper) of Royal Botanic Garden, Edinburgh.

Ballo'ta f. Greek name for black horehound, one of the species of this genus of sub-shrubs or perennial herbs. LABIATAE.

balsa'meus, -a, -um Similar to balsam; balsamic.

balsami'fer, -a, -um Balsam-bearing.

Balsa'mita f. Costmary, alecost. From L. *balsamum* balsam; presumably in allusion to the aromatic leaves. COMPOSITAE.

bal'ticus, -a, -um Of the area of the Baltic Sea.

Bambu'sa f. Bamboo. Latinized version of the Malayan vernacular name. GRAMINEAE.

bambusoi'des Resembling bamboo.

bana'ticus, -a, -um From the former Austria crownland of Banat, Roumania.

Banksia f. Australian honeysuckle. Named for Sir Joseph Banks (1743–1820), President of the Royal Society and virtual Director of Royal Botanic Gardens, Kew. Banks was a distinguished botanist in his own right as well as a wealthy and generous patron of science. A man of boundless interests, he made great contributions, both scholarly and material, to research. When he died, he left immense natural-history collections for the public good, now incorporated in those of the British Museum and the British Museum (Natural History), London. When Captain James Cook went to the Pacific in 1768 to observe the transit of Venus and investigate the possible existence of a great southern continent in the Pacific Ocean, Banks, as a young man, provided the scientific equipment of the expedition and engaged other trained naturalists and artists to accompany him on the voyage. He returned three years later with a rich harvest of plants, seeds, and general biological collections. He made other expeditions, also at his own expense, to Labrador, Newfoundland, and Iceland. Later (1787) he promoted a scheme for bringing the breadfruit tree from Tahiti to be established in the West Indian islands where food for the slaves had become a very difficult problem. Captain William Bligh (see *Blighia*) was selected to undertake the mission in the *Bounty*. It ended in the mutiny of the crew and the loss of the ship. A later expedition, also commanded by Bligh, successfully transported the breadfruit and thus made a contribution to West Indian food supplies. PROTEACEAE.

banks'iae Of Lady Dorothea Banks (1758–1828), wife of Sir Joseph.

Ba'phia f. Gr. *baphe*, a dye. The tree provides the bar or cam wood of commerce which yields a red dye. Violin bows are made from the wood. LEGUMINOSAE.

Bapti'sia f. Wild indigo of North America. From Gr. *bapto*, to dye. It has sometimes been used as a substitute for true indigo. LEGUMINOSAE.

barbaden'sis, -is, -e Of Barbados, West Indies.

Barba'rea f. Yellow rocket, winter or upland cress. Once generally known as herb of St. Barbara (*Herba Sanctae Barbarae*), patron saint of artillerymen and miners and protectress in thunderstorms. CRUCIFERAE.

bar'barus, -a, -um Foreign.

barba'tulus, -a, -um Somewhat bearded.

barba'tus, -a, -um Bearded; furnished with long weak hairs.

bar'biger, -a, -um Bearing barbs or beards.

barbiner'vis, -is, -e With veins barbed or bearded.

barbino'dis, -is, -e With beards at the nodes or joints.

barbula'tus, -a, -um Somewhat bearded or with a short beard.

barcinonen'sis, -is, -e Of the neighbourhood of Barcelona, Spain.

Barcla'ya f. In honour of Robert Barclay (1757–1830), English botanist and horticulturist. An earlier name is *Hydrostemma*. NYMPHAEACEAE.

Barklya f. A large Australian tree, named in honour of Sir Henry Barkly (1815–1898), a British colonial governor. LEGUMINOSAE.

Barle'ria Named in honour of Jacques Barrelier (1606–1673), Dominican monk, French botanist. ACANTHACEAE.

bar'ometz Specific epithet of *Cibotium barometz*, woolly fern or Scythian lamb. Barometz is from a Tartar word meaning lamb and refers to the woolly rootstocks, these being turned upside down with four leaf-stalks shortened to represent legs. This fern was commonly considered to be half animal and half plant.

Baros'ma f. Gr. *barys*, heavy; *ŏsmē*, odour; in reference to the powerful scent of the leaves of these evergreen shrubs. RUTACEAE.

Barringto'nia f. Evergreen trees and shrubs, named after the Hon. Daines Barrington (1727–1800), English jurist, antiquary and botanist, correspondent of Gilbert White, and close friend of Joseph Banks. LECYTHIDACEAE.

barysta'chys With heavy spikes; from Gr. *barys*, heavy in weight.

Basel'la f. Malabar nightshade. Latinized version of the vernacular name. The plant is edible and is cultivated in India and elsewhere in the tropics as a potherb. BASELLACEAE.

baselloi'des Resembling *Basella*.

basila'ris, -is, -e Pertaining to the base or bottom.

basi'licus, -a, -um Princely, royal. The classical and now the specific epithet of the herb sweet basil in allusion to its reputed healing qualities.

Bassia f. Named in honour of Ferdinando Bassi (1714–1774), Italian botanist, director of Bologna botanic garden. The name *Bassia* published by Allioni in 1766 has priority over *Kochia* of Roth (1800), invalidating *Bassia* Linnaeus (1771) for which *Madhuca* is the correct name.

bata'tas Specific epithet for the sweet potato, *Ipomoea batatas*. The word is the vernacular Carib (Haitian) Indian name for sweet potato. From it derives the English word potato.

Bateman'nia f. In honour of James Bateman (1811–1897), English orchid-grower, whose immense book *Orchidaceae of Mexico* (1837–1841) has some amusing illustrations by George Cruikshank. ORCHIDACEAE.

batesia'nus, -a, -um; batesii West African plants so-named commemorate George Latimer Bates (1863–1940), American ornithologist and botanist (formerly a school-teacher), from 1895 to 1928 in west tropical Africa. South African succulents commemorate John Thomas Bates (1882–1966), British Army sergeant in First World War, then London bus conductor, who became a very successful grower and keen student of succulent plants. His unrivalled collection went to the University of California from Hounslow in 1960.

bathyphyl'lus, -a, -um Thickly leaved.

batrachoi'des Resembling water-buttercup (*Ranunculus* sect. *Batrachium*).

battandi'eri In honour of Jules Aimé Battandier (1848–1922), French botanist, authority on Algerian plants.

Baue'ra f. Small evergreen shrubs named by Sir Joseph Banks after Franz (1758–1840) and Ferdinand (1760–1826) Bauer, Austrian brothers, both botanical artists, never excelled for their beautiful and accurate illustrations of plants. CUNONIACEAE.

bauera'nus, -a, -um Named for Ferdinand Bauer who was botanical artist on Captain Flinder's expedition to Australia.

Bauhi'nia f. Evergreen shrubs named after two illustrious Swiss botanists, the brothers Johann (Jean) and Caspar (Gaspard) Bauhin. Johann (1541–1613) was responsible for, among other works, the great *Historia Plantarum*, published nearly forty years after his death. Caspar (1560–1624) produced a valuable index of plant names and synonyms, *Pinax Theatri botanici* (1620). The two lobes of the leaf exemplify the two brothers. LEGUMINOSAE.

bava'ricus, -a, -um Bavarian.

Baxte'ria In honour of William Baxter (fl. 1820–1830), Scottish gardener, who collected plants in Western Australia between 1823 and 1829. LOMANDRACEAE.

beania'nus, -a, -um; beanii In honour of William Jackson Bean (1863–1947), curator of the Royal Botanic Gardens, Kew, which he served for 46 years, celebrated above all as a dendrologist.

Beaufor'tia f. Australian, heathlike shrubs named after Mary Somerset (*c.* 1630–1714), Duchess of Beaufort, patroness of botany early in the 18th century. MYRTACEAE.

Beaumon'tia f. Nepal trumpet-flower, herald's-trumpet. Named in honour of Lady Diana Beaumont (d. 1831) of Bretton Hall, Yorkshire. APOCYNACEAE.

beauverdia'nus, -a, -um In honour of Gustave Beauverd (1867–1940), Swiss botanist, for many years curator of the Barbey-Boissier herbarium at Geneva, and an accomplished botanical artist.

beesia'nus, -a, -um Commemorating the seed and nursery firm of Bees, founded by Arthur Killin Bulley (1861–1942) of Neston, Cheshire and based on a nickname for Bulley and his sister, the 'busy Bs'. Bulley sent George Forrest to China for the introduction of new plants.

Befaria See *Bejaria*.

Bego'nia f. A great group of cultivated ornamental plants named for Michel Bégon (1638–1710), Governor of French Canada and a patron of botany. BEGONIACEAE.

begonifo'lius, -a, -um Begonia-leaved, i.e. with markedly asymmetric leaves like a begonia.

Beja'ria f. Evergreen shrubs named in honour of José Bejar, 17th-century Spanish botanist. The name was erroneously published by Linnaeus as *Befaria*. ERICACEAE.

Belamcan'da f. Blackberry lily, leopard flower. Latinized version of the East Asiatic vernacular name, Malayalam *balamtandam*, Sanskrit *mālākanda*, for this herbaceous perennial. IRIDACEAE.

bel'gicus, -a, -um Of the Netherlands as a whole or of Belgium.

be'lladonna Italian word meaning beautiful lady. Specific epithet of *Atropa* and *Amaryllis.* Ladies used it to give brilliancy to the eyes—a property of the juice being to dilate the pupil. That vision was affected was probably not considered important.

Belleva'lia f. Small-flowered bulbous plants commemorating Pierre Richer de Belleval (1564–1632), founder in 1593 of the Montpellier botanic garden. LILIACEAE (HYACINTHACEAE).

bellidifo'lius, -a, -um With leaves like *Bellis*.

bellidifor'mis, -is, -e Daisy-like.

bellidioi'des Resembling *Bellium*.

Bel'lis f. Daisy. L. *bellus*, pretty. COMPOSITAE.

Bell'ium n. From *Bellis*, daisy. Only for technical reasons is it not included in *Bellis* which it much resembles. The English name 'Little Mary' (colloquial for stomach) is a pun on the Latin name *Bellium minutum* perpetrated by Clarence Elliott. COMPOSITAE.

belloi'des Resembling *Bellis* or daisy.

bel'lus, -a, -um Beautiful, handsome.

Belope'rone f. Gr. *bĕlŏs*, an arrow; *pĕrŏnē*, buckle, rivet; in allusion to the arrow-shaped connective, the part of the filament of the stamen connecting the anther lobes. ACANTHACEAE.

belophyl'lus With arrow-shaped leaves.

benedic'tus, -a, -um Blessed; well spoken of.

bengalen'sis, -is, -e (sometimes **benghalensis**) Of Bengal, India.

Beninca'sa f. Annual running squash-like herbs with edible fruits named for Count Giuseppe Benincasa (d. 1596), Italian botanist who founded the Botanic Garden in Pisa. CUCURBITACEAE.

benthamia'nus, -a, -um; bentha'mii In honour of the celebrated English botanist George Bentham (1800–1884), author of numerous scholarly and fundamental works including the *Flora Australiensis* and most of the *Genera Plantarum* of Bentham and Hooker.

Benthami'dia f. In honour of George Bentham (see above). CORNACEAE.

ben'zoin From an Arabic vernacular word meaning aromatic gum. See *Lindera*.

Berberido'psis f. From *Berberis*, barberry, and Gr. *opsis*, similar to. FLACOURTIACEAE.

Ber'beris f. Barberry. The Latinized form of the Arabian name for the fruit. This genus of shrubs comprises about 450 species. BERBERIDACEAE.

Berche'mia f. Deciduous shrubs named for M. Berchem, French botanist of the 17th century. RHAMNACEAE.

Berge'nia f. Decorative perennials named for Karl August von Bergen (1704–1760), professor at Frankfurt an der Oder. SAXIFRAGACEAE.

Bergerocac'tus m. Named for Alwyn Berger (1871–1931), German horticulturist and botanist with a special interest in succulents, for some years Superintendent of the Hanbury Garden, La Mortola, Italy. CACTACEAE.

bergianus, -a, -um Referring to the Hortus Bergianus near Stockholm, Sweden, commemorating Peter Jonas Bergius (1730–1790), Swedish botanist and physician.

Berkhe'ya f. In honour of Jan Le Francq van Berkhey (1729–1812), Dutch botanist. COMPOSITAE.

Berlandie'ra f. Named for J. L. Berlandier (d. 1851), Belgian botanist, who explored in Texas and New Mexico. COMPOSITAE.

bermudia'nus, -a, -um Of Bermuda.

berolinen'sis, -is, -e Of Berlin, Germany.

Bertero'a f. Dwarf herbs named for Carlo Giuseppe Bertero (1789–1831), Italian physician who botanized in Guadeloupe and other West Indian islands and later practised in Chile. He was drowned at sea when returning from Tahiti to Chile. CRUCIFERAE.

berthelotii Of Sabin Berthelot (1794–1880), French consul on Teneriffe and collaborator with P. B. Webb on the natural history of the Canary Islands.

Bertholle'tia f. Brazil-nut. Named for Claude-Louis Berthollet (1748–1822), French chemist, who discovered the bleaching qualities of chlorine. The Brazil-nut tree (*B. excelsa*) grows to great size. The globular fruits consist of a hard, woody casing packed with about twenty of the nuts of commerce, each in its own shell. LECYTHIDACEAE.

Bertolo'nia f. Named for Antonio Bertoloni (1775–1869), Italian botanist, professor at Bologna. MELASTOMATACEAE.

bessarab'icus, -a, -um Of Bessarabia, eastern Europe.

Bes'sera f. In honour of Wilibald S. J. Th. von Besser (1784–1842), Austrian-Polish botanist, professor at Kremenets. ALLIACEAE.

Be'ta f. Latin name. Cultivated from remote times, the genus includes the familiar red and the white sugar beet, both varieties of *B. vulgaris*, and Swiss chard (var. *cicla*). About half of the world's sugar supplies derive from beet. CHENOPODIACEAE.

beta'ceus, -a, -um Beetlike.

Beto'nica f. Variant of *Vettonica*, name of a plant which grew in Spain.

betonicifo'lius, -a, -um With leaves like betony, *Stachys officinalis*.

Be'tula f. Birch. The Latin name. The tree has special emotional appeal for the Northern peoples and is surrounded with legend and symbolism. Oil of birch gives to Russian leather its particular fragrance. BETULACEAE.

betulifo'lius, -a, -um Having leaves like birch.

betuli'nus, -a, -um; betuloi'des Resembling birch.

bi- Two.

bi'color Of two colours.

bicor'nis, -is, -e; bicornu'tus, -a, -um Two-horned; having two horns or hornlike spurs.

Bi'dens f. Tickseed, beggar's ticks. L. *bis*, twice; *dens*, a tooth; in reference to the two teeth on what is commonly called the seed but which, correctly, is the cypsela or dry one-seeded fruit. Other names are stick-tight and bur-marigold. COMPOSITAE.

bidenta'tus, -a, -um Having two teeth.

bidwi'llii In honour of John Carne Bidwill (1815–1853), English gardener and cattle-dealer in Australia and New Zealand, who collected many New Zealand plants.

bien'nis, -is, -e Biennial.

bi'fidus, -a, -um Cleft into two parts.

biflo'rus, -a, -um Twin-flowered.

bifo'lius, -a, -um Twin-leaved.

Bifrena'ria f. L. *bis*, twice; *frenum*, a bridle; in allusion to the two caudicles by which the pollen masses are connected with their gland in these epiphytic orchids. ORCHIDACEAE.

bifurca'tus, -a, -um Bifurcate; forked into two generally almost equal stems or branches.

Bigelo'via f. Plumed golden rod. A North American shrubby and herbaceous perennial named in honour of Dr. John M. Bigelow (1787–1879), Boston physician who assisted Engelmann (see *engelmannii*) on the U.S. – Mexican boundary mission. He collected regularly while visiting his patients in the Boston area on horseback. COMPOSITAE.

Bigno'nia f. Trumpet-creeper or cross vine. Named in honour of Abbé Jean Paul Bignon (1662–1743), Librarian to Louis XIV. BIGNONIACEAE.

bignonioi'des Resembling *Bignonia*.

bi'hai Tropical American vernacular name for the wild plantain.

biju'gus, -a, -um Double-yoked; two pairs joined.

Billardie'ra f. Apple berry. Evergreen climbers named in honour of Jacques Julien de La Billardière (1755–1834), French explorer and botanist. PITTOSPORACEAE.

Billber'gia f. Evergreen plants named in honour of Gustaf Johan Billberg (1772–1844), Swedish botanist. BROMELIACEAE.

binervatus, -a, -um; binervis, -e; binervosus, -a, -um With two prominent veins (nerves).

bipinna'tus, -a, -um Two-pinnated. A compound leaf is pinnate when the leaflets are arranged feather fashion. It is bipinnate when the leaflets are replaced by leaflets arranged feather fashion.

birma'nicus, -a, -um Of Burma.

Bischo'fia f. Named in honour of Gottlieb Wilhelm Bischoff (1797–1854), professor of botany at Heidelberg, author of important works on botanical terminology. EUPHORBIACEAE.

Biscutel'la f. L. *bis*, twice; *scutella*, a small flat dish; in allusion to the form of the fruits of these annual or perennial herbs. CRUCIFERAE.

bisec'tus, -a, -um Divided into two equal parts.

biserra'tus, -a, -um Double-toothed, i.e. with the teeth on the leaves being themselves toothed.

Bismar'ckia f. A palm, named in honour of Prince Otto von Bismarck (1815–1898), Prussian statesman, first Chancellor of the German Empire. PALMAE.

Bistor'ta f. Medieval plant name from *bis*, twice; *tortus*, twisted; referring to the twisted root. POLYGONACEAE.

biterna'tus, -a, -um Twice ternate. When leaflets are borne in threes, as in clover, they are described as ternate. A plant is biternate when the three divisions each bear three leaflets.

bithynicus, -a, -um From Bithynia, in ancient geography a region of north-west Asia Minor.

bitumino'sus, -a, -um Tarry; in some way resembling bitumen.

bival'vis, -is, -e Having two valves. The two halves of a pea-pod are the valves.

Bi'xa f. Latinized version of the South American vernacular name for this evergreen tree; the orange seeds are the source of annatto dye. BIXACEAE.

Blacksto'nia f. In honour of John Blackstone (d. 1753), English apothecary and botanist. GENTIANACEAE.

Blandfor'dia f. In honour of George Spencer-Churchill (1766–1840), Marquis of Blandford, then 5th Duke of Marlborough, who had a celebrated garden at Whiteknights, Reading. LILIACEAE (BLANDFORDIACEAE).

blan'dus, -a, -um Mild; not strong or bitter; pleasing, charming.

Blech'num n. From the classical Greek name *blēchnŏn* for a fern, probably not this one. BLECHNACEAE.

ble'o Brazilian vernacular name for *Pereskia bleo*.

blepharoca'lyx With fringed calyx.

blepharophyl'lus, -a, -um With leaves fringed like eyelashes, from Gr. *blĕpharon*, eye-lid; *phyllon*, leaf.

Bletil'la f. Diminutive of *Bletia*, a genus of terrestrial orchids of no horticultural interest which it closely resembles. Named in honour of Luis Blet, a Spanish apothecary who had a botanic garden in Algeciras towards the end of the 18th century. ORCHIDACEAE.

Bli'ghia f. Akee tree. A tropical tree with edible fruit named in honour of William Bligh (1754–1817), master of the *Resolution* on Cook's third voyage, and captain of the *Bounty*. Bligh is notorious for the mutiny against him on the *Bounty* and less well known for his extraordinary feat of navigating 4,000 miles across the Pacific from Tahiti to the island of Timor when cast adrift by the *Bounty* mutineers in an open boat with the loyal remnant of his crew, without the loss of a man. His mission on the *Bounty* was to bring the breadfruit tree from Tahiti to the West Indies, as planned by Sir Joseph Banks (see *Banksia*). Bligh succeeded in a later voyage and the breadfruit is now established in the West Indies. Later he became Governor of New South Wales where, in 1808, he was again involved in a mutiny. SAPINDACEAE.

blitoi'des Resembling *Blitum* (an old name for strawberry-blite), a coarse weed with a red fruit.

Bloome'ria f. Bulbous plants named in honour of Dr. H. G. Bloomer (1821–1874), pioneer Californian botanist. ALLIACEAE.

blossfeldianus, -a, -um In honour of Robert Blossfeld, nurseryman at Potsdam, Germany *c.* 1920–1940.

Blumenba'chia f. Plants covered with stinging hairs. Named in honour of Johann Friederich Blumenbach (1752–1840), professor of medicine at Göttingen, and the virtual founder of modern anthropology. LOASACEAE.

Bly'xa f. Apparently from Gr. *blyzo*, gush forth. HYDROCHARITACEAE.

Bocco'nia f. Plume poppy. Named in honour of Paolo Boccone (1633–1703), Italian monk and physician, who spent much of his life in Sicily. PAPAVERACEAE.

bodini'eri In honour of Emile Marie Bodinieri (1842–1901), French missionary in Kweichow (Guizhou), China, who made big plant collections.

bodnantense, -is Referring to Bodnant Garden, Talycafn, North Wales, largely created by 2nd Lord Aberconway (see *aberconwayi*) who gave it to the National Trust in 1949.

Boehme'ria f. China-grass. Named in honour of George Rudolf Boehmer (1723–1803), professor of botany and anatomy at Wittenberg, Germany. One of the species, *B. nivea*, is the source of ramie, a very fine fibre, taken from the inner bark. Ramie is one of the strongest fibres known but, owing to certain technical drawbacks, it is not widely used commercially. Its principal use is in the manufacture of gas mantles. URTICACEAE.

Boenninghause'nia f. After Clemens M. F. von Boenninghausen (1785–1864), doctor at Münster, Germany, author of a local flora. RUTACEAE.

boeo'ticus, -a, -um Of Boeotia, the district around Thebes, north-west of Athens, Greece.

bohemicus, -a, -um Of Bohemia, western Czechoslovakia, one time kingdom.

boissi'eri In honour of Edmond Boissier (1810–1885), celebrated Swiss botanist, describer of many species from Spain and the Near and Middle East, author of *Flora Orientalis*, etc. The great tragedy of Boissier's life was the death of his young wife Lucille (*née* Butini, 1822–1849) when accompanying him on a botanical expedition in Spain; *Chionodoxa luciliae* and *Omphalodes luciliae* commemorate her.

bolan'deri In honour of Henry Nicholson Bolander (1831–1897), Californian botanist.

Bolbi'tis f. From Gr. *bolbos*, bulb; with reference to the bulbously thickened veinlets of these ferns. ASPLENIACEAE.

bol'dus Latinized form of a Chilean vernacular name.

bolivia'nus, -a, -um Of Bolivia, South America.

Bol'lea f. In honour of Carl Bolle (1821–1909) of Berlin, German dendrologist and ornithologist who collected in the Canary and Cape Verde islands. ORCHIDACEAE.

Bolto'nia f. False chamomile. Herbaceous perennials and annuals, named in honour of James Bolton (d. 1799), a weaver at Halifax, who became a self-taught botanist and botanical artist, an authority on British ferns and fungi. COMPOSITAE.

Boma'rea f. Handsome flowering climbers, named after Jacques Christophe Valmont de Bomare (1731–1807), French patron of science. AMARYLLIDACEAE (ALSTROEMERIACEAE).

Bom'bax n. From Gr. *bombyx*, silk; in allusion to the fluffy, silky hairs which fill the seed capsule. This is the cotton-tree of the East Indies. BOMBACEAE.

bomby'cinus, -a, -um Silky.

bo'na-nox Good-night. Specific epithet for a night-flowering morning glory.

bonarien'sis, -is, -e Of Buenos Aires, Argentina.

bon'duc Arabic name for the hazel nut. Specific epithet of the leguminous *Guilandina bonduc*, with nut-like seeds.

Bongar'dia f. In honour of August Gustav Heinrich Bongard (1786–1839), German botanist who spent most of his life in Russia. BERBERIDACEAE.

bononien'sis, -is, -e Of Bologna (formerly Bononia), Italy.

bonus-henri'cus Literally, good Henry. Specific epithet of a *Chenopodium* called good-King-Henry and sometimes mercury. It is a hardy, rank vegetable which may be eaten like spinach.

boo'thii In honour of T. J. Booth (b. 1829), nephew of Thomas Nuttall, who collected in Assam in 1850.

Bora'go f. The herb borage. Possibly from L. *burra*, a hairy garment; in allusion to the hairy leaves. The leaves have a fragrance like cucumber and are used in such mixtures as claret-cup, giving a delicate flavour. BORAGINACEAE.

Borassus f. From Gr. *bŏrassŏs*, immature spadix of date palm. PALMAE.

Borbo'nia f. Evergreen shrubs named for Gaston de Bourbon (1608–1660), Duke of Orleans and son of Henry IV of France, a patron of botany. LEGUMINOSAE.

borbo'nicus, -a, -um

(*a*) From the island of Réunion in the Indian Ocean, once called Île Bourbon.

(*b*) In honour of the Bourbon kings of France.

borea'lis, -is, -e Northern.

borinque'nus, -a, -um Of Puerto Rico, called Borinquen by the Spanish.

borisii, *borisii-regis* In honour of King Boris (1894-1943) of Bulgaria; son of King Ferdinand; cf. *ferdinandi-coburgi.*

borneen'sis, -is, -e Of Borneo, East Indies.

bornmuel'leri In honour of Joseph Bornmüller (1862–1948), traveller and botanical collector especially in the Balkan Peninsula and Asia Minor, for many years curator of the Haussknecht Herbarium at Weimar.

Boro'nia f. Fragrant-flowered evergreen shrubs named by J. E. Smith for Francesco Borone (d. 1794 at age of 25), an Italian lad from Milan, engaged by J. E. Smith on his European tour in 1787, who accompanied Afzelius as an assistant in Sierra Leone and John Sibthorp in Greece. He was killed by an accidental fall from a window in Athens. Smith wrote: 'I think he had more acuteness in finding out specific differences of plants than anybody I ever knew.' RUTACEAE.

Borzicac'tus m. Named in honour of Antonino Borzi (1832–1921), Italian botanist. CACTACEAE.

Bossiae'a f. Evergreen shrubs, usually with yellow flowers, named in honour of de Boissieu La Martinière, doctor of medicine and botanist, on the voyage of La Pérouse, the noted French navigator, who was wrecked at Vanikoro in the Santa Cruz Islands in 1788. LEGUMINOSAE.

Boswel'lia f. Evergreen trees, source of the biblical resin, frankincense. Named in honour of James Boswell (1740–1795), friend and biographer of Dr. Samuel Johnson. BURSERACEAE.

Botry'chium n. Moonwort. Grape fern. From Gr. *botrys,* a bunch of grapes; a reference to the bunchlike formation of the spore-bearing organs of these deciduous ferns. OPHIOGLOSSACEAE.

botryoi'des Resembling a bunch of grapes.

Bougainvill'ea f. Perhaps the handsomest and most widely planted tropical vine. Named in honour of Louis Antoine de Bougainville (1729–1811), sailor and explorer, who sailed round the world 1767–1769, his name being commemorated in several South Pacific place names. He was a noted mathematician, scientist, lawyer, soldier and author, besides being a Fellow of the Royal Society of London. NYCTAGINACEAE.

bourgeaua'nus, -a, -um; bourgea'ui Named in honour of Eugène Bourgeau (1813–1877), French botanical traveller, who collected in France, Spain, Corsica, North Africa, Asia Minor and North America.

Boussingaul'tia f. Madeira-vine. Named for Jean-Baptiste Joseph Boussingault (1802–1887), noted French agricultural chemist. BASELLACEAE.

Bouvar'dia f. Named for Dr. Charles Bouvard (1572–1658), Superintendent of the Jardin du Roi, Paris. RUBIACEAE.

bowden'ii Epithet of a *Nerine* collected in South Africa and introduced into British gardens in 1902 by Athelstan Bowden-Cornish.

Bowie'a f. In honour of James Bowie (c. 1789–1869), Kew gardener who collected in Brazil and South Africa. LILIACEAE.

Bowke'ria f. Named for James Henry Bowker (d. 1900) and his sister Mrs. Mary Elizabeth Barber (d. 1899), South African botanists. *Aloe barberae* was named after the latter. SCROPHULARIACEAE.

bowlesia'nus, -a, -um In honour of Edward Augustus Bowles (1865–1954), distinguished English amateur horticulturist and botanist, an authority especially on *Crocus* and *Narcissus*.

Boyki'nia f. Perennial herbs named in honour of an American field botanist, Dr. Samuel Boykin (1786–1846) of Milledgeville, Georgia. SAXIFRAGACEAE.

Brabe'jum n. From Gr. *braběiŏn*, a prize. PROTEACEAE.

brachia'tus, -a, -um Branched at right angles; armlike.

brachy- In compound words signifying short. Thus:
brachyan'thus (with short flowers); **brachy'botrys** (short-clustered); **brachycar'pus** (with short fruits); **brachy'cerus** (short-horned); **brachype'talus** (short-petalled); **brachyphyl'lus** (with short leaves); **brachysi'phon** (with short tube).

Brachy'chiton m. Bottle tree. From Gr. *brachys*, short; *chitōn*, a tunic; a reference to the overlapping scales. STERCULIACEAE.

Brachy'come f. Swan River daisy. From Gr. *brachys*, short; *kŏmē*, hair; in allusion to the short bristles of the pappus. COMPOSITAE.

bractea'tus, -a, -um Having bracts. A bract is a leaflike and often a brightly coloured organ generally just below the flower cluster. The red 'flower' of the *Poinsettia* is, in fact, composed of bracts, the real flower being somewhat insignificant and generally yellow. Other showy bracts are the 'flowers' of dogwood and *Bougainvillea*.

bracteo'sus, -a, -um With well-developed or conspicuous bracts.

Bra'hea f. Fan-leaved palms, named in honour of Tycho Brahe (1546–1601), noted Danish astronomer, who was responsible for some of the great scientific advances of his day. PALMAE.

brandegea'nus, -a, -um In honour of Townsend Stith Brandegee (1843–1925), American civil engineer, noted for his studies of plants of California and Mexico.

Brase'nia f. Water-shield. The origin of the botanical name for this aquatic plant is obscure. CABOMBACEAE.

brasilien'sis, -is, -e Brazilian.

Brassa'vola f. A genus of epiphytic orchids named after Antonio Musa Brassavola (1500–1555), Italian botanist and physician, professor at Ferrara. ORCHIDACEAE.

Bras'sia f. A genus of epiphytic orchids named after William Brass (d. 1783), English botanist, who collected in West Africa for Sir Joseph Banks (see *Banksia*) and others in 1782–1783. ORCHIDACEAE.

Bras'sica f. The classical Latin name for cabbage. CRUCIFERAE.

brassicifo'lius, -a, -um Cabbage-leaved.

brau'nii In honour of Alexander Carl Heinrich Braun (1805–1877), successively professor of botany at Karlsruhe, Freiburg im Breisgau, Giessen and Berlin, cryptogamist and morphologist.

Bravo'a f. Mexican genus now included in *Polianthes*.

bre'vis, -is, -e Short. Also:
brevifol'lius, -a, -um (with short leaves); **brevipeduncula'tus, -a, -um** (with a short flower stalk); **brevisca'pus, -a, -um** (with a short scape); **brevisty'lis, -is, -e** (with a short style).

Brevoor'tia f. Floral firecracker. Named for J. C. Brevoort (1811–1887) of New York State University. The single species, *B. ida-maia*, is named for Ida May Burke, daughter of a California stage-coach driver, who brought it to the attention of Alphonso Wood (1810–1881), American botanist. LILIACEAE (ALLIACEAE).

bre'weri; breweria'nus, -a, -um In honour of William Henry Brewer (1828–1910), American botanist,professor of agriculture at Yale, pioneer botanical explorer in California.

Brey'nia f. Named in honour of Jacob Breyne (1637–1697), merchant in Danzig, and his son Johann Philipp Breyne (1680–1764), physician in Danzig, authors of works on rare and little-known plants. EUPHORBIACEAE.

Brickell'ia f. Named in honour of Dr. John Brickell (*c.* 1749–1809), Irish physician who settled in Georgia, U.S.A. COMPOSITAE.

Brimeu'ra f. In honour of Maria de Brimeur, a 16th-century flower-loving Dutch noblewoman. LILIACEAE (HYACINTHACEAE).

britan'nicus, -a, -um Of Great Britain.

Brittonas'trum n. In honour of Nathaniel Lord Britton (1859–1934), for many years director of the New York Botanical Garden and author of important floras of the eastern United States and various West Indian islands. LABIATAE.

Bri'za f. Quaking grass. The Greek name for one of the food grains, possibly rye. GRAMINEAE.

brizifor'mis, -is, -e; brizoi'des Resembling quaking grass or *Briza*.

Brodiae'a f. Named for James Brodie (1744–1824) of Brodie, Elgin, Scottish botanist, primarily interested in algae. ALLIACEAE.

Brome'lia f. Perennial herbs with stiff, pineapple-like leaves, named in honour of Olof Bromel (1629–1705), Swedish botanist. BROMELIACEAE.

Bromhea'dia f. In honour of Sir Edward Thomas ffrench Bromhead (1789–1855), High Steward of Lincoln. ORCHIDACEAE.

bromoi'des Resembling brome grass or *Bromus*.

Bro'mus m. Brome grass. From Gr. *brŏmŏs*, oats. GRAMINEAE.

bronchia'lis, -is, -e Useful in treating bronchitis.

Bro'simum n. From Gr. *brōsimŏs*, edible. The genus includes the West Indian bread-nut which also provides the beautifully marked snake-wood. Another species of the genus is the cow-tree, the *palo de vaca*, of South America, which has milky sap reputed to be as rich and wholesome as cow's milk. MORACEAE.

Broughto'nia f. Named for Arthur Broughton (d. 1796), English botanist and physician from Bristol who emigrated to Jamaica in 1783 and made drawings of its plants. ORCHIDACEAE.

Broussone'tia f. Named for Pierre Marie Auguste Broussonet (1761–1807), professor of botany at Montpellier, France. One species is the paper mulberry, *B. papyrifera*, the bark of which yields a kind of paper. MORACEAE.

Browal'lia f. Named in honour of Johan Browall (1707–1755), Bishop of Abo, a Swedish botanist, who in 1739 defended the sexual system of classification proposed by Linnaeus. SOLANACEAE.

Brow'nea f. Named for Patrick Browne (1720–1790), Irish physician and author of *A Civil and Natural History of Jamaica* (1756). LEGUMINOSAE.

Bruckentha'lia f. Spike heath. Named in honour of Samuel von Bruckenthal (1721–1803) and Michael von Bruckenthal (*fl.* 18th century), Austrian noblemen resident in Transylvania. ERICACEAE.

Brugman'sia f. Angel's trumpet. Named for Sebald Justin Brugmans (1763–1819), professor of natural history at Leiden. This genus of South American tropical shrubs is often included in *Datura*. SOLANACEAE.

bruma'lis, -is, -e Winter-flowering; literally, pertaining to the winter solstice.

Brunel'la f. Alternative spelling of *Prunella*, first used by Brunfels.

Brunfel'sia f. Evergreen, free-flowering shrubs named for Otto Brunfels (1489–1534), Carthusian monk and one of the earliest German botanists, who in 1530 published the first good figures of plants. SOLANACEAE.

brunneifo'lius, -a, -um Brown-leaved.

Brun'nera f. Perennial herbs named for Samuel Brunner (1790–1844), Swiss botanist who collected in Italy, the Crimea and west tropical Africa. BORAGINACEAE.

bru'nneus, -a, -um Brown.

Bruno'nia f. In honour of Robert Brown (1773–1858), Scottish botanist, who circumnavigated Australia on Flinders' voyage in 1801–1803 and later became keeper of botany in the British Museum, London. Among his many important discoveries was the Brownian movement of minute particles. The epithets *brownii*, *brunonianus* and *brunonis* commemorate him and also this Australian herbaceous perennial. BRUNONIACEAE.

Brunsdon'na f. Hybrid between *Amaryllis* (*Belladonna*) and *Brunsvigia*. AMARYLLIDACEAE.

Brunsvi'gia f. In honour of Carl Wilhelm Ferdinand (1713–1780), Duke of Brunswick-Lüneburg, an enlightened German prince who encouraged the arts and sciences in his domains. This is Josephine's lily of South Africa. It is very close to *Amaryllis*. AMARYLLIDACEAE.

brutius, -a, -um Of Brutium (Bruttium), southern Italy, modern Calabria.

Bry'a f. In honour of J. T. de Bry (1564–1617), engraver. LEGUMINOSAE.

bryoi'des Resembling moss.

Bryo'nia f. Bryony. Gr. name *bryōnia* used by Dioscorides. CUCURBITACEAE.

Bryophyl'lum n. From Gr. *bryo*, to sprout; *phyllon*, a leaf; in allusion to the vegetative buds on the edges of the leaves of these succulents which need only to be laid on damp sand to start new plants. Their sturdy constitution and ease of propagation make them popular house plants, with such English names as air plant, life-plant, and floppers. CRASSULACEAE.

buccinator'ius, -a, -um; buccina'tus, -a, -um Shaped like a crooked trumpet or horn.

bucha'ricus, -a, -um Of the vicinity of Bokhara, Turkistan.

Buddle'ja f. Butterfly bush. Ornamental small trees and shrubs. Named in honour of the Reverend Adam Buddle (1660–1715), English botanist and vicar of Farmbridge in Essex. Often spelled *Buddleia* but spelled *Buddleja* by Linnaeus. LOGANIACEAE.

buddleoi'des Resembling *Buddleja*.

buer'geri In honour of Heinrich Bürger (1806–1858), German Jewish pharmacist and naturalist in Dutch employ as a botanical and zoological collector in Japan.

bufo'nius, -a, -um Pertaining to toads; growing in damp places.

bul'bifer, bulbi'fera, -um Bulb-bearing; often used of plants with flowers replaced by bulbils.

bulbifor'mis, -is, -e Shaped like a bulb.

Bulbi'ne f. Gr. *bŏlbinē*, name of a bulbous plant, from Gr. *bŏlbŏs*, a bulb. LILIACEAE (ASPHODELACEAE).

Bulbinel'la f. Diminutive of *Bulbine*. LILIACEAE (ASPHODELACEAE).

Bulboco'dium n. Gr. *bŏlbŏs*, a bulb; *kodion*, wool.

Bulbophyl'lum n. From Gr. *bŏlbŏs*, bulb; *phyllon*, a leaf. The leaves of these epiphytic orchids grow from a pseudobulb. ORCHIDACEAE.

bulbo'sus, -a -um Bulbous; may be applied to any swollen underground stem, not necessarily a bulb.

bulga'ricus, -a, -um Bulgarian.

bulla'tus, -a, -um Bullate; blistered or puckered; usually applying to foliage.

Bumal'da f. After Ovidio Montalbani (1601–1671), of Bologna, Latinized name Bumaldus. Epithet for a *Staphylea*.

Bume'lia f. From Gr. *boumĕlios*, Greek name for the ash tree; reason for application to these small American trees and shrubs is obscure. SAPOTACEAE.

bungea'nus, -a, -um; bun'gei In honour of Alexander von Bunge (1803–1890), Russian botanist, professor of botany at Dorpat (Tartu), Estonian monographer of *Astragalus*.

Bu'nium n. Gr. *bouniŏn*, name of the earth-nut. UMBELLIFERAE.

Buphthal'mum n. From Gr. *bous*, an ox; *ophthalmos*, an eye; an allusion to the appearance of the flower. COMPOSITAE.

Bupleu'rum n. Hare's ear. From the Gr. *boupleuros*, meaning ox-rib, a name for another plant. UMBELLIFERAE.

Burchel'lia f. Evergreen shrubs. In honour of William John Burchell (1781–1863), English traveller in Africa and Brazil. RUBIACEAE.

burea'vii In honour of Edouard Bureau (1830–1918), French botanist, authority on *Bignoniaceae*.

burkwoo'dii In honour of the brothers Albert and Arthur Burkwood, English nurserymen and hybridists at Kingston upon Thames.

burma'nicus, -a, -um Of Burma.

bur'sa-pasto'ris Old generic name, now the specific epithet for the weed shepherd's purse, which is the literal translation.

Bursa'ria f. L. *bursa*, a purse; in allusion to the shape of the reddish fruits. PITTOSPORACEAE.

Burse'ra f. Named after Joachim Burser (1583–1649), German physician and botanist, maker of a celebrated herbarium studied by Linnaeus and now at Uppsala, Sweden. BURSERACEAE.

Bu'tea f. Named for John Stuart, third Earl of Bute (1713–1792), for whom *Stewartia* (*Stuartia*) was also named, British Prime Minister 1762–1763 but a better plantsman than politician, the unofficial director of Kew before Sir Joseph Banks. *B. monosperma*, 'Flame of the Forest', with red flowers is one of the most conspicuous of Indian trees. LEGUMINOSAE.

Bu'tia f. Vernacular name in Brazil for a species of palm (*B. capitata*) with numerous varieties. PALMAE.

Bu'tomus m. Gr. *boutŏmŏs.* from, *bous*, ox; *tŏmŏs*, cutting, from *temnein* to cut; in allusion to the sharp leaf margins which make the plant quite unsuitable for fodder. BUTOMACEAE.

Butyrosper'mum n. From Gr. *boutyron*, butter; *sperma*, seed; in allusion to the fatty seeds. Now treated as a synonym of *Vitellaria*. SAPOTACEAE.

buxbau'mii In honour of Johann Christian Buxbaum (1693–1730), German botanist who collected plants in the Near East.

buxifo'lius, -a, -um Box-leaved.

Bu'xus f. Box. The classical Latin name. BUXACEAE.

byzanti'nus, -a, -um Of Istanbul (Constantinople), classical Byzantium.

C

Cabom'ba f. Fanwort, fish grass. Latinized version of the native Guiana name for these perennial aquatics. CABOMBACEAE.

cacaliifo'lius, -a, -um With leaves like *Cacalia*, a genus of virtually no garden interest.

caca'o Aztec name for the cocoa tree and still the name both for the tree and the chocolate in many countries. *Theobroma cacao* is the botanical name.

cachemi'ricus, -a, -um Of Kashmir.

Cac'tus m. Name applied by the ancient Greeks to some spiny plant in no way connected with these American plants. Strictly *Cactus* was the name of one genus only, but it has been adopted loosely as a vernacular name for the whole family *Cactaceae*, comprising about 85 genera and 2,000 species.

cad'micus, -a, -um Metallic; like tin in appearance.

caerules'cens Becoming blue, bluish. Often incorrectly spelled 'coerulescens'.

caeru'leus, -a, -um Dark blue. Often incorrectly spelled 'coeruleus'.

Caesalpi'nia f. Ornamental trees and shrubs named in honour of Andrea Cesalpini (1524/25–1603), Italian botanist, philosopher, and physician to Pope Clement VIII, author of *De Plantis* (1583) and other works. LEGUMINOSAE.

caesiglaucus, -a, -um Blue and glaucous.

cae'sius, -a, -um Light blue.

caespito'sus, -a, -um Growing in dense clumps; cespitose.

caf'fer, caf'fra, caffrum; caffro'rum Of South Africa; from Arabic *kafir*, unbeliever, pagan.

caini'to West Indian vernacular for the star apple (*Chrysophyllum cainito*).

Caio'phora f. Now known as *Blumenbachia*, *q.v.* Gr. *kaios*, a burn; *phoreo*, to bear; in allusion to the stinging hairs of these usually climbing or trailing plants. LOASACEAE.

cai'ricus, -a, -um From Cairo, Egypt.

ca'jan Adaptation of Malay vernacular name *katjang* for the pigeon pea.

cajanifo'lius, -a, -um With leaves like the pigeon pea.

Caja'nus m. Latinized version of the vernacular name *katjang* for the nutritious pigeon pea which is widely grown throughout the tropics. In India it is called 'dall', and is cooked into a tasty potage to be eaten with curries. LEGUMINOSAE.

cala'ba West Indian vernacular name of *Calophyllum calaba*.

calab'ricus, -a, -um Of Calabria, southern Italy.

Cala'dium n. Latinized form of Malay plant-name *kaladi* used by Rumphius. ARACEAE.

calamifo'lius, -a, -um With reedlike leaves.

Calamin'tha f. Gr. *kalaminthē*, Greek name for savory, from Gr. *kalos*, beautiful, and *minthē*, mint. LABIATAE.

Calamophyl'lum n. From *kalamos*, reed, and *phyllon*, leaf, with reference to the slender terete leaves. AIZOACEAE.

Ca'lamus m. Cane and rattan palms. Gr. *kalamos*, a reed. PALMAE.

Calandri'nia f. Rock purslane. Named in honour of Jean Louis Calandrini (1703–1758), professor of mathematics and philosophy at Geneva. PORTULACACEAE.

Calan'the f. Gr. *kalos*, beautiful; *anthos*, a flower. Chiefly terrestrial orchids. ORCHIDACEAE.

Calathe'a f. Gr. *kalathos*, a basket; in allusion to the flower cluster of these perennial foliage plants, which looks like flowers in a basket. MARANTACEAE.

calathi'nus, -a, -um Basketlike.

calcara'tus, -a, -um Spurred.

calca'reus, -a, -um Pertaining to lime.

Calceola'ria f. Slipper-wort. L. *calceolus*, a slipper; in allusion to the form of the flower. SCROPHULARIACEAE.

calce'olus Diminutive of *calceus*, shoe.

calci'cola Dweller on limy soil.

calda'sii In honour of Francisco José de Caldas (1771–1816), Spanish American botanist executed for his revolutionary activities as a Colombian patriot.

Caldcluvia f. In honour of Alexander Caldcleugh. (fl. 1803–1858), traveller in South America. CUNONIACEAE.

Calen'dula f. Marigold. L. *calendae*, the first day of the month and also the day on which interest must be paid. The allusion is to the long flowering of some of the species. *C. officinalis* is the pot-marigold which was used in English country cooking to flavour thick soups. A single head thrown into the pan imparts a pleasing taste. COMPOSITAE.

calendula'ceus, -a, -um Like *Calendula* in colour.

califor'nicus, -a, -um Of California.

Cal'la f. Presumably from Gr. *kallos*, beauty. *C. palustris*, the water-arum, is the only species. It has nothing to do with the popularly named calla-lily which belongs to the genus *Zantedeschia*. ARACEAE.

Callian'dra f. Gr. *kalli-*, beautiful; *aner*, *andros*, male, hence a stamen; in allusion to the numerous conspicuous stamens. LEGUMINOSAE.

callian'thus, -a, -um Having beautiful flowers.

Callicar'pa f. Gr. *kalli-*, beautiful; *karpos*, fruit. The shrubs are sometimes called beauty-berry. VERBENACEAE.

callicar'pus, -a, -um With beautiful fruit.

Calli'coma f. Gr. *kalli-*, beautiful; *kŏmē*, hair; in allusion to the tufted heads of flowers. CUNONIACEAE.

callimor'phus, -a, -um Beautifully shaped; of beautiful form.

Callir'hoe f. Poppy-mallow. Named in honour of the daughter of a minor Greek deity, Achelous, a river god. MALVACEAE.

Callis'ia f. From Gr. *kallŏs*, beauty. COMMELINACEAE.

callista'chyus, -a, -um With beautiful spikes.

callistegioi'des Resembling *Calystegia*, a genus of *Convolvulaceae* which includes *C. sepium*, the vicious bindweed; from *calyx*, calyx; *stĕgon*, cover; in allusion to the large bracts covering the calyx.

Calliste'mon m. Bottlebrush tree. Gr. *kalli-*, beautiful; *stēmōn*, a stamen; in allusion to the beauty of the flowers of these ornamental shrubs. MYRTACEAE.

Calli'stephus m. China or annual aster. From Gr. *kalli-*, beautiful; *stĕphŏs*, a crown; from the appearance of the flowers. COMPOSITAE.

callis'tus, -a, -um Very beautiful.

Calli'triche f. Starwort. From Gr. *kalli-*, beautiful; *trix*, *trichos*, hair. CALLITRICHACEAE.

Calli'tris f. Cypress pine. From Gr. *kalli-*, beautiful; *treis*, three; in allusion to the beautiful three-fold arrangement of its parts, i.e. the leaves and cone-scales are in threes. CUPRESSACEAE.

callizo'nus, -a, -um Beautifully banded or zoned.

callo'sus, -a, -um Thick-skinned; with callouses.

Callu'na f. Heather or ling. From Gr. *kalluno*, to cleanse to adorn; in allusion to the use of heather for making brooms. ERICACEAE.

Calocar'pum n. See *Achras*.

Caloced'rus f. Gr. *kalos*, beautiful; *cedrus*, cedar. CUPRESSACEAE.

Caloce'phalus f. Gr. *kalos*, beautiful; *cĕphalē*, a head; an allusion to the flower. COMPOSITAE.

caloce'phalus, -a, -um With beautiful head.

Calochor'tus m. Mariposa lily. Gr. *kalos*, beautiful; *chortos*, grass. LILIACEAE.

Caloden'drum n. Cape chestnut. Gr. *kalos*, beautiful; *dĕndrŏn*, a tree. RUTACEAE.

Calome'ria f. From Gr. *kalos*, beautiful, good; *mĕris*, a part; in compliment to Napoleon Bonaparte. See *Humea*. COMPOSITAE.

Calonyc'tion n. Moonflower. From Gr. *kalos*, beautiful; *nyx*, *nyktos*, night; in allusion to the beauty of these night-blooming plants. This group is now included in *Ipomoea*. CONVOLVULACEAE.

Calo'phaca f. From Gr. *kalos*, beautiful; *phake*, a lentil. These are leguminous shrubs not unlike laburnum. LEGUMINACEAE.

Calophyl'lum n. Calaba tree. From Gr. *kalos*, beautiful; *phyllon*, a leaf. The leaves are beautifully veined. GUTTIFERAE.

calophyl'lus, -a, -um With beautiful leaves.

calophy'tum Beautiful plant.

Calopo'gon m. Gr. *kalos*, beautiful; *pōgōn*, a beard; in allusion to the fringed lip of this native North American bog-orchid. ORCHIDACEAE.

calostro'tus, -a, -um Beautifully covered; with a beautiful cover.

Calotham'nus m. Gr. *kalos*, beautiful; *thamnos*, a shrub. MYRTACEAE.

Calo'tropis f. Gr. *kalos*, beautiful; *trŏpis*, ship. ASCLEPIADACEAE.

caloxan'thus, -a, -um Of beautiful yellow.

Calpur'nia f. African trees and shrubs named by the learned Ernst Meyer after T. Julius Calpurnius (1st cent. A.D.), a second-rate Latin poet who imitated Virgil. The genus is closely related to and may be said to imitate the genus *Virgilia*. LEGUMINOSAE.

Cal'tha f. In England, marsh-marigold or king-cup; in America, cowslip. From the Latin name for a plant with a yellow flower, probably *Calendula*. RANUNCULACEAE.

cal'vus, -a, -um Bald; hairless; naked.

calycan'thema Varietal epithet of the *Campanula* known as the cup-and-saucer Canterbury bell. Literally, blossoming in the calyx.

Calycan'thus m. Carolina allspice. Gr. *kalyx*, calyx; *anthos*, a flower. The calyx and the petals are the same colour in the flowers of these North American aromatic deciduous shrubs. CALYCANTHACEAE.

calyci'nus, -a, -um Calyx-like.

Calyco'tome Gr. *kalyx*, calyx; *tŏmē*, a part left after cutting; a tree stump. In these deciduous spiny shrubs the upper part of the calyx drops after the flower opens. LEGUMINOSAE.

Caly'pso f. Bog-orchid named in honour of the nymph Calypso, daughter of Atlas, who entertained Odysseus for seven years but, even with the promise of immortality, could not overcome his longing to return home. In the end, Zeus had to send Hermes to bid her release him. ORCHIDACEAE.

calyptra'tus, -a, -um Bearing a calyptra, any caplike covering of a flower or fruit.

Calyste'gia f. From Gr. *kalyx*, calyx; and *stĕgŏn*, cover; with reference to the large bracts concealing the calyx. CONVOLVULACEAE.

Caly'thrix f. Gr. *kalyx*, calyx; *thrix*, a hair. In these heathlike shrubs, the divisions of the calyx end in long bristling hairs. MYRTACEAE.

camaldul'ensis, -is, -e Of the Camaldoli garden near Naples, Italy.

ca'mara South American vernacular name for a species of *Lantana*.

Camaro'tis f. Gr. *kamarotos*, an arch; a reference to the arched shape of the lip of the flower of these epiphytic orchids. ORCHIDACEAE.

Camas'sia f. Camas or bear-grass. From Nootka Chinook *kamas*, variant of *Quamash*, used by western North American Indians, who ate the bulbs. LILIACEAE.

cambessede'sii Named for Jacques Cambessèdes (1799–1863), a French botanist who published a flora of the Balearic Islands, where his peony occurs.

cam'bricus, -a, -um Of Wales (Cambria).

Camel'lia f. Evergreen flowering trees and shrubs named for Georg Josef Kamel (1661–1706), Jesuit pharmacist, born at Brno, Moravia; he botanized from 1688 onwards in Luzon in the Philippines and wrote an account of the plants there which was published in 1704 by his English correspondent John Ray under his Latinized name, Camĕllus. Commercially the most important species is the tea plant, *Camellia sinensis*. THEACEAE.

camelliflo'rus, -a, -um With flowers like *Camellia*.

Camoen'sia f. Shrubby climbers named in honour of Luis de Camoens (1524–1580), celebrated Portuguese epic poet and author of *The Lusiads*. LEGUMINOSAE.

campaniflo'rus, -a, -um With bell-shaped flowers.

Campa'nula f. Canterbury-bell. Bellflower. Harebell. Diminutive of late L. *campana*, a bell; in allusion to the form of the flower. CAMPANULACEAE.

campanula'rius, -a, -um; campanula'tus, -a, -um Bell-shaped.

campanuloi'des Resembling *Campanula*.

campbel'lii Named in honour of Dr. Archibald Campbell (1805–1874), Superintendent of Darjeeling and Political Agent to Sikkim, who accompanied Joseph Hooker on his celebrated Sikkim journey in 1849.

campes'tris, -is, -e Of the fields or open plains.

camphora'tus, -a, -um Pertaining to or resembling camphor.

Camphoros'ma f. Gr. *kamphora*, camphor; *ŏsmĕ*, an odour. The shrub smells like camphor. CHENOPODIACEAE.

Campsi'dium n. From its likeness to *Campsis*. BIGNONIACEAE.

Cam'psis f. Trumpet creeper. Gr. *kampe*, something bent. The stamens of these deciduous climbing plants are curved. BIGNONIACEAE.

Camptoso'rus m. Gr. *kamptos*, curved; *sorus*, a group of fern spore cases. ASPLENIACEAE.

campylocar'pus, -a, -um With curved or bent fruit.

campylo'gynus, -a, -um With curved or bent ovary.

camtschatcen'sis, -is, -e; camtscha'ticus, -a, -um Of the Kamchatka Peninsula on the Siberian coast.

canaden'sis, -is, -e Canadian, but used by early writers also to cover the north-eastern United States.

canalicula'tus, -a, -um Channelled; grooved.

Canan'ga f. Version of the Malayan vernacular name *Kenanga*, one species yields cananga or macassar oil. ANNONACEAE.

canarien'sis, -is, -e Of the Canary Islands.

Canari'na f. The type species comes from the Canary Islands. CAMPANULACEAE.

Canava'lia f. Jack-bean. Latinized version of the Malabar vernacular name *kanavali* for these climbing herbs or sub-shrubs. LEGUMINOSAE.

can'byi In honour of William Marriott Canby (b. 1831), American botanist.

cancella'tus, -a, -um Cross-barred; latticed.

candelab'rum Like a branched candlestick.

can'dicans Shining or woolly-white.
candidis'simus, -a, -um Very white.
can'didus, -a, -um Shining or pure white.
Candol'lea f. In honour of Augustin Pyramus de Candolle (1778–1841), professor of botany first at Montpellier, then Geneva, Switzerland, founder of a widely adopted system of classification, author of many important botanical works, notably the *Prodromus Systematis naturalis*. STYLIDIACEAE.
Canel'la f. Wild cinnamon. From L. *canna*, a reed, in allusion to the rolled bark. CANELLACEAE.
canes'cens With hairs off-white or ashy-grey colour.
cani'nus, -a, -um Pertaining to dogs; applied metaphorically to an inferior kind, e.g. a scentless as opposed to a scented species.
Canis'trum f. L. *canistrum*, basket; the narrow bract-covered inflorescence with a cluster of flowers emerging at the top suggested a flower basket. BROMELIACEAE.
Can'na f. Gr. *kanna*, a reed. Tropical plants with spectacular flowers, from America and Asia. CANNACEAE.
cannabi'nus, -a, -um Like *Cannabis* or hemp.
Can'nabis f. Hemp. *kannabis*, L. *cannabis*, derived from a language of Central Asia or the Near East and giving rise to both English 'canvas' and 'hemp', the latter resulting from the substitution of *h* for *k* and of *f* or *p* for *b* in Teutonic languages, a process called the Gothonic sound-shift. The female plant, besides producing fibre for ropes etc. and hemp seed for varnishes and bird food, is the source of the drugs hashish and marijuana. CANNABACEAE.
cannifo'lius, -a, -um With leaves like a *Canna*.
cantab'ricus, -a, -um Of Cantabria, a district of northern Spain, or the Cantabrian mountains.
cantabrigien'sis, -is, -e Relating to Cambridge, England; to be distinguished from *cantabricus*, *cantianus*, and *cantuariensis* (qq. v.)
cantia'nus, -a, -um Relating to Kent, England.
Can'tua f. Latinized version of the Peruvian name. POLEMONIACEAE.
cantuarien'sis, -is, -e Relating to Canterbury, Kent, England.
ca'nus, -a, -um Off-white; ash-coloured.
capen'sis, -is, -e Of the Cape of Good Hope, South Africa.
capera'tus, -a, -um Wrinkled.
capilla'ceus, -a, -um; capilla'ris, -is, -e Resembling hair; very slender.
capilla'tus, -a, -um Furnished with fine hairs.
capillifor'mis, -is, -e Like hair.
capilli'pes Slender-footed.
capil'lus-ve'neris Venus' hair. Specific epithet of *Adiantum*.

capita'tus, -a, -um Growing in a dense head—referring to the flowers, the fruit, or the whole plant.

capitella'tus, -a, -um; capitel'lus, -a, -um; capitula'tus, -a, -um Having a small head.

cappado'cicus, -a, -um Of eastern Asia Minor, the ancient Province of Cappadocia.

Cap'paris f. Caper. The ancient Greek name for these evergreen shrubs. The pickled flowerbuds of *C. spinosa* are known as capers. CAPPARACEAE.

capreola'tus, -a, -um Provided with tendrils.

cap'reus, -a, -um Pertaining to goats.

capricor'nis, -is, -e

(*a*) Like a goat's horn.

(*b*) From or below the Tropic of Capricorn in the Southern Hemisphere.

caprifo'lium From L. *caper*, goat; *folium*, leaf.

Cap'sicum n. Peppers, both hot and sweet. From Gr. *kapto*, to bite. The true peppercorn, however, is *Piper nigrum*. SOLANACEAE.

capsula'ris, -is, -e Having capsules.

caracasa'nus, -a, -um Of Caracas, Venezuela.

Caraga'na f. Latinized version of the Mongolian name for a species of the genus. LEGUMINOSAE.

Caralluma f. Latinized from the Indian (Telinga) name *car-allum*. ASCLEPIADACEAE.

Carda'mine f. Bitter cress, cuckoo flower, lady's smock. From the Greek name for a plant of the cress family. CRUCIFERAE.

cardaminifo'lius, -a, -um With leaves like *Cardamine*.

Cardian'dra f. Gr. *kardia*, a heart; *aner*, *andros*, man, hence anther; in allusion to the shape of the anthers. HYDRANGEACEAE.

cardina'lis, -is, -e Scarlet; cardinal red.

cardiobasis With heart-shaped base.

Cardiocri'num n. Gr. *kardia*, heart; *krinon*, lily; referring to the heart-shaped leaves. LILIACEAE.

cardiope'talus, -a, -um With heart-shaped petals.

cardiophyl'lus, -a, -um With heart-shaped leaves.

Cardiosper'mum n. Gr. *kardia*, heart; *sperma*, a seed. The black seed of these climbing herbs has a heart-shaped spot. SAPINDACEAE.

cardua'ceus, -a, -um Resembling a thistle.

Carduncellus m. Diminutive of *Carduus*.

cardun'culus, -a, -um Resembling a small thistle.

carducho'rum Of the Kurds, eastern Turkey and Iraq.

Ca'rduus m. Thistle. The classical Latin name. COMPOSITAE.

Car'ex f. Sedge. The classical Latin name. CYPERACEAE.

caribae'us, -a, -um Of the Caribbean area.

Ca'rica f. Papaya, paw paw. The L. name *carica*, Gr. *karike*, a kind of fig, was transferred to the papaya on account of its fig-like leaves. The melon-like fruit, with its delicate and distinctive flavour, is a natural source of pepsin and is therefore a tenderizer. Tough meat wrapped in the rind overnight is much improved. (See *carica.*) CARICACEAE.

car'ica Specific epithet for fig, *Ficus carica*. The word also refers to Caria, a district in Asia Minor, where figs seem to have been extensively cultivated.

caricifo'lius, -a, -um Sedge-leaved.

carici'nus, -a, -um Resembling sedge or *Carex*.

carina'tus, -a, -um Having a keel.

carinthia'cus, -a, -um Of Carinthia in Austria.

Caris'sa f. Latinized version of the Indian vernacular name for these spiny shrubs with edible fruits. APOCYNACEAE.

carle'sii In honour of William Richard Carles (*c*. 1867–1900) of the British consular service in China, who collected plants in Korea between 1883 and 1885.

Carli'na f. Carline thistle. Medieval plant name derived from Carolus (Charles) and referring to the legend that Charlemagne, having been shown the root of this thistle by an angel, used it successfully as a remedy for the plague which prevailed in his army. COMPOSITAE.

Carludovi'ca f. Palmlike plants named in honour of Charles IV (1748–1819) and his queen, Louise (1751–1819), of Spain. CYCLANTHACEAE.

Carmichae'lia f. This genus, confined to New Zealand and Lord Howe Island, was discovered in 1769 by Joseph Banks and Daniel Solander. It remained unpublished until 1825 when Robert Brown named it in honour of Captain Dugald Carmichael (1722–1827), Scottish army officer and botanist who collected plants at the Cape of Good Hope and in Mauritius, Tristan da Cunha and India, but not New Zealand. LEGUMINOSAE.

carmichae'lii When applied to *Aconitum* commemorates Dr. J. R. Carmichael (d. 1877), medical man and plant collector in China 1863–1877.

carmineus, -a, -um Carmine.

Carnegi'ea f. Giant cactus or saguaro, state flower of Arizona. Named in honour of Andrew Carnegie (1835–1919), steel magnate and philanthropist. It is the only species in the genus. It was named *Cereus giganteus* by Engelmann in 1848 but placed in a genus of its own as *Carnegiea gigantea* by Britton and Rose in 1908. CACTACEAE.

carnerosa'nus, -a, -um Of the Carnerosa Pass in Mexico.

car'neus, -a, -um; carni'color Flesh-coloured; deep pink.

carnio'licus, -a, -um Of Carniola in what was north-western Jugoslavia.

carno'sulus, -a, -um Somewhat fleshy.

carno'sus, -a, -um Fleshy.

caroli-alexandri Named for the reigning Grandduke Carl Alexander of Sachsen-Weimar-Eisenach (1818–1900) with Weimar, Germany, his capital.

carolinia'nus, -a, -um; carolinen'sis, -is, -e; caroli'nus, -a, -um From North Carolina or South Carolina.

Caro'ta f. Variant of Gr. *karŏtŏn*, carrot.

carpa'ticus, -a, -um (sometimes **carpathicus**) Of the Carpathian Mountains between Czechoslovakia and Poland.

Carpente'ria f. An evergreen flowering shrub named in honour of Professor William M. Carpenter (1811–1848), Louisiana physician. HYDRANGACEAE.

carpeta'nus, -a, -um Of Carpētania, in classical times a large region of central Spain inhabited by the Carpetani, whose territory included Sierra de Guadarrama and whose chief city was Toletum (Toledo).

carpinifo'lius, -a, -um With leaves like hornbeam.

Carpi'nus f. Hornbeam. The classical Latin name. The English name is said to derive from the use of the wood for ox-yokes. BETULACEAE.

Carpobro'tus m. Hottentot fig. Gr. *karpos*, fruit; *brōtŏs*, edible. These succulents were formerly included in *Mesembryanthemum*. AIZOACEAE.

Carpo'detus m. Gr. *karpos*, fruit; *detos*, bound; from the fact that the fruit is contracted in the middle. GROSSULARIACEAE.

carpus, -a, -um In Gr. compounds means -fruited, e.g. *melanocarpus*, black-fruited; *xanthocarpus*, yellow-fruited.

Carrie'rea f. Deciduous trees with attractive foliage and flowers; named in honour of Elie Abel Carrière (1816–1896), French botanist and horticulturist, for many years editor of *Revue Horticole*. FLACOURTIACEAE.

Car'thamus m. Distaff thistle, false saffron. From an Arabic word meaning to paint—an allusion to the brilliant colour yielded by the flowers. COMPOSITAE.

carthusiano'rum Of the monks of the Carthusian Monastery of Grande Chartreuse near Grenoble, France.

carthusianus, -a, -um See previous entry.

cartilagi'neus, -a, -um Resembling cartilage.

Ca'rum n. Latin form of Greek name *karŏn*. Caraway, *C. carvi*, from Caria, the name of the district in Asia Minor where it was much grown, is one of the few aromatics growing widely throughout the northern temperate zone where it is much used to flavour cakes and breads. It yields an aromatic oil. UMBELLIFERAE.

Ca'rya f. Hickory. From Gr. *karya*, a walnut tree, the fruit of which was known as *karyon*, a word also applied to other nuts. According to legend, Carya, daughter of the King of Laconia, was changed by Bacchus into a walnut tree. JUGLANDACEAE.

caryophyl'lus, -a, -um From Gr. *karya*, walnut; *phyllŏn*, a leaf; referring to the aromatic smell of walnut leaves, which led to the use of the name for the clove and thence to the clove pink (*Dianthus caryophyllus*).

caryopteridifo'lius, -a, -um With leaves like *Caryopteris*.

Caryop'teris f. Blue spiraea. From Gr. *karyon*, a nut; *pteron*, a wing. The fruit is winged. VERBENACEAE.

Caryo'ta f. Gr. *karyon*, nut. *C. urens* is the toddy palm, widely grown throughout India and the East generally. It produces sap freely—as much as three gallons daily from a mature tree. This palm 'wine' quickly ferments and very soon turns into the highly intoxicating toddy, looking and tasting rather like inferior gasoline. PALMAE.

caryoti'deus, -a, -um *Caryota*-like.

caryotifo'lius, -a, -um With leaves like *Caryota* or fishtail palm.

cashmeria'nus, -a, -um; cashmiriensis, -is, -e Of Kashmir, western Himalaya.

Casimiro'a f. Mexican evergreen trees, named for Casimiro Gomez de Ortega (1740–1818), Spanish physician and botanist, director of the Madrid botanic garden from 1771 to 1801. RUTACEAE.

cas'picus, -a, -um; cas'pius, -a, -um From the shores of the Caspian Sea.

Cassan'dra f. Trojan princess with the gift of prophecy, whose warnings of calamity were disregarded but came true. The custom of naming genera of *Ericaceae* after persons in Greek mythology (e.g. *Andromeda*, *Cassiope*) led to its use for this genus earlier called *Chamaedaphne*. ERICACEAE.

Cas'sia f. Senna. The Greek name for a genus of leguminous plants which provide the senna leaves and pods important in pharmacy. Not to be confused with the cassia of commerce which is *Cinnamomum cassia*, an adulterant for true cinnamon. It is cheaper and looks the same, though coarser in flavour. LEGUMINOSAE.

cas'sia Specific epithet of cinnamon. (See *Cassia*.)

cas'sine North American Indian name for dahoon, *Ilex cassine*.

Cassi'nia f. Named after Count Henri de Cassini (1781–1832), French botanist specializing in the study of *Compositae*. COMPOSITAE.

cassinoi'des Resembling *Ilex cassine*.

Cassi'ope f. In Greek mythology the mother of Andromeda; hence the use of the name for a genus related to the genus *Andromeda*. ERICACEAE.

Casta'nea f. Sweet chestnut. The Latin name for these trees, after the town of Castania in Thessaly famous for them. FAGACEAE.

casta'neus, -a, -um Chestnut-coloured.

Castano'psis f. Gr., resembling chestnut; indicating its relationship to the genus *Castanea*. FAGACEAE.

Castanosper'mum n. Moreton Bay chestnut. From *Castanea*, chestnut; Gr. *sperma*, a seed; in allusion to the form and size of the seed. LEGUMINOSAE.

Castille'ja f. Indian paint brush. Named for an 18th-century botanist, Domingo Castillejo, of Cadiz, Spain. SCROPHULARIACEAE.

Castillo'a f. In honour of Juan Castillo y Lopez (d. 1794), Spanish botanist. One of these species, *C. elastica*, yields a milky juice which can be a commercial source of rubber when the market is high enough. This plant yielded the rubber balls which astonished Columbus. MORACEAE.

cas'tus, -a, -um Spotless; chaste.

Casuari'na f. So named because the long, drooping branches of this Pacific genus are supposed to resemble the feathers of a cassowary (*Casuarinus*). Called beefwood, sheoke, and Australian pine, these trees are features of many tropical beaches where they alone seem to grow well. CASUARINACEAE.

catacos'mus, -a, -um Gr. *katakŏsmus*, adorned.

Catal'pa f. North American Indian name. BIGNONIACEAE.

catalpifo'lius, -a, -um With leaves like catalpa.

Catanan'che f. Cupid's dart. Greek name for these herbs, traditionally the basis of love potions. The Greek name means a strong incentive. COMPOSITAE.

catap'pa f. Malayan name for the Indian almond (*Terminalia catappa*).

cata'ria f. Pertaining to cats.

catarrac'tae; catarrac'tarum Of a waterfall or waterfalls.

Catase'tum n. Gr. *kata*, downward; *seta*, a bristle. A reference to the two horns of the column or structure formed by the junction of stamens and pistils in some species of these epiphytic orchids. ORCHIDACEAE.

catawbien'sis, -is, -e Of the region of the Catawba River in the Blue Ridge Mountains in the south-eastern United States.

ca'techu Vernacular name of various Asiatic plants including betel palm (*Areca catechu*), apparently from Malayan *caccu*.

Catesbae'a f. Lily thorn. In honour of Mark Catesby (1682–1749), of Sudbury, England, author of *A Natural History of Carolina*. RUBIACEAE.

Ca'tha f. Latinized version of the Arabian name of this evergreen shrub with edible leaves, *qat* or *khat*. CELASTRACEAE.

cathar'ticus, -a, -um Purgative; cathartic.

cathaya'nus, -a, -um; cathayen'sis, -is, -e Of China.

Cathcar'tia f. A poppy named in honour of John Fergusson Cathcart (1802–1851), Calcutta judge and amateur botanist. PAPAVERACEAE.

cat'jang Malayan vernacular name for *Vigna cylindrica*.

Cat'tleya f. In honour of William Cattley (d. 1832) of Barnet, England, a wealthy patron of botany and an ardent collector of rare plants, who employed young John Lindley to describe and draw his plants and had this genus named for him in return. ORCHIDACEAE.

caucas'icus, -a, -um Of the Caucasus; Caucasian.

cauda'tus, -a, -um Having a tail; caudate.

caules'cens Having a stem.

cauliala'tus, -a, -um Having a winged stem.

cauliflo'rus, -a, -um Bearing flowers on the stem or trunk.

-caulis In Latin compounds means -stemmed, e.g. *atrocaulis*, red-stemmed; *crassicaulis*, thick-stemmed.

Caulophyl'lum f. Blue cohosh. From Gr. *kaulos*, a stem; *phyllon*, a leaf. The stem of this perennial seems to form a stalk for a single large leaf. BERBERIDACEAE.

caulora'pus, -a, -um With a stem like a turnip, as in kohlrabi.

caus'ticus, -a, -um Burning or caustic to taste.

cauti'cola Dweller on rocks or cliffs.

Cautle'ya f. In honour of Sir Proby Thomas Cautley (1802–1871), English military engineer, geologist and palaeontologist, responsible for large irrigation works in India. ZINGIBERACEAE.

cautleoides Resembling *Cautleya*.

cav'us, -a, -um Hollow.

cazorlensis, -is, -e Of the Sierra de Cazorla, south-east Spain.

Ceano'thus f. New Jersey tea. From the Greek name for a spiny plant, not this one. This genus of shrubs is more appreciated abroad than in its native North America. RHAMNACEAE.

Cecro'pia f. Milky-juiced trees named for C̄ecrŏps, mythical Egyptian founder of Athens. CECROPIACEAE.

Ced'rela f. Diminutive of L. *cedrus*, a cedar; from a similarity in the appearance and fragrance of the wood. The genus includes West Indian cedar, much in demand for making cigar boxes. MELIACEAE.

Cedronel'la f. Diminutive of L. *cedrus*, cedar; probably from the fragrance of *C. triphylla* (balm of Gilead) which gives out a very sweet scent when crushed. LABIATAE.

Ced'rus f. Cedar. The Latin name. The genus includes such true cedars as Atlas cedar, the cedar of Lebanon and deodar. The L. *c̄edrus* and Gr. *kĕdros* may originally have been applied to species of *Juniperus*. The trees popularly called 'cedar' in the United States are generally *Juniperus* and *Thuja*. The name *citrus* is a variant of *cedrus*. PINACEAE.

Cei'ba f. Latinized version of the South American name for the silk cotton tree which is the source of kapok, used for stuffing lifebelts and upholstery. BOMBACACEAE.

celastri'nus, -a, -um Resembling *Celastrus*.

Celas'trus f. Bittersweet. From the Greek name *k̄elastrŏs* for an evergreen tree. CELASTRACEAE.

cele'bicus, -a, -um Of the island of Celebes (Sulawesi), Indonesia.

Celmi'sia f. New Zealand daisy. Named after Celmisios, son of the Greek nymph Alciope, for whom a related genus was named. COMPOSITAE.

Celo'sia f. Cock's comb. Gr. *k̄elĕŏs*, burning; in allusion to the brilliant colour of some of the flowers. AMARANTHACEAE.

Cel'sia f. Named in honour of Olof Celsius (1670–1756), professor of theology at Uppsala and a supporter of Linnaeus in his impecunious student days. Now included in *Verbascum*. SCROPHULARIACEAE.

Cel'tis f. Hackberry. Gr. name for another tree. ULMACEAE.

cem'bra Italian name of the Swiss stone pine, *Pinus cembra*.

ceni'sius, -a, -um Of the area of Mont Cenis between France and Italy.

Centau'rea f. Knapweed, bachelor's button. From Gr. *kentauros*, centaur. The genus includes sweet sultan and dusty miller. COMPOSITAE.

Centaur'ium n. Centaury. From Gr. *kentauros*, centaur, the half-man half-horse of Greek mythology. This herb was supposed to have excellent medicinal qualities, especially in healing wounds. Chiron the Centaur, famous for his knowledge of the uses of plants, is said to have used centaury to heal a wound in his foot, caused by one of the arrows of Hercules. GENTIANACEAE.

centifo'lius, -a, -um Many-leaved; with a hundred leaves or petals.

Centrade'nia f. Gr. *kĕntrŏn*, a spur; *adēn*, a gland. The anthers of these flowering evergreen shrubs have a spurlike gland. MELASTOMACEAE.

centra'lis, -is, -e Central, in the middle.

centranthifo'lius, -a, -um With leaves like *Centranthus* or red valerian.

Centran'thus m. Red valerian. From Gr. *kĕntrŏn*, a spur; *anthŏs*, a flower; from the spurred flower. Often incorrectly spelled *Kentranthus*. VALERIANACEAE.

Centropo'gon m. Gr. *kĕntrŏn*, a spur; *pōgōn*, a beard; in allusion to the fringed stigma of these shrubs and sub-shrubs. CAMPANULACEAE.

Centrose'ma n. Butterfly pea. From Gr. *kĕntrŏn*, a spur; *sēma*, a standard. The standard has a short spur behind it. LEGUMINOSAE.

ce'pa f. Latin for onion.

Cephae'lis f. Gr. *kĕphalē*, a head; from the close arrangement of the flowers of these shrubs and herbs, from one of which (*C. ipecacuanha*) ipecacuanha is produced. RUBIACEAE.

Cephalan'thus m. Button bush. Gr. *kĕphalē*, a head; *anthŏs*, a flower. The flowers of these shrubs are closely packed in globose clusters. RUBIACEAE.

Cephala'ria f. Giant scabious. Gr. *kĕphalē*, a head. The flowers of these herbs are in round clusters. DIPSACACEAE.

Cephaloce'reus m. Old-man cactus. From Gr. *kĕphalē*, a head; *Cereus*, a genus of cactus. A woolly head is produced when the plant is of flowering size. CACTACEAE.

cephalo'nicus, -a, -um Of the Grecian island of Cephalonia (Kefallinia) in the Ionian Sea.

-cephalus In Gr. compounds means headed, e.g. *macrocephalus*, large-headed.

Cephalosta'chyum n. Gr. *kĕphalē*, a head; *stachys*, ear of corn, hence a spike. GRAMINEAE.

Cephalota'xus f. Plum yew. From Gr. *kĕphalē*, a head; *Taxus*, yew; a reference to the appearance of the trees which resemble yews. CEPHALOTAXACEAE.

cephalo'tes Resembling a small head.

cepifo'lius, -a, -um With leaves resembling onion.

cera'ceus, -a, -um Waxy.

cera'micus, -a, -um Resembling pottery; or from Ceram or Serang, an island in the Moluccas.

cera'sifer, cerasi'fera, -um Bearing cherries or cherry-like fruit.

cerasifor'mis, -is, -e Shaped like a cherry.

cera'sinus, -a, -um Cherry-red.

cerastioi'des Resembling *Cerastium* or chickweed.

Ceras'tium n. Mouse-ear chickweed. From Gr. *kĕrastēs*, horned; in allusion to the horned seed capsule. CARYOPHYLLACEAE.

Ce'rasus f. Latin name for cherry, and used when the cherries are put in a genus of their own, but also now the specific epithet of sour cherry, *Prunus cerasus*. ROSACEAE.

ceratocau'lis, -is, -e With a stalk like a horn.

Cerato'nia f. Carob. From Gr. *kĕratōnia*, the name of the carob, from Gr. *kĕras*, horn, in allusion to the pods. The seed pods, sometimes sold for food, are called St. John's bread and are supposed to be the locusts which John the Baptist ate with wild honey in the wilderness. The seeds were used as weights and from them derives the carat, the jewellers' weight for gold and precious stones. LEGUMINOSAE.

Ceratope'talum n. Gr. *kĕrăs*, a horn; *petalon*, a petal. In one species the petals look like stag's horns. CUNONIACEAE.

Ceratophyl'lum n. Hornwort. From Gr. *kărăs*, a horn; *phyllon*, a leaf; from the antler-like appearance of the leaves of these submerged plants. CERATOPHYLLACEAE.

Ceratop'teris f. Gr. *kĕrăs*, a horn; *pteris*, a fern; from the horned appearance of this water fern. PARKERIACEAE.

Ceratostig'ma n. Gr. *kĕrăs*, a horn and *stigma*; from the hornlike excrescence on the stigma of the flower. PLUMBAGINACEAE.

Ceratoza'mia f. Gr. *kĕrăs*, a horn; *Zamia*, the name of another genus, which it resembles except for the fact that here the cones of these Mexican foliage plants have horned scales. ZAMIACEAE.

Cercidiphyl'lum n. Gr. *kerkis*, redbud or Judas tree; *phyllon*, a leaf; from the resemblance of the leaves to those of *Cercis siliquastrum*. CERCIDIPHYLLACEAE.

Cerci'dium n. Resembling *Cercis* or redbud.

Cer'cis f. Redbud, Judas tree. The Greek name *kerkis* for a European species, probably a poplar, but also applied to *Cercis siliquastrum*, which through confusion between Judas and Judaea is now traditionally the tree on which

Judas hanged himself. The flowers are produced along the branches and trunk before the leaves appear. LEGUMINOSAE.

Cercocar'pus m. Gr. *kerkos*, a tail; *karpos*, a fruit. The fruit of these trees and shrubs has a tail-like plume. ROSACEAE.

cerea'lis, -is, -e Pertaining to agriculture. Ceres was the goddess of farming.

cerefo'lius, -a, -um Waxen-leaved.

Ce'reus m. L. *cereus*, a wax taper; in allusion to the shape of some of the species of this cactus. CACTACEAE.

ce'reus, -a, -um Waxy.

ce'rifer, ceri'fera, -um Wax-bearing.

Cerin'the f. Honeywort. From Gr. *kērinos*, waxen. It was once supposed that bees obtained wax from these flowers. BORAGINACEAE.

cerinthoi'des Resembling *Cerinthe* or honeywort.

ceri'nus, -a, -um Waxy.

cer'nuus, -a, -um Drooping; nodding.

Cerope'gia f. Gr. *kērŏs*, wax; *pēgē*, a fountain; in allusion to the waxen appearance of the flower clusters. ASCLEPIADACEAE.

Cero'pteris f. A synonym of *Pityrogramma*.

Ceroxy'lon n. Wax palm. From Gr. *kērŏs*, wax; *xylon*, wood. The trunks of these feather palms are coated with wax. PALMAE.

cerris Classical Latin name for the Turkey or mossy-cup oak, and now the specific epithet (*Quercus cerris*).

Ces'trum n. Bastard jasmine. Greek name for some other plant than these ornamental shrubs. SOLANACEAE.

Chaenome'les f. Japanese quince. From Gr. *chaino*, to gape; *mēlŏn*, an apple; an allusion to the erroneous belief that the fruit splits. ROSACEAE.

Chaenorhi'num n. From Gr. *chaino*, to gape, and *Antirrhinum*; an allusion to the open mouth of the corolla which distinguishes the genus from *Antirrhinum* and *Linaria*. SCROPHULARIACEAE.

Chaeno'stoma n. Gr. *chaino*, to gape; *stŏma*, a mouth; an allusion to the corolla. SCROPHULARIACEAE.

Chaerophyl'lum n. Chervil. From Gr. *chairo*, to please, rejoice; *phyllon*, a leaf. The foliage is fragrant. UMBELLIFERAE.

chaetomal'lus, -a, -um From Gr. *chaite*, loose flowing hair; *mallŏs*, flock of wool.

chalcedo'nicus, -a, -um From Chalcedon, the classical name for what is now Kadekoy, a district in Asia Minor opposite Istanbul.

chalmersia'nus, -a, -um In honour of James B. Chalmers (1841–1901), traveller and author of works on New Guinea where he was killed by cannibals.

chamae- n. Gr. prefix indicating dwarfness or low growth, from *chamai*, on the ground.

Chamaeba'tia f. From Gr. *chaimaibatos*, a dwarf bramble. ROSACEAE.

Chamaebu'xus f. Gr. *chamai*, on the ground; *Buxus*, box; from its box-like leaves and dwarf creeping habit. POLYGALACEAE.

Chamaece'reus m. Gr. *chamai*, on the ground; *Cereus*, a genus of cactus; an allusion to its prostrate habit. CACTACEAE.

Chamaecy'paris f. False cypress. From Gr. *chamai*, dwarf; *kuparissos*, cypress. Often sold as *Retinospora*, the genus includes such specimen and timber trees as Port Orford and Alaska cedar, Hinoki cypress, and southern white cedar. CUPRESSACEAE.

Chamaedaph'ne f. Gr. *chamai*, on the ground, dwarf; *daphne*, laurel. Also known by the later name *Cassandra*. ERICACEAE.

Chamaedo'rea f. Gr. *chamai*, on the ground; *dorea*, a gift. The shining bright-coloured fruits of these palms are very easily reached. PALMAE.

chamaedrifo'lius, -a, -um With leaves like *Chamaedrys* or ground oak.

chamae'drys Gr. *chamai* on the ground, dwarf; *drys*, oak; used by Theophrastus, for a low-growing plant with oak-like leaves.

chamae'iris Specific epithet for an iris meaning dwarf iris.

Chamaeli'rium n. Devil's bit. This North American perennial is also sometimes called blazing star. From Gr. *chamai*, dwarf; *lirion*, a lily. LILIACEAE.

Chamaepericly'menum n. Gr. *chamai*, on the ground, dwarf; *pěriklyměnon*, honeysuckle; from the resemblance of their red berries. CORNACEAE.

Chamae'rops f. Dwarf fan-palm. From Gr. *chamai*, dwarf; *rhōps*, a bush. PALMAE.

Chambeyro'nia f. Palms named for Captain Chambeyron, who commanded the ship in which the French botanist and doctor of medicine, Eugene Vieillard (1819–1896), explored the coasts of New Caledonia where this genus was found. PALMAE.

Chamelau'cium n. Meaning and derivation of the name for these Australian evergreen shrubs is obscure. MYRTACEAE.

Chameran'themum n. Gr. *chamai*, dwarf; *Eranthemum*. ACANTHACEAE.

chapma'nii In honour of Alvin Wentworth Chapman (1809–1899), American botanist, author of *Flora of the Southern United States*.

charan'tia A pre-Linnaean name and now the specific epithet for *Momordica charantia*.

Chardi'nia f. In honour of Jean Chardin, later Sir John Chardin (1643–1713), French jeweller, traveller in Asia Minor, Iran, Caucasus and India, Huguenot refugee in England where in 1681 Charles II knighted him. COMPOSITAE.

charian'thus, -a, -um With elegant flowers.

Chasman'the f. From Gr. *chasme*, yawning, gaping; *chasmamai*, to yawn; *anthos*, flower. A South African genus. IRIDACEAE.

chasman'thus, -a, -um With wide-open or gaping flowers.

chatha'micus, -a, -um Of the Chatham Islands in the South Pacific.

Cheilan'thes f. Lip ferns. From Gr. *cheilos*, a lip; *anthos*, a flower; in allusion to the form of the membranous covering (indusium) of the spore-bearing parts. ADIANTACEAE.

cheilan'thus, -a, -um With a flower furnished with a lip.

Cheiran'thus m. Wallflower. The meaning is obscure although some suppose that it may derive from Arabic *keiri* and others from Gr. *cheir*, a hand; *anthos*, a flower; and thus refer to the custom of carrying these old-fashioned, sweetly scented favourites in the hand as a bouquet. CRUCIFERAE.

chelidonioi'des Resembling *Chelidonium* or celandine.

Chelido'nium n. Greater celandine. From Gr. *chĕlidōn*, a swallow. Supposedly because its first flowering and the arrival of swallows sometimes coincide. PAPAVERACEAE.

Chelo'ne f. Turtlehead. Greek word *chĕlōnē* for a tortoise; in allusion to the flower of these North American perennial herbs. SCROPHULARIACEAE.

Chenopo'dium n. Goosefoot. From Gr. *chēn*, a goose; *pŏdion*, a little foot; from the shape of the leaf of these plants, mostly weeds but including some edible species, such as Good-King-Henry. CHENOPODIACEAE.

Chiapa'sia f. Named after the state of Chiapas, Mexico, where this cactus grows. CACTACEAE.

Chiastophyl'lum n. From Gr. *chiastos*, arranged diagonally, crosswise; *phyllon*, leaf; referring to the pairs of opposite leaves. CRASSULACEAE.

chilen'sis, -is, -e Of Chile, South America.

Chilio'trichum n. From Gr. *chilii*, thousand; *trix*, hair, referring to the numerous bristles of the pappus. COMPOSITAE.

chiloen'sis, -is, -e Of the island of Chiloe off the Chilean coast.

Chilo'psis f. Gr. *cheilos*, a lip; *opsis*, resembling. The calyx of this low growing tree or shrub has a distinct lip. BIGNONIACEAE.

chimae'ra Specific epithet of uncertain application. The word means a monster; and, figuratively, a wild fancy.

Chima'phila f. Pipsissewa. From Gr. *cheima*, winter weather; *phileo*, to love. These small perennials are evergreen. PYROLACEAE.

Chimonan'thus m. Wintersweet. Gr. *cheimōn*, winter; *anthos*, flower. These shrubs flower in winter. CALYCANTHACEAE.

Chimonobambu'sa f. From Gr. *cheimōn*, winter; *Bambusa*, bamboo. GRAMINEAE.

chinen'sis, -is, -e Chinese; also spelled *sinensis*, which see.

Chiococ'ca f. Snowberry. Gr. *chiōn*, snow; *kŏkkŏs*, a berry. The berries of one species are snow-white. RUBIACEAE.

Chio'genes f. Creeping snowberry. Gr. *chiōn*, snow; *genos*, offspring; in allusion to the snow-white berries. ERICACEAE.

Chionan'thus m. Gr. *chiōn*, snow; *anthos*, a flower. The type species (*C. virginicus*) has snowy white flowers. OLEACEAE.

chionan'thus, -a, -um Having snow-white flowers.

Chionodo'xa f. Gr. *chiōn*, snow; *doxa*, glory. Among the very earliest of spring flowers, they often bloom when snow is still on the ground. The type-species, discovered by Boissier, was flowering among melting snow high in the Tmolus mountains of Asia Minor and named *C. luciliae* in honour of his wife. LILIACEAE (HYACINTHACEAE).

Chiono'phila f. Gr. *chiōn*, snow; *phileo*, to love. It grows at high elevations in the Rocky Mountains. SCROPHULARIACEAE.

Chionoscil'la f. Hybrids between *Chionodoxa* and *Scilla*.

chiotil'la Mexican name for a cactus, *Escontria chiotilla*.

Chiri'ta f. Nepalese vernacular name. GESNERIACEAE.

Chiro'nia f. South African plants named for Chiron, a centaur, son of Phyllira and Saturn. One of the beings half-man and half-horse, he was skilled in the medicinal use of plants. GENTIANACEAE.

chirophyl'lus, -a, -um With leaves resembling a hand.

Chlidan'thus m. From Gr. *chlidē*, luxury, costly ornament; *anthos*, flower. AMARYLLIDACEAE.

chloo'des Grass-green.

Chlor'acca Green tip; epithet of *Paedranassa*.

chloracan'thus, -a, -um With green spines.

Chlorae'a f. Gr. *chloros*, green. Some species of this genus of terrestrial orchids have green flowers. ORCHIDACEAE.

chloran'thus, -a, -um With green flowers.

chlorifo'lius, -a, -um With leaves like *Chlora* (i.e. *Blackstonia*), a genus of little horticultural interest.

Chlo'ris f. Finger-grass. Ornamental flowering grasses named in honour of Chloris, the Greek goddess of flowers. GRAMINEAE.

chlorochi'lon With a green lip.

Chloro'galum n. Gr. *chlōrŏs*, green; *gala*, milk; the sap of these bulbous plants from California is green. LILIACEAE (HYACINTHACEAE).

Chloro'phora f. Gr. *chlōrŏs*, green; *phoreo*, to bear. One species (*C. tinctoria*) of these trees produces a green or yellow dye. MORACEAE.

Chloro'phytum n. From Gr. *chlōrŏs*, green; *phyton*, a plant. Some species are grown for their foliage. LILIACEAE (ANTHERICACEAE).

chlo'rops With a green eye.

Choisy'a f. Mexican orange. Named in honour of Jacques Denis Choisy (1799–1859), Swiss botanist and professor of philosophy at Geneva. RUTACEAE.

Chordospar'tium n. From Gr. *chŏrdē*, string; *Spartium*; in allusion to the slender branches. LEGUMINOSAE.

chorda'tus, -a, -um Cordlike.

Chori'sia f. Floss silk tree. Named in honour of Ludwig (Louis) Choris

(1795–1828), a botanical artist who accompanied the noted Russian navigator Otto von Kotzebue on his second scientific expedition to the Pacific, 1823–1826. BOMBACACEAE.

Chorize'ma n. Contraction of *chorizonema* from Gr. *chōris*, separate; *nēma*, thread; referring to the separate filaments of the stamens. This genus was discovered and named by the French botanist Labillardière in Australia in 1792. LEGUMINUSAE.

Chrozo'phora f. From Gr. *chrōzō*, to tinge, dye, stain; *phoros*, carrying; in allusion to importance as source of turnsole dye. EUPHORBIACEAE.

Chrysalidocar'pus m. Gr. *chrysos*, gold; *karpos*, fruit; in allusion to the golden fruit of one species. *C. lutescens* is the *Areca* of florists, properly cane palm. PALMAE.

Chrysan'themum n. Gr. *chrysos*, gold; *anthĕmŏn* flower. According to modern classification this formerly large genus has now been reduced to two Mediterranean annual species known in classical times, *C. coronarium* and *C. segetum*, both yellow-flowered. The other species have been put in some 10 segregated genera, of which horticulturally the most important are *Dendranthema* (including the florist's chrysanthemums in their great variety), *Leucanthemum* (including the Shasta daisy) and *Tanacetum* (including the garden Pyrethrums). COMPOSITAE.

chrysanthoi'des f. Resembling *Chrysanthemum*.

chrysan'thus, -a, -um With golden flowers.

chry'seus, -a, -um Golden.

Chrysoba'lanus f. Gr. *chrysos*, gold; *balanos*, an acorn; in allusion to the yellow fruit of some species. CHRYSOBALANACEAE.

chrysocar'pus, -a, -um With golden fruit.

Chryso'coma f. Gr. *chrysos*, gold; *kŏmē*, hair. COMPOSITAE.

chryso'comus, -a, -um With golden hairs.

chrysodo'ron A golden gift.

Chryso'gonum n. Golden star. Gr. *chrysos*, gold; *gonu*, a knee. COMPOSITAE.

chrysographes Written on with gold; with golden markings.

chryso'lepis, -is, -e With golden scales.

chrysoleu'cus, -a, -um Gold-and-white.

chryso'lobus, -a, -um Golden-lobed.

chrysophyl'lus, -a, -um With golden leaves.

chrys'ops Gr. *chrysos*, golden; *ōps*, eye, face.

Chrys'opsis f. Golden aster. Gr. *chrysos*, gold; *opsis*, appearance. The flowers are yellow. COMPOSITAE.

chryso'stomus, -a, -um Golden-mouthed.

Chrysotham'nus m. Gr. *chrysos*, gold; *thamnos*, shrub. COMPOSITAE.

Chus'quea f. A local South American name for these bamboos. GRAMINEAE.

Chy'sis f. From Gr. *chysis*, melting. The pollen masses in these orchids appear to be fused into a whole. ORCHIDACEAE.

Cibo'tium n. Scythian lamb (see *barometz*). Gr. *kibōtŏs*, a small box; from the appearance of the sori of these tree ferns. THYRSOPTERIDACEAE.
Ci'cer f. Chick pea or garbanzo. The classical Latin name. LEGUMINOSAE.
Cicer'bita. f. Italian name, apparently of medieval origin, for sow-thistle, but now for a genus of blue-flowered composites related to *Lactuca*. COMPOSITAE.
cichoria'ceus, -a, -um Resembling chicory or *Cichorium*.
Cicho'rium n. Chicory, endive. The Latinized version of the Arabic name for one species. COMPOSITAE.
Cicu'ta f. Water hemlock. The Latin name for poison hemlock (*Conium maculatum*), the official state poison of ancient Athens used for the execution of Socrates. UMBELLIFERAE.
cicuta'rius, -a, -um Resembling water hemlock or *Cicuta*.
cicutifo'lius, -a, -um With leaves like water hemlock or *Cicuta*.
cilia'ris, -is, -e; cilia'tus, -a, -um Fringed with hairs on many leaves and petals; ciliate; from L. *cilium*, eyelid, then eyelash.
cili'cicus, -a, -um Of Cilicia, a classical region of southern Asia Minor.
ciliica'lyx With a fringed calyx.
cilio'sus, -a, -um Fringed or ciliate.
Cimici'fuga f. Bugbane. Black cohosh. From L. *cimex*, a bug; *fugo*, to drive away; from the use of *C. foetida* as an insect repellent. RANUNCULACEAE.
Cincho'na f. Named in honour of Countess de Chinchon, wife of the Viceroy of Peru, who, legend has it, was cured in 1638 of malaria by use of what was then called Peruvian bark. It is the source of quinine, the commercial production of which is now centred in Java. During World War II it was an important part of Japanese strategy to cut off the world from supplies of quinine. RUBIACEAE.
cin'ctus, -a, -um Girdled; girt around.
Cinera'ria f. From L. *cinereus*, ash-coloured; from the grey down on the leaves of these herbs and sub-shrubs. COMPOSITAE.
cinerariifo'lius, -a, -um With leaves like *Cineraria*.
cineras'cens Becoming ashy-grey.
cine'reus, -a, -um Ash-coloured.
cinnabari'nus, -a, -um Cinnabar-red; vermilion.
cinnamo'meus, -a, -um Brown like cinnamon.
cinnamomifo'lius, -a, -um With leaves like cinnamon.
Cinnamo'mum n. From the classical Greek name for cinnamon. The genus includes the camphor tree (*C. camphora*) and also cassia (*C. aromaticum*) which is widely grown throughout the Far East as an adulterant of true cinnamon (*C. verum*). LAURACEAE.
Cipu'ra f. Derivation of the name of these South American bulbous plants is unknown. IRIDACEAE.

circina'lis, -is, -e; circina'tus, -a, -um Coiled.

cirra'tus, -a, -um; cirrho'sus, -a, -um Equipped with tendrils.

Cirrha'ea f. L. *cirrus* or *cirrhus*, tendril; in allusion to the prolonged tendril-like rostellum of this orchid. ORCHIDACEAE.

Cirrhope'talum n. Presumably from *cirrus* or *cirrhus*, curl, tendril; *petalum*, petal. In allusion to the long lateral segments of the flowers of these orchids. ORCHIDACEAE.

Cir'sium n. Plumed thistle. From the Greek for a kind of thistle. The genus includes a great many pernicious weeds and very few plants of value to the gardener. COMPOSITAE.

cismonta'nus, -a, -um On this side of the mountain—meaning the south side of the Alps, the side on which Rome stands.

cisplati'nus, -a, -um From the neighbourhood of the River Plate in South America.

Cis'sus f. Gr. *kissos*, ivy; an allusion to the climbing habits of these lianes. *C. antarctica* is the kangaroo vine. VITACEAE.

cistiflo'rus, -a, -um *Cistus*-flowered.

cistifo'lius, -a, -um With leaves like *Cistus*.

Cis'tus m. Rock-rose. From the Greek name for the plant. CISTACEAE.

Cithare'xylum n. Gr. *kithara*, a lyre; *xylon*, wood; a Latinization of the Jamaican name 'fiddlewood' from its former use (not a corruption of *bois fidèle*). VERBENACEAE.

citra'tus, -a, -um Resembling *Citrus*.

citrifo'lius -a, -um With leaves resembling *Citrus*.

citriniflo'rus, -a, -um With lemon-yellow flowers.

citri'nus, -a, -um Lemon-coloured or resembling citron.

citriodo'rus, -a, -um Lemon-scented.

citroi'des Resembling *Citrus*.

Citro'psis f. *Citrus* plus Gr. *opsis*, resembling. These small trees are somewhat like *Citrus*, and used as stocks. RUTACEAE.

Citrul'lus m. From *Citrus*, from the appearance of the fruit. Watermelon is *C. lanatus*. CUCURBITACEAE.

Cit'rus f. Latin name for the citron (*Citrus medica*), the fruit of which was substituted in ancient Jewish ritual for the cone of the cedar (*Cedrus libani*) which had detested associations with Bacchus (Dionysos), but applied by Linnaeus and earlier authors to the whole genus which includes oranges, lemons, limes, and grapefruit. RUTACEAE.

Cladan'thus m. From Gr. *klados*, a branch; *anthos*, a flower. The flowers of this annual herb occur at the ends of the branches. COMPOSITAE.

Cla'dium n. Gr. *kladion*, a small branch. CYPERACEAE.

Cladas'tris f. Kentucky yellow-wood. Gr. *klados*, a branch; *thraustos*, fragile; from the brittle nature of the twigs. LEGUMINOSAE.

clandesti'nus, -a, -um Hidden.

Clar'kia f. North American annuals named in honour of Captain William Clark (1770–1838), who, with Captain Meriwether Lewis, made the first transcontinental expedition, crossing the Rocky Mountains in 1806. ONAGRACEAE.

Clause'na f. Named in honour of Peder Claussøn (1545–1614), a Danish priest who wrote on topography and natural history. RUTACEAE.

clau'sus, -a, -um Closed; shut.

cla'va-her'culis Club of Hercules. A specific epithet of *Zanthoxylum*.

clava'tus, -a, -um Club-shaped.

clavella'tus, -a, -um Shaped like a small club.

clave'nae In honour of Niccolo Chiavema (d. 1617), Italian apothecary.

Clavi'ja f. Named in honour of Don Jose de Viera y Clavijo (1731–1813), who translated the works of Buffon into Spanish. THEOPHRASTACEAE.

Clayto'nia f. Named in honour of John Clayton (1686–1773), who came to Virginia from England in 1705. He corresponded with the botanical great of the day—Linnaeus, Gronovius, Kalm, and John Bartram—as well as with Benjamin Franklin and Thomas Jefferson. Collinson (see *Collinsonia*), the English Quaker botanist, described him as the greatest botanist in America. The common spring-beauty, sometimes called good-morning spring (*C. virginica*), is a species of the genus. PORTULACEAE.

Cleistocac'tus m. From Gr. *kleistos*, closed; plus *cactus*. The flowers hardly open. CACTACEAE.

clemati'deus, -a, -um Like *Clematis*.

Cle'matis f. Greek name *klēmatis* for various climbing plants. RANUNCULACEAE.

Clematocle'thra f. From the generic names *Clematis* and *Clethra*. ACTINIDIACEAE.

clementi'nae In honour of Clementine, wife of the plant-collector George Forrest.

Cleo'me f. Derivation unknown. Spider flower is *C. hassleriana* or *C. spinosa*. CAPPARACEAE.

Cleroden'drum n. Gr. *klēros*, chance; *dendron*, a tree; supposed to be an allusion to the variable medicinal qualities of these shrubs, trees, and climbers. VERBENACEAE.

Cle'thra f. Sweet pepper bush. White alder. *Klēthra*, the Greek name for alder, the leaves of which are similar to those of *Clethra*. CLETHRACEAE.

clethroi'des Resembling *Clethra*.

Cley'era f. In honour of Andreas Cleyer (d. 1697 or 1698), German doctor and soldier in service of Dutch East India Company in Java and Japan. THEACEAE.

Clian'thus m. Glory pea. Gr. *klĕŏs*, glory; *anthos*, a flower; in allusion to the brilliantly coloured flowers. LEGUMINOSAE.

Cliffor'tia f. In honour of George Clifford (1685–1760), Dutch merchant-banker of English descent, the plants of whose magnificent garden at De

Hartekamp were described by Linnaeus in his *Hortus Cliffortianus* (1738). ROSACEAE.

Clifto'nia f. An evergreen shrub named in honour of William Clifton, lawyer, Attorney-General of Georgia in 1759. CYRILLACEAE.

Clinto'nia f. Bear-tongue. Named for Governor De Witt Clinton (1769–1828), of New York, who promoted the construction of the Erie Canal. LILIACEAE.

Clito'ria f. Butterfly pea. From L. *clitoris*, clitoris, small female genital appendage; an allusion to a characteristic of the flower. LEGUMINOSAE.

Cli'via f. Kaffir lily. Bulbous evergreen plants named after Lady Charlotte Florentina Clive, Duchess of Northumberland (d. 1866), granddaughter of Robert Clive, and the first to flower the type-species. AMARYLLIDACEAE.

clivo'rum Of the hills.

clusia'nus, -a, -um; clu'sii In honour of Carolus Clusius (Charles de l'Ecluse, 1526–1609), celebrated Flemish botanist and polymath, a much travelled, highly observant man who after many misfortunes ended his days happily as a professor at the newly founded University of Leiden.

clypea'tus, -a, -um Resembling the small round Roman shield (*clipeus*).

clypeola'tus, -a, -um Somewhat shield-shaped.

Clyto'stoma n. Gr. *klytos*, beautiful; *stŏma*, a mouth. BIGNONIACEAE.

Cneo'rum n. Spurge olive. From the Greek name *knēōron* for some shrub resembling olive. CNEORACEAE.

Cni'cus m. Blessed thistle. From Gr. *knēkos*, a thistle. COMPOSITAE.

coarcta'tus, -a, -um Pressed or crowded together.

Coba'ea f. Tendril-climbers named in honour of a Jesuit Father, Bernardo Cobo (1572–1659), Spanish missionary and naturalist in Mexico and Peru. POLEMONIACEAE.

coc'cifer, cocci'fera, -um Berry-bearing; being a host to coccoid insects, the gravid females of which resemble berries.

cocci'neus, -a, -um Scarlet.

Cocci'nia f. L. *coccineus*, scarlet; from the colour of the fruits. CUCURBITACEAE.

Cocco'loba f. Seaside grape. From Gr. *kokkŏlŏbis*, ancient name of a kind of grape also called *balisca* and transferred to these plants in allusion to the fruit. POLYGONACEAE.

Coccothri'nax f. From Gr. *kŏkkŏs*, a berry, and *Thrinax*, a related genus of fan palms. They have berrylike fruits. PALMAE.

Coc'culus m. Moonseed, snail-seed, coral beads. From the diminutive of Gr. *kŏkkŏs*, a berry; in allusion to the fruits of these evergreen climbers or shrubs. MENISPERMACEAE.

Cochemi'ea f. From the name of an Indian tribe which once lived in lower California, home of these plants. CACTACEAE.

cochenilli'fera Cochineal-bearing. Specific epithet of the cochineal fig,

Nopalea cochenillifera. Cochineal is a dyestuff made from the dried bodies of the females of a scale insect found in Mexico and Central America. It is used as a reddish or purple colouring for foodstuffs and is the source of carmine.

Cochlea'ria f. Scurvy grass. Gr. *kŏchlarion*, L. *cochlear*, a spoon; from the shape of the basal leaves of these herbs. CRUCIFERAE.

cochlearifo'lius, -a, -um With leaves like *Cochlearia.*

cochlea'ris, -is, -e Spoon-shaped.

cochlea'tus, -a, -um Spirally twisted like a snail-shell (*cŏchlĕa*).

Cochleoste'ma n. From Gr. *kŏchlŏs*, spiral shell; *stēma*, penis, stamen; in allusion to the spirally twisted anthers. COMMELINACEAE.

Cochlio'da f. From Gr. *kochlŏĕides*, like a snail-shell; in allusion to the snail-shell-like thickening on the lip of the type-species of this genus of orchids. ORCHIDACEAE.

Cochlosper'mum n. From Gr. *kŏchlŏs*, snail-shell or other spiral shell; *sperma*, seed; in allusion to the shape of the seeds of these tropical trees. BIXACEAE.

cockburnia'nus, -a, -um In honour of members of the Cockburn family domiciled in China.

Co'cos f. The coconut palm has become so widespread through human activity over many centuries that its original home, probably in the Malaysian region, is uncertain. It has spread itself by natural means over the sea coasts of the tropics wherever its seedlings could find a footing on a sandy shore unmolested by wild pigs, rats and crabs. One of the important sources of vegetable oil, it is an indispensable article of diet in many places. Sailing vessels in many parts of the world are rigged with coconut-fibre ropes. The fibre is also used for doormats, matting cord and peat substitute in composts. The nut shells are used as drinking and other cups, the leaves are used for thatching, while the trunks make a handsome timber commercially called porcupine wood. The name comes from Portuguese *coco*, mask. PALMAE.

Codia'eum m. Croton. Latinized version of the Ternate vernacular name, *kodiho*.

Codono'psis f. From Gr. *kōdōn*, a bell; *ŏpsis*, resembling; from the shape of the corolla of these herbs. CAMPANULACEAE.

coelesti'nus, -a, -um; coeles'tis, -is, -e Sky-blue.

Coe'lia f. Gr. *kŏilŏs*, hollow; from an erroneous concept that the pollen masses of this epiphytic orchid were hollow. ORCHIDACEAE.

coe'licus, -a, -um Celestial; heavenly.

coeli-ro'sa Specific epithet meaning rose of the sky.

Coeloglos'sum n. From Gr. *kŏilŏs*, hollow; *glōssa*, tongue, referring to the lip of these orchids. ORCHIDACEAE.

Coelo'gyne f. Gr. *kŏilŏs*, hollow; *gynē*, female; referring to the depressed stigma of these epiphytic orchids. ORCHIDACEAE.

coerules'cens; coeruleus, -a, -um The first, bluish, becoming blue; the second, blue.

Co'ffea f. From *kahwah*, the Arabic name of the beverage. The coffee we drink is prepared from the seeds which are enclosed in a bright scarlet berry. RUBIACEAE.

coffea'tus, -a, -um Coffee brown like the roasted seeds.

coggygria A false rendering of the Gr. *kŏkkugia*, name for the smoke-tree (see *Cotinus*).

cogna'tus, -a, -um Closely related to.

Co'ix f. Job's tears. From the Greek name for a reedlike plant. GRAMINEAE.

Co'la f. From the African vernacular name. STERCULIACEAE.

Col'chicum n. Autumn crocus. Said by ancient authors to be especially abundant in Colchis, the Black Sea region of Georgia, Caucasus. LILIACEAE.

col'chicus, -a, -um From Colchis.

Coleone'ma n. From Gr. *kŏlĕŏs*, a sheath; *nema*, a thread; referring to the filaments of the stamens which are folded in the petals of these South African evergreen shrubs. RUTACEAE.

Coleotry'pe f. From Gr. *kŏlĕŏs*, a sheath; *trypē*, hole; in allusion to the young growths breaking through a hole in the leaf-sheath instead of emerging from the top of the sheath. COMMELINACEAE.

Co'leus m. Gr. *kŏlĕŏs*, a sheath; an allusion to the manner in which the stamens are enclosed. This genus is considered insufficiently distinct from *Plectranthus*. An alternative name is *Solenostemon*. LABIATAE.

Colle'tia. f. Spiny shrubs named in honour of Philibert Collet (1643–1718), French botanist. RHAMNACEAE.

collettia'nus, -a, -um; collettii. In honour of General Sir Henry Collett (1836–1901) of the Indian Army, author of *Flora Simlensis* (1902).

Collin'sia f. Blue-eyed Mary. Annuals named in honour of Zaccheus Collins (1764–1831), Vice-President of the Philadelphia Academy of Natural Sciences. SCROPHULARIACEAE.

Collinso'nia f. Horse balm, horse weed. Coarse herbs named in honour of Peter Collinson, 18th-century English Quaker merchant, who corresponded widely with men such as Linnaeus, John Bartram and John Clayton (see *Claytonia*) in America, and helped to introduce many North American plants into British gardens by way of his garden at Mill Hill. LABIATAE.

colli'nus, -a, -um Pertaining to hills.

Collo'mia f. From Gr. *kŏlla*, glue. The seeds of these herbs are mucilaginous when wet. POLEMONIACEAE.

Coloca'sia f. West Indian kale, taro root, elephant's ear. From the Gr. *kŏlŏkasia*, used for the root of *Nelumbo nucifera*, according to Dioscorides. ARACEAE.

colora'tus, -a, -um Coloured.

Colquhou'nia f. Named in honour of Sir Robert Colquhoun (d. 1838), British Resident in Nepal in 1819, patron of the Calcutta Botanic Garden in the early part of the 19th century. The name is pronounced *ko-hoo-nia*. LABIATAE.

colubri'nus, -a, -um Snakelike; shaped like a snake.

columba'rius, -a, -um Dovelike or pertaining to doves.

columbia'nus, -a, -um Of British Columbia or the Columbia River separating Washington and Oregon, U.S.A.

columel'laris, -is, -e Pertaining to a small pillar or pedestal.

colum'naris, -is, -e Columnar; in the shape of a column.

Colum'nea f. Named in honour of Fabius Columna (Fabio Colonna, 1567–1640), author of the first botanical book with copper-plate illustrations, published in Naples 1592. GESNERIACEAE.

colur'na Classical name, now the specific epithet of the hazelnut.

Colu'tea f. Bladder senna. From the Greek name *kŏlutea* for these shrubs. LEGUMINOSAE.

colvi'lei Epithet of a Himalayan buddleja named for Sir James William Colvile (1801–1880), Scottish lawyer, judge in Calcutta (1845–1859).

Colvil'lea f. A red-flowered Madagascan tree named in honour of Sir Charles Colville (1770–1843), Scottish soldier, governor of Mauritius 1828–1834. LEGUMINOSAE.

Coma'rum n. From Gr. *kŏmaros*, a name of the strawberry-tree, transferred to *Potentilla comarum* on account of the strawberry-like appearance of the fruiting heads. ROSACEAE.

coma'tus, -a, -um Furnished with a tuft, sometimes of hair.

com'beri In honour of Harold F. Comber (1898–1969), English gardener and plant collector, who settled in U.S.A. as lily grower.

Combre'tum n. Latin name for a climbing plant, but not of this genus. COMBRETACEAE.

Comesper'ma n. From Gr. *kŏmē*, hair; *sperma*, a seed; in allusion to the tufts of hair on the seeds. POLYGONACEAE.

Commeli'na f. Day-flower. Named for the Dutch botanists Commelin (Commelijn), Johan (1629–1692) and his nephew Caspar (1667–1731), by Plumier and adopted by Linnaeus, who while in Holland evidently learned of Caspar's son Caspar, the author of a Latin poem in 1715 but otherwise unknown. To quote Linnaeus, 'Commelina has flowers with three petals, two of which are showy, while the third is not conspicuous, from the two botanists called Commelin, for the third died before accomplishing anything in botany.' COMMELINACEAE.

Commerso'nia f. In honour of Philibert Commerson (1727–1773), French doctor and naturalist on Bougainville's voyage of (1766–1769) . Commerson remained in Mauritius to make further biological investigations and never returned to France. Among the new plants collected by him was *Bougainvillea*. STERCULIACEAE.

commix'tus, -a, -um Mixed together; mingled.

commu'nis, -is, -e Common; general; growing in company.

commuta'tus, -a, -um Changed or changing; used for a species close to one already known.

como'sus, -a, -um Furnished with a tuft, sometimes of sterile flowers, as in *Muscari comosum*, or conspicuous bracts, as in *Eucomis comosa*.

compac'tus, -a, -um Compact; dense.

Comparet'tia f. In honour of Andrea Comparetti (1745 or 1746–1801 or 1802), Italian botanist, professor of medicine at Padua. ORCHIDACEAE.

complana'tus, -a, -um Flattened; levelled.

complex'us, -a, -um Encircled; embraced; complex.

complica'tus, -a, -um Complicated; complex.

compo'situs, -a, -um Compound.

compres'sus, -a, -um Compressed; flattened.

Compto'nia f. Sweet fern. Named for Henry Compton (1632–1713), Bishop of London, dendrologist and patron of botany. Compton began his career as an officer of the Life Guards. As Bishop of London he was suspended by King James II for his strongly expressed protestant religious beliefs and was one of the aristocrats who invited William of Orange to invade England. At the time of the Revolution of 1688, he girded on sword and pistols and personally escorted Princess Anne, later to be Queen, out of London in his own coach. MYRICACEAE.

comp'tus, -a, -um Adorned; ornamented.

Conan'dron n. From Gr. *kōnŏs*, a cone; *anēr*, *andrŏs*, man, hence stamen, the appendages of the anthers of this herb being arranged as a cone around the style. GESNERIACEAE.

conca'tenans Linking together.

conca'vus, -a, -um Hollowed out; basin-shaped.

conchifo'lius, -a, -um With leaves like sea shells.

concin'nus, -a, -um Neat; elegant; well-made.

con'color Of the same colour throughout.

condensa'tus, -a, -um; conden'sus, -a, -um Crowded together.

confertiflo'rus, -a, -um With the flowers crowded together.

confer'tus, -a, -um Crowded.

confor'mis, -is, -e Conforming to the type; similar in shape or other ways to related species.

confu'sus, -a, -um Confused; uncertain; apt to be taken for another species.

conges'tus, -a, -um Arranged very closely together; congested.

conglomera'tus, -a, -um Crowded together, conglomerate.
congola'nus, -a, -um Of the Congo.
Conico'sia f. From Gr. *kōnicŏs*, cone-shaped, conical. A genus of S. African plants of the *Mesembryanthemum* group. AIZOACEAE.
co'nifer, coni'fera, -um Cone-bearing.
Coniogram'me f. Gr. *kŏniŏs*, dusty; *grammē*, a line; referring to the spore cases being arranged along the veins. ADIANTACEAE.
Coni'um n. Poison hemlock. Gr. *kōneion*, L. *cōnīum* for both the plant and the poison derived from it, which was administered to those condemned in Athens to die for certain offences against the state. Plato's description of Socrates taking the poison remains one of the most moving passages in literature (see *Cicuta*). UMBELLIFERAE.
conjuga'tus, -a, -um; conjuga'lis -is, -e Joined in pairs; wedded.
conjunc'tus, -a, -um Joined.
conna'tus, -a, -um United; twin; having opposite leaves joined together at their bases.
conoi'deus, -a, -um Conelike.
Conophy'tum n. Cone-plant. From Gr. *kŏnŏs*, a cone; *phytŏn*, a plant; an allusion to the shape of these succulents. AIZOACEAE.
cono'pseus, -a, -um Gnat-like, from Gr. *kōnōps*, gnat, mosquito.
conradi'nae In honour of Conradine, wife of the German botanist B. A. E. Koehne (1848–1918); see *helenae*.
consangui'neus, -a, -um Related.
Conso'lida f. Medieval name of a wound-healing herb, *Symphytum*, from L. *consolido*, make firm, but transferred to these annual delphiniums. RANUNCULACEAE.
conso'lidus, -a, -um Solid; stable.
consper'sus, -a, -um Scattered.
conspi'cuus, -a, -um Conspicuous.
constantinopolita'nus, -a, -um Of Constantinople (now Istanbul), Turkey.
constric'tus, -a, -um Constricted.
conti'guus, -a, -um Near together—so close as to touch one another. Closely related.
continenta'lis, -is, -e Continental.
contor'tus, -a, -um Twisted; bent irregularly; contorted.
contrac'tus, -a, -um Drawn together; contracted.
controver'sus, -a, -um Doubtful; controversial.
Convalla'ria f. Lily-of-the-valley. From L. *convallis*, a valley, the late medieval name being *Lilium convallium* taken from the Vulgate Bible translation of the Song of Solomon. LILIACEAE (CONVALLARIACEAE).
convallaroi'des Resembling lily-of-the-valley.

convalliodo'rus, -a, -um Scented like lily-of-the-valley.
convolvula'ceus, -a, -um Similar to *Convolvulus*.
Convol'vulus m. Bindweed. From L. *convolvo*, to twine around. CONVOLVULACEAE.
conyzoi'des Resembling *Conyza*, a genus of no garden interest.
coo'peri In honour of Thomas Cooper (1815–1913), English gardener, who collected in South Africa from 1859 to 1862, introducing many notable plants; father-in-law of the Kew botanist N. E. Brown (1849–1934) who also worked in his youth for W. W. Saunders and thereby became interested in South African plants.
Coope'ria f. Evening-star. Night-flowering bulbous plants named in honour of Joseph Cooper, gardener to Earl FitzWilliam at Wentworth, Yorkshire, *c.* 1830. AMARYLLIDACEAE.
Copai'fera f. From *copaiba*, the Brazilian name for a balsam, plus L. *fero*, to bear. The trees yield copal, a commercial gum used in making varnish. LEGUMINOSAE.
copalli'nus, -a, -um Gummy; resinous.
Coperni'cia f. Carnanba palm, named for Copernicus (Nicolaus Koppernigk, 1473–1543), Polish astronomer, physician, economist, and theologian, most famous for having propounded and proved the heliocentric theory which placed the sun as the centre of our solar system around which the planets, including the earth, revolve. Thus, once and for all, he upset the Ptolemaic system which had been accepted for 1,500 years. PALMAE.
Copiapo'a f. From the province of Copiapo, Chile, where these cacti grow. CACTACEAE.
Copros'ma f. Gr. *kŏprŏs*, dung; *ŏsmē*, a smell; in allusion to the fetid smell of the type species. RUBIACEAE.
Cop'tis f. Gold thread. Mouth root. From Gr. *kopto*, to cut; a reference to the deeply cut leaves. RANUNCULACEAE.
coralliflo'rus, -a, -um With flowers of coral-red.
coralli'nus, -a, -um Coral-red.
coralloden'dron Specific epithet of the coral-tree (*Erythrina*).
Cor'chorus m. Greek name *kŏrchŏros* of obscure derivation. Jute is obtained from *C. capsularis* and *C. olitorius*. TILIACEAE.
corda'tus, -a, -um Heart-shaped.
Cor'dia f. Named in honour of Euricius Cordus (1486–1535) and his son Valerius (1515–1544), German botanists and pharmacists. BORAGINACEAE.
cordifo'lius, -a, -um With heart-shaped leaves.
cordifor'mis, -is, -e Heart-shaped.
Cordy'line f. Gr. *kŏrdylē*, a club. Often offered as *Dracaena* to which this genus is closely related. AGAVACEAE.

corea'nus, -a, -um Of Korea.

Core'ma n. Gr. *korēma*, a broom; from the habit of these heathlike shrubs. EMPETRACEAE.

Coreo'psis f. Tickseed. From Gr. *kŏris*, a bug; *opsis*, like. The seed (cypsela) of these herbs looks like a bug or tick. COMPOSITAE.

coria'ceus, -a, -um Thick and tough; leathery.

Corian'drum n. Coriander. The Greek name, *kŏriandrŏn*, from *kŏris*, referring to the unpleasant smell of the unripe fruits which disappears when they are ripe and dry. The plant has a long history of use as a flavouring. Today coriander seed is used in such diverse things as gin, confectionery, bread, curry powder, and so on. UMBELLIFERAE.

Coria'ria f. L. *corium*, leather. Some species of these shrubs and herbs are used in tanning. CORIARIACEAE.

coria'rius, -a, -um Like leather.

coridifo'lius, -a, -um With leaves like *Coris*, a plant of no garden interest.

corifo'lius, -a, -um With leathery leaves.

coriophyl'lus, -a, -um With leaves like *Coris*.

coris Specific epithet for a St. John's Wort (*Hypericum*), from a similarity to *Coris*.

cor'neus, -a, -um Horny.

cornicula'tus, -a, -um With small horns.

cor'nifer, corni'fera, -um; cor'niger, corni'gera, -um Bearing or being furnished with a horn.

cornubien'sis, -is, -e Of Cornwall, south-west England.

cornuco'piae Like a cornucopia or horn of plenty.

Cor'nus f. Dogwood or cornel, bunchberry. The Latin name for the cornelian cherry (*Cornus mas*). This genus is easily broken up into smaller groups treated as genera by many botanists, e.g. *Benthamidia*, *Cornus* proper (*Macrocarpium*) and *Swida* (*Thelycrania*). CORNACEAE.

cornu'tus, -a, -um Horned or horn-shaped.

Coro'kia f. From the Maori name. CORNACEAE.

corolla'tus, -a, -um Like a corolla.

coromandelia'nus, -a, -um Of the Coromandel Coast, the south-eastern part of the Indian peninsula south of the Kistna River.

cor'onans; corona'tus, -a, -um Crowned.

corona'rius, -a, -um Used for garlands, or pertaining to garlands.

Coronil'la f. Crown vetch. Diminutive of L. *corona*, a crown; in reference to the umbel. LEGUMINOSAE.

coronopifo'lius, -a, -um With leaves like *Cŏrōnŏpus*, a rather weedy plant.

Coro'zo f. Native South American name for this palm. PALMAE.

Corpuscula'ria f. From Latin *corpusculum*, a little body. AIZOACEAE.

Cor'rea f. Australian shrubs named in honour of José Francesco Correa da Serra (1751–1823), Portuguese botanist. RUTACEAE.

corruga'tus, -a, -um Corrugated; wrinkled.

cor'sicus, -a, -um Of Corsica.

Cortade'ria f. Pampas grass. From the Argentinian name. GRAMINEAE.

cortico'sus, -a, -um With bark like a cork tree; heavily barked.

Cortu'sa f. Perennial herbs named in honour of Jacobi Antonio Cortusi (1513–1593), director of the Padua botanic garden. PRIMULACEAE.

cortusoi'des Resembling *Cortusa.*

corus'cans Glittering; coruscating.

Coryan'thes f. Helmet flower, bucket orchid. From Gr. *kŏrys*, a helmet; *anthos*, a flower. ORCHIDACEAE.

corya'nus, -a, -um In honour of Reginald R. Cory (1817–1934), coal-owner, benefactor by bequest of the Cambridge Botanic Garden and the Lindley Library, Royal Horticultural Society.

Cory'dalis f. Greek word meaning a lark; the flowers have spurs like those of larks. FUMARIACEAE.

corylifo'lius, -a, -um With leaves like *Corylus* or hazel.

Corylo'psis f. Gr. *kŏrylŏs*, hazel; *ŏpsis*, like; from the resemblance of the leaves of these flowering shrubs to *Corylus* or hazel. HAMAMELIDACEAE.

Co'rylus f. Hazel. The Greek name. Now the name for a genus including cob-nut and filbert. BETULACEAE.

corym'bifer, corymbi'fera, -um Corymb-bearing. A corymb is a cluster of flowers in which the outer stalks or pedicels are longer than the inner ones. The blossom of the sweet cherry is an example.

corymbiflo'rus, -a, -um With flowers produced in a corymb.

corymbo'sus, -a, -um Provided with corymbs; corymbose.

coryne'phorus, -a, -um Carrying a club-shaped organ of some sort.

coryno'calyx With a calyx like a club.

Corynocar'pus m. New Zealand laurel. From Gr. *kŏrynē*, a club; *karpos*, fruit; from the shape of the fruit of this tree. CORYNOCARPACEAE.

Co'rypha f. Gr. *koryphe*, a summit or hilltop; with reference to the terminal crown of leaves of this palm. PALMAE

coryphae'us, -a, -um Gr. *koryphaios*, leading.

Coryphan'tha f. Gr. *koryphe*, a summit; *anthos*, a flower. The flowers are on the top of these globular or cylindrical cacti. CACTACEAE.

cosmophyl'lus, -a, -um With leaves like *Cosmos.*

Cos'mos m. Cosmea. Gr. *kosmos*, beautiful. COMPOSITAE.

costa'tus, -a, -um Ribbed.

Cos'tus m. Spiral-flag. The Latin name derived from an Oriental name for an imported aromatic root. ZINGIBERACEAE.

cotinifo'lius, -a, -um With leaves like *Cotinus* or smoke-tree.
cotinoi'des Resembling *Cotinus* or smoke-tree.
Co'tinus m. Smoke-tree. From Gr. *kŏtinŏs* for the wild-olive but of uncertain application here. ANACARDIACEAE.
Cotonea'ster m. L. *cotoneum*, quince; *-aster*, suffix indicating incomplete resemblance, hence applied to a wild or inferior kind; from a similarity to quince *Cydonia* in some species of this small-fruited genus. ROSACEAE.
Co'tula f. Greek word meaning a small cup. The bases of the leaves of these small herbs form cups. COMPOSITAE.
Cotyle'don f. Gr. *kŏtylēdōn*, from *kŏtylĕ*, a cavity or small cup; in allusion to the cuplike leaves of some species of these succulent plants. CRASSULACEAE.
coul'teri In honour of Thomas Coulter (1793–1843), Irish botanist who collected in Mexico and California.
coum Of the island of Kos in the Aegean Sea.
Couta'rea f. Latinized version of the vernacular name in Guyana for these evergreen trees and shrubs. RUBIACEAE.
Cowa'nia f. Named in honour of James Cowan (d. 1823), London merchant who travelled in Mexico and Peru and introduced many plants. ROSACEAE.
cowania'nus, -a, -um In honour of John Macqueen Cowan (1892–1960), Scottish forester and botanist, authority on the genus *Rhododendron*.
Cram'be f. Seakale. The Greek name for cabbage. CRUCIFERAE.
Craspe'dia f. Gr. *kraspĕdon*, a fringe; in allusion to the feathery pappus. COMPOSITAE.
crassicau'lis, -is, -e Thick-stemmed.
crassifo'lius, -a, -um Thick-leaved.
cras'sipes Thick-footed or thick-stemmed.
crassius'culus, -a, -um Somewhat thick.
Cras'sula f. Diminutive of L. *crassus*, thick; from the somewhat thick leaves of these shrubs and herbs. CRASSULACEAE.
cras'sus, -a, -um Thick; fleshy.
crataegifo'lius, -a, -um With leaves like *Crataegus* or hawthorn.
crataegoi'des Resembling hawthorn.
Crataegomes'pilus f. Graft hybrids of *Crataegus* and *Mespilus*.
Cratae'gus f. Hawthorn. The Greek name for the tree. From *kratos*, strength; an allusion to the strength and hardness of the wood. ROSACEAE.
Crataemes'pilus f. Sexual hybrids of *Crataegus* and *Mespilus*.
cre'ber, cre'bra, -um Thickly clustered; close; frequent.
crebiflo'rus, -a, -um With thickly clustered flowers.
Cremantho'dium n. From Gr. *krĕmao*, hang, and *anthodium*, flower-head or capitulum (from *anthōdes*, flower-like), referring to the drooping flower-heads. COMPOSITAE.

crenatiflo'rus, -a, -um With flowers cut in rounded scallops.
crena'tus, -a, -um Cut in rounded scallops; crenate.
crenula'tus, -a, -um Somewhat scalloped.
crepida'tus, -a, -um Shaped like a sandal or slipper.
Cre'pis f. Hawk's-beard. A large genus of herbs with dandelion-like flowers. Gr. *krēpis*, a boot; but why is not clear. COMPOSITAE.
cre'pitans Rustling; crackling.
Crescen'tia f. The Calabash tree, bearing the big fruits from which bowls and dippers are made in hot countries. Named in honour of Pietro Crescenzi (1230–1321), Italian author of a work on country life. BIGNONIACEAE.
creta'ceus, -a, -um Pertaining to chalk; chalky.
cretensis, -is, -e; cre'ticus, -a, -um Of the island of Crete.
cri'niger, crini'gera, -um Bearing hairs; from L. *crinis*, hair.
crini'tus, -a, -um Furnished with long, generally weak, hairs.
Crinoden'dron n. From Gr. *krinŏn*, lily; *dendron*, tree; in allusion to the flowers. ELAEOCARPACEAE.
Crinodon'na f. Generic name given to a hybrid group between *Crinum* and *Amaryllis* (*Belladonna*). AMARYLLIDACEAE.
Cri'num n. Gr. *krinon*, a lily. Bulbous plants. AMARYLLIDACEAE.
crispa'tus, -a, -um; cris'pus-, -a, -um Finely waved; closely curled.
cris'ta-gal'li A specific epithet meaning cock's comb.
crista'tus, -a, -um Having tassel-like tips, crested.
crithmifo'lius, -a, -um With leaves like *Crithmum*.
Crith'mum n. Sea samphire. Gr. *krithe*, barley; in allusion to the fruit ('seed'). UMBELLIFERAE.
croca'tus, -a, -um; cro'ceus, -a, -um Saffron-coloured; yellow.
Crocos'mia f. Gr. *krŏkŏs*, saffron; *ŏsmē*, a smell. The dried flowers placed in warm water have a strong saffron smell. IRIDACEAE.
crocosmiiflo'rus, -a, -um *Crocosmia*-flowered.
Cro'cus m. Gr. *krŏkŏs*, saffron, derived from Semitic *karkom*, and one of the most ancient plant names. The genus includes both the common and the saffron crocus. The saffron crocus, *C. sativus*, is widely grown in Mediterranean countries and elsewhere for the stigmas and part of the styles which are dried to produce a yellow dye and a spice used in cookery. IRIDACEAE.
Crossan'dra f. Gr. *krŏssos*, a fringe; *aner*, *andros*, male. The anthers are fringed with hairs. ACANTHACEAE.
Crotala'ria f. Rattle-box. Gr. *krŏtalon*, a rattle, clapper; with reference to the seeds rattling in the inflated pods. LEGUMINOSAE.
Cro'ton m. Gr. *krŏtōn*, a tick, from the appearance of the seeds. *C. tiglium* is the source of croton oil, a drastic purgative. EUPHORBIACEAE.
crotonifo'lius, -a, -um With leaves like *Croton*.

Cro'wea f. In honour of James Crowe (1750–1807), an English surgeon, born at Norwich, who studied mosses, fungi, and willows, of which he had a large living collection. RUTACEAE.

Crucianel'la f. Crosswort. From the diminutive of L. *crux*, a cross; the leaves of these herbs being crosswise. RUBIACEAE.

crucia'tus, -a, -um In the form of a cross.

cru'cifer, cruci'fera, -um Bearing a cross.

cruen'tus, -a, -um Blood-coloured, gory.

Crupi'na f. Latinized from a Belgian or Dutch vernacular name. COMPOSITAE.

crus-gal'li A specific epithet meaning cock's spur.

crusta'tus, -a, -um Encrusted.

crux-an'drae Specific epithet meaning St. Andrew's cross.

crux-maltae Specific epithet meaning Maltese cross.

Cryptan'tha f. From Gr. *krypto*, to hide; *anthos*, a flower. BORAGINACEAE.

Cryptan'thus m. From Gr. *krypto*, to hide; *anthos*, a flower; the lower part of the flower is covered by the sheathing leaves. BROMELIACEAE.

Cryptoco'ryne f. From Gr. *krypto*, to hide; *kŏrynē*, club, the club-shaped spadix being completely hidden within the spathe. ARACEAE.

Cryptogram'ma n. Rockbrake. From Gr. *krypto*, to hide; *gramma*, writing. In these ferns the spore cases (sori) are concealed beneath the reflexed leaf margin. ADIANTACEAE.

Cryptome'ria f. Japanese cedar. Gr. *krypto*, to hide; *meris*, a part. All flower parts of this tree are concealed. TAXODIACEAE.

Cryptoste'gia f. Rubber vine. From Gr. *krypto*, to hide; *stĕgo*, to cover. The 5 scales are enclosed within the corolla tube. ASCLEPIADACEAE.

Cryptostem'ma n. Gr. *krypto*, to hide; *stĕmma*, a crown. The woolly hairs of the 'seeds' (cypselae) cover the pappus. COMPOSITAE.

crystal'linus, -a, -um Crystalline; with crystal-like papilloae.

Ctenan'the f. From Gr. *kteis*, *ktĕnos*, a comb; *anthos*, a flower; from the arrangement of the bracts. MARANTACEAE.

ctenoi'des Resembling a comb.

cube'ba Specific epithet for *Piper cubeba* from the pre-Linnaean name for cubeb.

Cucuba'lus m. Latin plant name mentioned by Pliny. CARYOPHYLLACEAE.

cuculla'ris, -is, -e; cuculla'tus, -a, -um Hoodlike; having sides or apex curved inwards to resemble a hood.

cucumerifo'lius, -a, -um Having leaves like cucumber.

cucumeri'nus, -a, -um Resembling cucumber.

Cu'cumis m. Cucumber, melon. The ancient Latin name. CUCURBITACEAE.

Cucur'bita f. Latin name for a gourd. The genus also includes such plants as squash and pumpkin. CUCURBITACEAE.

Culcita f. Latin for mattress, cushion, pillow; with reference to the woolly rootstock of this fern being used in Madeira for stuffing cushions. THYSOPTERIDACEAE.

culto'rum. Of cultivated land, gardens, plantations.

cultra'tus, -a, -um; cultrifor'mis, -is, -e Shaped like a knife blade.

Cumi'num n. Cumin. Greek name for this aromatic herb of ancient use. UMBELLIFERAE.

cunea'tus, -a, -um Wedge-shaped, generally with the narrow end down.

cuneifo'lius, -a, -um With wedge-shaped leaves.

cuneifor'mis, -is, -e In the form of a wedge.

Cu'nila f. American dittany. Latin name for a mint. LABIATAE.

Cunningha'mia f. Decorative coniferous evergreens named for James Cunninghame (d. *c.* 1709), East India Company surgeon at Amoy, China, who, between 1698 and 1702, sent home large collections of plants and plant drawings from China, Malacca, etc. TAXODIACEAE.

Cupa'nia f. Ornamental evergreen trees, resembling *Blighia*, named in honour of Francesco Cupani (1657–1711), Italian monk and author of works on Sicilian plants. SAPINDACEAE.

Cu'phea f. Gr. *kyphos*, curved; from the shape of the seed capsule. LYTHRACEAE.

cuprea'tus, -a, -um; cupreus, -a, -um Coppery; copper-coloured.

cupressifo'lius, -a, -um With leaves like cypress.

cupressifor'mis, -is, -e; cupressi'nus, -a, -um; cupressoi'des Resembling cypress.

Cupressocy'paris f. Hybrid between *Chamaecyparis* and *Cupressus*.

Cupres'sus f. Cypress. The Latin name for the Italian cypress tree (*C. sempervirens*). CUPRESSACEAE.

curassa'vicus, -a, -um Of Curaçao, Dutch Antilles, Caribbean Sea.

Curcu'ligo f. L. *curculio*, a weevil. The ovary is beaked like a weevil. HYPOXIDACEAE.

Curcu'ma f. Turmeric. The Latinized version of the Arabic name. The stout rootstocks are the source of turmeric which is both a dyestuff and a condiment important in curry to which it imparts the yellow colour. The name crocus has the same derivation. ZINGIBERACEAE.

Curto'nus m. From Gr. *kyrtos*, swelling, hunch-backed, bent. IRIDACEAE.

cur'tus, -a, -um Shortened.

curva'tus, -a, -um Curved.

curvifo'lius, -a, -um With curved leaves.

Cuscu'ta f. Medieval Latin name for dodder. CONVOLVULACEAE.

cuscutifor'mis, -is, -e Resembling *Cuscuta* or dodder.

cuspida'tus, -a, -um With a sharp stiff point; cuspidate.

cuspidifo'lius, -a, -um With leaves having a sharp stiff point.

Cyanan'thus m. Gr. *kyanos*, blue; *anthos*, a flower; from the colour of the flower. CAMPANULACEAE.

cyanan'thus, -a, -um With blue flowers.
Cyanel'la f. Gr. *kyanos*, blue. TECOPHILACEAE.
cya'neus, -a, -um Blue.
cyanocar'pus, -a, -um With blue fruit.
cyanophyl'lus, -a, -um With blue leaves.
Cyano'tis f. Gr. *kyanos*, blue; *ous*, an ear; an allusion to the colour and form of the petals. COMMELINACEAE.
cy'anus Old name of a dark blue substance now the specific epithet for the cornflower.
Cya'thea f. Tree fern. From Gr. *kyatheion*, a little cup; in allusion to the spore cases. CYATHEACEAE.
cyatheoi'des Resembling *Cyathea*.
Cyatho'des f. From Gr. *kyathōdēs*, cup-like; in allusion to the cup-shaped, toothed disc of these evergreen shrubs. EPACRIDACEAE.
Cybiste'tes m. Gr. *kybistētēs*, one who tumbles; referring to the dry mature fruiting heads being tumbled and blown about by the wind. AMARYLLIDACEAE.
Cy'cas f. Greek name for a kind of palm. Although palmlike these are not palms. Sago-palm is *C. revoluta*. CYCADACEAE.
Cyc'lamen n. Sowbread. The Greek name presumably from *kylos*, circle; referring to the rounded tubers. They are regarded as a favourite food for swine in the South of France, Sicily, and Italy. PRIMULACEAE.
cyclami'neus, -a, -um Resembling *Cyclamen*.
Cyclanthe'ra f. Gr. *kyklos*, a circle; *anthera*, anther; from the arrangement of the anther cells into rings . CUCURBITACEAE.
Cyclan'thus m. Gr. *kyklos*, a circle; *anthos*, a flower; from the arrangement of the flowers of this herb. CYCLANTHACEAE.
cyclocar'pus, -a, -um With fruit arranged in a circle.
Cyclo'pia f. From Gr. *kyklŏs*, circle; *pous*, foot; referring to the circular base of the calyx. LEGUMINOSAE.
cyc'lops Gigantic, like the huge one-eyed Cyclops of mythology.
Cydis'ta f. From Gr. *kydistos*, most glorious; in reference to the spectacular flowers of these ornamental vines. BIGNONIACEAE.
Cydo'nia f. Quince. The Latin name as also *cotonea* for quince, derived from the town of Cydon (now Khania) in Crete. ROSACEAE.
cylindra'ceus, -a, -um; cylindricus, -a, -um Long and round, cylindrical.
cylindrosta'chyus, -a, -um With a cylindrical spike.
Cymba'laria f. Ivy-leaved toadflax. From Gr. *kymbalon*, L. *cymbalum*, cymbal, referring to the leaf shape of some species. SCROPHULARIACEAE.
Cymbi'dium Gr. *kymbē*, a boat. There is a hollow recess in the lip. ORCHIDACEAE.
cymbifor'mis, -is, -e Boat-shaped.

Cymbopo'gon m. Gr. *kymbē*, a boat; *pōgōn*, a beard; in allusion to the spikelets of these oil-producing grasses. GRAMINEAE.

cymo'sus, -a, -um Furnished with cymes or flower clusters in which the flower in the centre opens first, the remainder of the flowers open in succession outwards to the periphery (as in *Phlox*); from L and Gr. *cyma*, a young sprout.

Cynanast'rum n. Gr. *kyanos*, blue; *-astrum*, incompletely resembling. The name was apparently suggested by resemblance to the genus *Cyanella*. TECOPHILACEAE.

Cynan'chum n. Gr. *kyōn*, a dog; *ancho*, to strangle. Most species of these herbs and sub-shrubs are poisonous. ASCLEPIADACEAE.

Cy'nara f. Artichoke and cardoon. The Latin name. COMPOSITAE.

cynaroi'des Resembling *Cynara*.

Cyno'ches n. From Gr. *kyknŏs*, swan; *auchē*, neck; referring to the arched slender column of the male flowers. ORCHIDACEAE.

Cy'nodon n. Bermuda grass. From Gr. *kyōn*, a dog; *ŏdŏus*, a tooth. GRAMINEAE.

Cynoglos'sum n. Hound's tongue. From Gr. *kyōn*, a dog; *glōssa*, a tongue; in allusion to the leaves. BORAGINACEAE.

cyparis'sias f. Latin name of a kind of spurge, from Gr. *kyparissos*, cypress.

Cypel'la f. Gr. *kypellon*, a goblet; from the form of the flowers. IRIDACEAE.

Cype'rus m. Galingale. Greek word meaning sedge. The genus includes *Papyrus*, from which the Egyptians made a writing material by cutting the pithlike tissues into strips, laying them crosswise and uniting them under pressure. CYPERACEAE.

Cyphoman'dra f. Gr. *kyphos*, a tumour; *aner*, *andros*, male; with reference to a thickening on the connective of the anther. SOLANACEAE.

Cypripe'dium n. Lady's slipper, moccasin flower. From Gr. *Kypris*, Latin *Cypria*, Venus (Aphrodite), who was worshipped especially on Cyprus; *pĕdilon*, a slipper, unfortunately Latinized as *pedium* by Linnaeus, from the shape of the flower. ORCHIDACEAE.

cyp'rius, -a, -um Cyprian, relating to the island of Cyprus.

Cyril'la f. Leatherwood. Named in honour of Dominico Cirillo (1734–1790), physician and professor of botany at Naples, executed by the Bourbons for his liberal opinions. CYRILLACEAE.

Cyrtanthe'ra f. From Gr. *kyrtos*, arched; in allusion to the curved anthers. ACANTHACEAE.

Cyrtan'thus m. Gr. *kyrtos*, arched; *anthos*, a flower; referring to the curved flower tube. AMARYLLIDACEAE.

Cyrto'mium n. Gr. *kyrtos*, arch; with reference to the veining of the leaves. ASPLENIACEAE.

Cyrtopo'dium n. Gr. *kyrtos*, arched; *pous*, a foot; from the form of the lip of the flower of these terrestrial orchids. ORCHIDACEAE.

Cyrto'stachys m. Gr. *kyrtos*, arched; *stachys*, ear of corn, a spike. The flower spikes of these palms are curved. PALMAE.

Cysto'pteris f. Bladder fern. From Gr. *kystis*, a bladder; *ptĕris*, a fern with reference to the subglobos indusium. ASPLENIACEAE.

cythe'reus, -a, -um From the Greek island of Kythera.

cytisoi'des Resembling *Cytisus* or broom.

Cy'tisus m. Broom. Gr. *kytisos*, used by the Greeks for several kinds of woody *Leguminosae*. LEGUMINOSAE.

D

Daboe'cia f. St. Dabeoc's heath. Named after a rather obscure Irish saint. Misspelled *Daboecii* by Linnaeus. ERICACEAE.

dacrydioi'des Resembling *Dacrydium*.

Dacry'dium n. Gr. *dakrydion*, a small tear, diminutive of *dacry*, tear. The trees, called pines in New Zealand, exude drops of resin. PODOCARPACEAE.

dacty'lifer, dactyli'fera, -um Fingerlike; furnished with fingers.

Dac'tylis f. Cock's foot grass. From Gr. *dactylŏs*, a finger; referring to the finger-like appearance of the inflorescence. GRAMINEAE.

dactyloi'des Resembling fingers.

Dactylorhi'za f. From Gr. *daktylŏs*, finger; *rhiza*, root; in allusion to the finger-like tubers of these orchids which contrast with the rounded tubers of *Orchis* proper. ORCHIDACEAE.

Daemo'norops f. From Gr. *daimon*, Latin *daemon*, evil spirit, demon; *rhōps*, a shrub; in allusion to the stem of these rattans. They climb by means of sharp recurved hooks on the main stem and leaf stalks. PALMAE.

Da'hlia f. Pronunciation is largely a matter of taste; probably the most convenient course would be to use 'day-lia' when referring to the garden dahlias in the English-speaking world but 'dah-lia' when referring to species of the genus *Dahlia*, from which the garden forms have been derived by hybridization, thus avoiding confusion with the genus *Dalea*. These perennials from Mexico were named in honour of Dr. Anders Dahl (1751–1789), Swedish botanist and pupil of Linnaeus, who sought unsuccessfully to retain the Linnaean collections in Sweden. COMPOSITAE.

dahu'ricus, -a, -um; dau'ricus, -a, -um; davu'ricus, -a, -um Of Dahuria (Dauria, Davuria), region of south-east Siberia and north-east Mongolia.

Da'is f. Greek word meaning a torch; an allusion to the form of inflorescence. THYMELAEACEAE.

Dalber'gia f. Named for the brothers Nils Dahlberg (1736–1820), Swedish botanist and royal physician, and Carl Gustav Dahlberg (*fl.* 1754–1775),

Swedish officer, who owned an estate in Surinam, whence he sent specimens to Linnaeus. LEGUMINOSAE.

Da'lea f. Herbs and small shrubs named in honour of Dr. Samuel Dale (1659–1739), English botanist, apothecary and physician, at Braintree, Essex. LEGUMINOSAE.

dalecar'licus, -a, -um Of the province of Dalecarlia, or Dalarna, in Sweden.

Dalecham'pia f. Named for Jacques Dalechamps (1513–1588), French physician and botanist, author of *Historia generalis Plantarum* (1587). EUPHORBIACEAE.

dalhou'siae In honour of Lady Dalhousie, wife of James A. B. Ramsay (1812–1860), 1st Marquess of Dalhousie, Governor-General of India.

Dalibar'da f. An herbaceous perennial named in honour of Thomas François Dalibard (1703–1779), French botanist, author of a *Flora Parisiensis Prodromus* (1749). ROSACEAE.

dalma'ticus, -a, -um Of Dalmatia, on the Adriatic side of Balkan Peninsula.

damasce'nus, -a, -um Of Damascus, Syria.

Damaso'nium n. Gr. name for water-plantain (*Alisma*). ALISMATACEAE

dam'meri. Named for Carl Lebrecht Udo Dammer (1860–1920), German botanist for many years at the Botanical Museum, Berlin-Dahlem.

Danaë f. Alexandrian laurel. Named for Dănāē, in Greek mythology the daughter of Acrisius, King of Argos. Having been warned that she would bear a son who would eventually kill him, he shut her up in a brazen tower only to have Zeus rescue her with the result that she gave birth to Perseus. Acrisius placed both in a wooden box and threw them into the sea. By one of those twists of fate which make up Greek tragic plots, Perseus grew up to kill his grandfather accidentally when practising with the discus. LILIACEAE (RUSCACEAE).

danaeifo'lius, -a, -um With leaves like *Danaë*.

Da'phne f. Greek name for the bay tree or laurel (*Laurus nobilis*) later transferred to the present genus. According to mythology it was named after a nymph changed by the gods into a bay tree to save her from pursuit by Apollo, but the name itself may be derived from an Indo-European root meaning 'odour'. THYMELEAECEAE.

Daphniphyl'lum n. Gr. *daphnē*, laurel; *phyllon*, a leaf. The leaves of these evergreen shrubs and small trees resemble laurel. DAPHNIPHYLLACEAE.

daphnoi'des Resembling *Daphne*.

Darlingto'nia f. Named in honour of Dr. William Darlington (1782–1863), Philadelphia physician and botanist who published the posthumous botanical work of his friend and fellow physician, William Baldwin (1779–1819), who died on an exploring trip up the Missouri River. SARRACENIACEAE.

Dar'mera f. In honour of Karl Darmer, 19th-century Berlin horticulturist. A replacement name for *Peltiphyllum* (Engler) Engler (1891), not *Peltophyllum* Gardner (1841). SAXIFRAGACEAE.

Darwi'nia f. Named in honour of Dr. Erasmus Darwin (1731–1802), physician, man of science and writer of poems on scientific subjects, grandfather of Charles Darwin. MYRTACEAE.

darwinii In honour of Charles Darwin (1809–1882) geologist, zoologist and botanist who collected plants on the 5-year voyage of *The Beagle* lastingly celebrated for his theory of evolution by means of natural selection. See *Fitzroya*.

dasy- A prefix in many Greek compound words meaning hairy or thick. Thus: **dasyacan'thus** (thick-spined); **dasyan'thus** (shaggy-flowered); **dasycar'pus** (with hairy fruit); **dasy'cladus** (shaggy-branched); **dasyphyl'lus** (shaggy-leaved); **dasyste'mon** (with hairy stamens).

Dasyli'rion n. Gr. *dasys*, thick; *lirion*, lily; in allusion to the thick stems and liliaceous flowers. AGAVACEAE.

Datis'ca f. The derivation of the name of these herbaceous perennials is obscure. DATISCACEAE.

Datu'ra f. Thorn-apple. From an Indian vernacular name. Some of the species are violently narcotic and poisonous, including Jimson weed (*D. stramonium*), after wilting. *D. metel* is common in India and provides a convenient source of poison. The shrubby species (Angel's trumpet) are now placed in the genus *Brugsmania*. SOLANACEAE.

Daubento'nia f. Herbs or shrubs named for a French naturalist, Louis Jean Marie Daubenton (1716–1799). LEGUMINOSAE.

daucifo'lius, -a, -um With leaves like carrot or *Daucus*.

daucoi'des Resembling carrot or *Daucus*.

Dau'cus m. Carrot. The Latin name. UMBELLIFERAE.

Daval'lia f. Hare's-foot fern. Named for Edmond Davall (1763–1798), Swiss botanist of English origin, a friend of James Edward Smith, to whom he bequeathed his herbarium. DAVALLIACEAE

Davi'dia f. Dove-tree, ghost-tree, handkerchief-tree. Named in honour of Abbé Armand David (1826–1900), French missionary in China, 1862–1873, who collected many plants and animals. He discovered what is now known as Père David's Deer (*Elaphurus davidianus*) in the grounds of the Imperial Summer Palace at Peking; it survives only in captivity. NYSSACEAE.

davi'dii In honour of Armand David.

Davie'sia f. In honour of Hugh Davies (*c.* 1739–1821), Welsh clergyman, author of *Welsh Botanology* (1813). LEGUMINOSAE.

davi'sii In honour of Peter Hadland Davis (1918–1992), editor and chief author of *Flora of Turkey and the East Aegean Islands* (1965–1988).

dawsonia'nus, -a, -um Species named for Jackson T. Dawson (1841–1916), first superintendent of the Arnold Arboretum, Boston, Massachusetts.

dealbatus, -a, -um Whitened; covered with opaque white powder.

Dea'mia f. Cactus named for Charles Clemon Deam (1865–1953), American druggist and botanist in Indiana. CACTACEAE.

deb'ilis, -is, -e Weak, frail.

Debregea'sia f. Named after Prosper Justin de Brégeas (b. 1807), French naval officer, who commanded the corvette *La Bonite* on a voyage of exploration to the Far East (1836–1837). URTICACEAE.

Decais'nea f. Named in honour of Joseph Decaisne (1807–1882), director of the Jardin des Plantes, Paris, eminent French botanist and horticulturist of Belgian origin. LARDIZABALACEAE.

decan'drus, -a, -um With ten stamens.

decape'talus, -a, -um With ten petals.

decaphyl'lus, -a, -um With ten leaves.

deci'duus, -a, -um Deciduous.

deci'piens Deceptive; cheating.

declina'tus, -a, -um Bent downwards.

De'codon n. Swamp loosestrife. Gr. *deka*, ten; *ŏdŏus*, a tooth. The calyx has ten teeth. LYTHRACEAE.

decolo'rans Discolouring; staining.

decompo'situs, -a,-um More than once divided, usually much divided.

decora'tus, -a, -um; deco'rus, -a, -um Decorative; becoming; comely.

decor'ticans Shedding bark.

decuma'nus, -a, -um Very large; immense; literally the tenth part, but acquiring its meaning of importance from association with the main entrance of Roman camps, called the Decumana Porta, because the tenth part of a Roman division was quartered near it.

Decuma'ria f. L. *decimus*, ten; from the number of parts of the flower. HYDRANGEACEAE.

decum'bens Trailing with tips upright; decumbent.

decur'rens Running down the stem; decurrent; e.g. a leaf base that forms a flange or wing that merges with the stalk below it.

Deerin'gia f. Named for George Charles Deering (*c.* 1695–1749), physician, who was born in Saxony, studied at Leiden and Paris and practised medicine in London and later Nottingham. He was the author of a catalogue of plants found at Nottingham. AMARANTHACEAE.

defl'exus, -a, -um Bent abruptly downward.

defor'mis, -is, -e Misshapen, deformed.

dehis'cens Dehiscent, namely the splitting or opening of a seed-pod for the release of seeds or the opening of an anther to discharge pollen. The function can sometimes be relatively violent and both visible and audible as in the popping of broom pods on a hot day. In the case of the castor-oil plant the seeds are thrown several feet and it is important to harvest them in time.

Deinan'the f. From Gr. *dĕinŏs*, wondrous; *anthē*, flower. HYDRANGEACEAE.
dejec'tus, -a, -um Debased.
delec'tus, -a, -um Chosen.
delicatis'simus, -a, -um Most delicate.
delica'tus, -a, -um Delicate.
delicio'sus, -a, -um Delicious.
De'lonix f. Royal poinciana, flamboyant, gul mohur, fancy-anna, possibly the most gorgeous tree in cultivation. From Gr. *dēlŏs*, evident, conspicuous; *ŏnux*, a claw; referring to the long-clawed petals. LEGUMINOSAE.
Delosper'ma n. From Gr. *dēlŏs*, evident; *sperma*, seed; in allusion to the seeds exposed in the unroofed chambers of the capsule. A genus formerly included in *Mesembryanthemum*. AIZOACEAE.
Delo'stoma n. Gr. *dēlos*, evident; *stŏma*, a mouth. the flower has a wide mouth. BIGNONIACEAE.
del'phicus, -a, -um Of Delphi, Greece.
delphinan'thus, -a, -um With flowers like *Delphinium*.
delphinen'sis, -is, -e Of Dauphiné, France.
delphinifo'lius, -a, -um With leaves like *Delphinium*.
Delphin'ium n. Larkspur. The Greek name, *dĕlphiniŏn*, derived from the word *dĕlphis* for a dolphin in allusion to the form of the flower of annual species (now placed in *Consolida*) found in Greece. RANUNCULACEAE.
deltoi'des; deltoi'deus, -a, -um Triangular, like the Greek letter Δ(delta).
Dema'zeria (incorrectly, **Desmazeria**) f. A genus of grasses named after Jean Baptiste Joseph Henri Desmazières (1796–1862), French botanist. GRAMINEAE.
demer'sus, -a, -um Living under water; submerged.
demis'sus, -a, -um Hanging down; weak.
Dendran'thema f. From Gr. *dĕndrŏn*, tree; *anthĕmŏn*, flower; with reference to the woody flower stems of these ornamental garden plants usually included in the genus *Chrysanthemum*. COMPOSITAE.
dendri'cola A dweller on trees.
Dendro'bium n. Gr. *dĕndrŏn*, a tree; *bios*, life. These tree-perching orchids are, next to the cattleyas, the most popular greenhouse orchids. ORCHIDACEAE.
Dendroca'lamus m. Giant bamboo. From Gr. *dĕndrŏn*, a tree; *calamus*, a reed. Some of the species grow to a height of 100 feet. Among them is the male bamboo from which surf-casting and split-cane fishing rods are made. GRAMINEAE.
Dendrochi'lum n. Gr. *dĕndrŏn*, a tree; *cheilos*, a lip. The flowers of these tree-perching orchids have conspicuous lips. ORCHIDACEAE.
dendroi'deus, -a, -um Treelike.
Dendrome'con f. Tree-poppy. From Gr. *dĕndrŏn*, a tree; *mēkōn*, a poppy; in allusion to its shrubby habit. PAPAVERACEAE.

Dendro'panax From Gr. *dĕndrŏn*, tree; *Panax*, which see. ARALIACEAE.

dendro'philus, -a, -um Tree-loving.

Denmo'za f. Name an anagram of Mendoza, the province of Argentina to which this cactus is native. CACTACEAE.

Dennstae'dtia f. Hayscented fern. Named for August Wilhelm Dennstedt, German botanist (1776–1826). DENNSTAEDTIACEAE.

densa'tus, -a, -um; densus, -a, -um Compact; dense.

dens-ca'nis Specific epithet meaning dog's tooth.

densiflor'us, -a, -um Densely flowered.

densifo'lius, -a, -um Densely leaved.

Denta'ria f. Toothwort. From L. *dens*, a tooth; in allusion to the toothlike scales on the rhizome which led to the supposition that it might be good for toothache. This is often included in *Cardamine*. CRUCIFERAE.

denta'tus, -a, -um; den'tifer, denti'fera, -um; dento'sus, -a, -um Toothed; furnished with teeth.

denticula'tus, -a, -um Slightly toothed.

denuda'tus, -a, -um Bare; naked.

deoda'ra Specific epithet from the North Indian name for the deodar (*Cedrus deodara*).

depaupera'tus, -a, -um Imperfectly developed; dwarfed.

depen'dens Hanging down.

depres'sus, -a, -um Flattened; pressed down.

Der'ris f. Gr., a leather covering; in allusion to the tough seed pods. The insecticide, derris, is made from the powdered tuberous root. In the East Indies the powder is used to poison fish by stupefying them. LEGUMINOSAE.

Descham'psia f. In honour of Louis August Deschamps (1765–1842), French surgeon-naturalist, who investigated the natural history of Java. GRAMINAE.

Descurai'nia f. In honour of François Descourain (1658–1740), French pharmacist. CRUCIFERAE.

deser'ti Of the desert.

deserto'rum Of the deserts.

Desfontai'nia f. A Chilean evergreen shrub named for René Louiche Desfontaines (1750–1833), French botanist, professor at the Jardin des Plantes, Paris; see *Fontanesia*. LOGANIACEAE.

Desman'thus m. Gr. *dĕsmē*, a bundle; *anthos*, a flower. The flowers of these herbs or shrubs are collected together in spikes resembling bundles. LEGUMINOSAE.

Desmo'dium n. Tick-trefoil. From Gr. *dĕsmos*, a band or chain; with reference to the jointed pods. LEGUMINOSAE.

desmoncoi'des Resembling the genus *Desmoncus*.

Desmon'cus m. Gr. *dĕsmos*, a band or chain; *onkos*, a hook. The ends of the leaves of these tropical American spiny palms are furnished with hook-like tips. PALMAE.

desquama'tus, -a, -um With scales removed; scoured; bereft of scales.

deter'sus, -a, -um Wiped clean.

deton'sus, -a, -um Bare; shorn.

deu'stus, -a, -um Burned.

Deut'zia f. Very ornamental shrubs. After Johan van der Deutz (1743–1788), Dutch friend and patron of Carl Thunberg, Swedish botanist and pupil of Linnaeus. HYDRANGEACEAE.

diabo'licus, -a, -um Devilish; applied, for example, to a plant with a two-horned fruit.

diacan'thus, -a, -um Furnished with two spines.

diac'ritus, -a, -um Separated; excellent.

Dia'crium n. From Gr. *di*, double; *akris*, peak, point; referring to the two horn-like points on the lip. ORCHIDACEAE.

diade'ma Gr. *dĭadēma*, royal head-dress, crown, diadem.

dian'drus, -a, -um Furnished with two or twin stamens.

Dianel'la f. Flax lily. Diminutive of Diana, goddess of the chase. LILIACEAE.

dianthiflo'rus, -a, -um *Dianthus*-flowered.

Dian'thus m. Pink, carnation. From Gr. *Di*, of Zeus or Jove; *anthos*, flower. CARYOPHYLLACEAE.

Diapen'sia f. Classical Greek plant name adopted by Linnaeus. DIAPENSIACEAE.

dia'phanus, -a, -um Transparent.

diap'repes Distinguished.

Dia'scia f. Gr. *di*, two; *askos*, a sac. The flowers have two spurs. SCROPHULARIACEAE.

Dicen'tra f. From Gr. *dis*, twice; *kentron*, a spur. The flowers of this genus, which includes bleeding heart, have two spurs. FUMARIACEAE.

Dichon'dra f. Gr. *di*, two; *chŏndrŏs*, lump, grain of corn; with reference to the two capsules. CONVOLVULACEAE.

Dichorisan'dra f. Gr. *dis*, twice; *chōri*, separate; *aner*, *andros*, male; in allusion to two lateral stamens which spread outwards and away from the converging four other stamens. COMMELINACEAE.

dicho'tomus, -a, -um Forked in pairs; repeatedly dividing into two branches.

Di'chroa f. Gr. *dis*, twice; *chrŏa*, colour. HYDRANGEACEAE.

dichroan'thus, -a, -um With flowers of two distinct colours.

dichro'mus, -a, -um; di'chrous,-a, -um Of two distinct colours.

Dickso'nia f. Tree ferns named after James Dickson (1738–1822), British botanist and nurseryman. DICKSONIACEAE.

Dicli'ptera f. Gr. *diklis*, double-folding; *pteron*, wing; in allusion to the two wing-like divisions of the valves of the capsule. ACANTHACEAE.

dicoc'cus, -a, -um Furnished with two berries or nuts.

Dicranostig'ma n. From Gr. *dikranos*, two-branched; plus *stigma*. PAPAVERACEAE.

Dictam'nus m. Greek name for a Cretan origanum, probably named after Mount Dikte, later transferred to this perennial herb which is variously called dittany, gas plant, burning bush, and fraxinella. The leaves and stem, which have a strong odour, give out a volatile oil which, on a hot still day, can be ignited with a match. RUTACEAE.

dictyophyl'lus, -a, -um With leaves having a conspicuous network of veins and veinlets. Similarly, *dictyocarpus*, with netted fruit.

Dictyosper'ma n. From Gr. *dictyon*, a net; *sperma*, seed; in a technical allusion to the seed of these *Areca*-like palms. PALMAE.

dictyo'tus, -a, -um Made in net-fashion; netted; latticed.

Didis'cus m. Gr. *di*, two; *diskos*, disc. UMBELLIFERAE.

Didymocar'pus m. From Gr. *didymos*, double; *karpos*, fruit; the capsule breaking into two parts. GESNERIACEAE.

di'dymus, -a, -um Twin or in pairs; two-fold.

Dieffenba'chia f. Dumb cane. Named in honour of J. F. Dieffenbach (1790–1863) in charge of the gardens of the royal palace of Schönbrunn at Vienna around 1830. ARACEAE.

dieffenba'chii Specific epithet of a curious monotypic umbelliferous genus, *Coxella*, confined to the Chatham Islands where the first botanical collections were made by Ernst Dieffenbach, naturalist to the New Zealand Company, 1839–1841.

dielsia'nus, -a, -um; dielsii In honour of Friedrich Ludwig Emil Diels (1874–1945), director of Berlin-Dahlem botanic garden and museum, especially noted for his study of Australian plants.

Diera'ma n. Wandflower. Gr. *dĭĕrāma*, a funnel; from the shape of the flower. IRIDACEAE.

Diervil'la f. Bush honeysuckle. Named for M. Dierville, a French surgeon who travelled in Canada, 1699–1700, and who introduced *D. lonicera* into European cultivation. CAPRIFOLIACEAE.

diffor'mis, -is, -e Of unusual form in relation to the normal run of the genus.

diffu'sus, -a, -um Spreading; diffuse.

digitaliflo'rus, -a, -um With foxglove-like flowers.

Digita'lis f. Foxglove. A latinization of the German name 'Fingerhut'. From L. *digitus*, a finger. The flower is like the finger of a glove. *D. purpurea* is a source of the digitalin used in cardiac medicine. SCROPHULARIACEAE.

digita'tus, -a, -um Shaped like an open hand; digitate.

dilata'tus, -a, -um; dila'tus, -a, -um Spread out; expanded.

Dille'nia f. Named by Linnaeus for Johann Jacob (John James) Dillenius (1684–1747), German botanist and physician. He came to England in 1721 from Giessen to help William Sherard, produced a magnificent book *Hortus*

Elthamensis (1732), illustrating and describing the plants of James Sherard's garden at Eltham near London, and became Sherardian professor of botany at Oxford in 1734 where in 1741 he published his celebrated *Historia Muscorum*. To quote Linnaeus, '*Dillenia* of all plants has the showiest flower and fruit, even as Dillenius made a brilliant show among botanists.' DILLENIACEAE.

Dillwy'nia f. Evergreen flowering shrubs named in honour of Lewis Weston Dillwyn (1778–1855), British botanist and porcelain manufacturer at Swansea, Wales. LEGUMINOSAE.

dimidia'tus, -a, -um Divided into two dissimilar or unequal parts.

Dimorphothe'ca f. Cape marigold. From Gr. *dis*, twice; *morphe*, a shape; *thēka*, 'fruit'; in allusion to the two kinds of achenes (*cypselae*) found in one and the same fruiting head; those of the disc and ray differ in shape. COMPOSITAE.

dimor'phus, -a, -um Having two forms of leaf, flower, or fruit on the same plant.

Dinteran'thus m. In honour of Professor Kurt Dinter (1868–1945), German botanist who made extensive collections in South-west Africa from 1897 to 1906. *Juttadintera* commemorates his wife. AIZOACEAE.

di'odon Having two teeth.

dioi'cus, -a, -um Dioecious, i.e. having the male reproductive organs borne on one plant and the female on another.

Diona'ea f. Venus flytrap. One of the Greek names for Venus. Found only in North and South Carolina, the plant has leaves each divided into two hinged, valvelike segments, the inner faces of which have three sensitive 'trigger' hairs. When an insect lands inside and touches any two of the hairs, the valves close like jaws, the interlocking stiff hairs around the outside edge of the leaves holding the victim. The trap reopens when the insect is digested. DROSERACEAE.

Diony'sia f. From *Dĭōnȳsŏs*, Greek name of Bacchus. PRIMULACEAE.

Dio'on n. From Gr. *dis*, twice; *ōŏn*, an egg; with reference to the paired seeds; originally spelled *Dion*. ZAMIACEAE.

Diosco'rea f. Yam. Named in honour of Pedanios Dioscorides, 1st-century Greek physician and herbalist who compiled *Materia Medica* which was the most important work on medical herbs throughout the Middle Ages. DIOSCOREACEAE.

Dios'ma f. Gr. *dios*, divine; *ŏsme*, fragrance. The leaves when crushed are very fragrant. RUTACEAE.

diosmifo'lius, -a, -um With leaves like *Diosma*.

Dios'pyros f. Persimmon, ebony. Gr. *dĭŏspyrŏs* from *dios*, divine; *pyros*, wheat; a name transferred to this genus with edible fruits. EBENACEAE.

Dipca'di n. A Turkish name (not a South African one) for the musk hyacinth but transferred to this predominantly African genus. HYACINTHACEAE.

Dipel'ta f. Gr. *di*, two; *pelte*, a shield; referring to the two conspicuous shield-like bracts at the base of the flower. CAPRIFOLIACEAE.

dipe'talus, -a, -um Having two petals.

Diphylle'ia f. Gr. *di*, two; *phyllon*, a leaf, with reference to the two leaves on the stem. BERBERIDACEAE.

diphyl'lus, -a, -um Having two leaves or leaflets.

Diplade'nia f. Gr. *diplŏŏs*, double; *aden*, a gland. There are two glands on the ovary. APOCYNACEAE.

Diplarrhe'na f. From Gr. *diplŏŏs*, double; *arrēn*, male; referring to the two fertile stamens. IRIDACEAE.

Dipla'zium n. Gr. *diplasios*, double. The indusium, the covering over the spores, is double in these ferns. ASPLENIACEAE.

Diploglot'tis f. Gr. *diplŏŏs*, double; *glottis*, a tongue. The inner scale of the petals is doubled on this Australian tree. SAPINDACEAE.

diplostephioi'des Resembling *Diplostephium*, a genus of little garden interest.

diplo'trichus, -a, -um Gr. *diplŏŏs*, double; *thrix*, a hair; with two kinds of hair, that is, long and short hairs intermixed.

dipsa'ceus, -a, -um Resembling the teasel.

Dip'sacus m. Teasel. Gr. *dipsakos*, from *dipsa*, thirst; the united bases of two opposite leaves form a little basin holding rain water. The plants are cultivated for the fruiting heads which are covered with stiff, hooked points and are used in the woollen industry to raise the nap on cloth. So far, no man-made device has been found as effective. DIPSACACEAE.

Dipteracan'thus m. Gr. *dipteras*, two-winged; *Acanthus*; in allusion to the two-leaved peduncle. ACANTHACEAE.

dipterocar'pus, -a, -um Having two-winged fruit.

Diptero'nia f. Gr. *di*, two; *pteron*, a wing; in allusion to the winged fruit. ACERACEAE.

dip'terus, -a, -um Two-winged.

dipyre'nus, -a, -um Having two seeds or kernels.

Dir'ca f. Leatherwood. From Gr. *dirke*, a fountain. These early blooming North American shrubs grow in moist places. THYMELAEACEAE.

Di'sa f. In allusion to the mythical Queen Disa of Sweden who was commanded to come before the King of the Sveas neither naked nor clothed and who accordingly came wrapped in a fishing net; the upper lip of the flower of the type-species has a somewhat netted appearance. ORCHIDACEAE.

Disan'thus m. Gr. *dis*, twice; *anthos* flower; in allusion to the paired flowers. HAMAMELIDACEAE.

Disca'ria f. Gr. *diskos*, a disc. The flower has a large fleshy disc. RHAMNACEAE.

discifor'mis, -is, -e Disc-shaped.

Discocac'tus m. Gr. *diskos*, a disc, and *cactus*; from the shape of these plants which are globe-shaped or flattened endwise. CACTACEAE.

discoi'deus, -a, -um Discoid; without rays.
disco'lor Of two different and, usually, distinct colours.
Disocac'tus m. From Gr. *dis*, twice; *ĭsŏs*, equal; *cactus*; the sepals and petals of these cacti are equal in number. CACTACEAE.
dis'par Unequal; dissimilar from the normal state of the genus.
disper'sus, -a, -um Scattered.
Dis'porum n. Fairybells. From Gr. *dis*, two; *spora*, seed; in allusion to the two ovules in each chamber of the ovary. LILIACEAE.
dissec'tus, -a, -um Deeply cut; divided into deep lobes or segments.
dissi'milis, -is, -e Unlike the normal for the genus.
dissitiflo'rus, -a, -um With flowers in loose heads, not compact.
dis'situs, -a, -um Lying apart, well-spaced.
dista'chyus, -a, -um With two spikes.
dis'tans Widely separated.
distichophyl'lus, -a, -um with leaves arranged in two ranks.
dis'tichus, -a, -um In two ranks.
Dis'tictis f. Gr. *dis*, twice; *stiktos*, spotted; the flattened seeds look like two rows of spots in the capsule. BIGNONIACEAE.
distor'tus, -a, -um Misshapen; of grotesque form.
Disty'lium n. Gr. *dis*, two; *stylos*, a style; there are two styles. HAMAMELIDACEAE.
disty'lus, -a, -um Having two styles.
Diu'ris f. From Gr. *di-* two; *oura*, tail; referring to the two tail-like lateral sepals. ORCHIDACEAE.
diur'nus, -a, -um Day-flowering.
divarica'tus, -a, -um Spreading; growing in a straggling manner.
diver'gens Spreading out widely from the centre.
diversi'color Diversely coloured. Also:
diversiflo'rus, -a, -um (diversely flowered); **diversifo'lius, -a, -um** (diversely leaved); **diversifor'mis, -is, -e** (of diverse forms).
divi'sus, -a, -um Divided.
di'vus, -a, -um Belonging to the gods.
Docy'nia f. Anagram of *Cydonia*. These trees are closely related to *Cydonia*, quince. ROSACEAE.
dodecan'drus, -a, -um Having twelve stamens.
Dodeca'theon n. Shooting star, American cowslip. From Gr. *dodeka*, twelve; *thĕŏs*, god; the name given by Pliny to a spring flower, believed to be the primrose, which was under the care of the twelve principal gods. PRIMULACEAE.
Dodonae'a f. Named in honour of Rembert Dodoens (1517–1585), celebrated Flemish royal physician and herbalist, who became professor of medicine at Leiden in 1582. SAPINDACEAE.
dodonaeifo'lius, -a, -um *Dodonaea*-leaved.

dolabra'tus, -a, -um; dolabrifor'mis, -is, -e Hatchet-shaped.

Dolichan'dra f. Gr. *dŏlichŏs*, long; *anēr*, *andros*, male; in allusion to the long anthers of this climbing shrub. BIGNONIACEAE.

Do'lichos m. Greek for long; used by the ancient Greeks for long-podded beans, whence transferred to these climbing plants closely related to beans. LEGUMINOSAE.

dolichoste'mon Gr. *dŏlichŏs*, long; *stēmŏn*, thread, stamen.

Dolichothe'le f. Gr. *dŏlichŏs*, long; *thēlē*, a nipple; an allusion to the elongated tubercles or small above-ground tubers. CACTACEAE.

dolo'sus, -a, -um Deceitful; appearing like some other plant.

Dombey'a f. Ornamental evergreen trees or shrubs named in honour of Joseph Dombey (1742–1794), French botanist, who travelled in Chile and Peru with Ruiz and Pavon. STERCULIACEAE.

domes'ticus, -a, -um Frequently used as a house plant; domesticated.

Do'nax m. Gr. *dŏnax*, a kind of reed. MARANTACEAE.

Doo'dia f. Named in honour of Samuel Doody (1656–1706), London apothecary, curator of the Chelsea Physic Garden from 1691. BLECHNACEAE.

Dore'ma n. Gr. *dōrēma*, gift; in allusion to the gum ammoniacum given by one species. UMBELLIFERAE.

Dori'tis f. From Gr. *dŏry*, a spear; the lip being spear-shaped. ORCHIDACEAE.

doronicoi'des Resembling *Doronicum*.

Doro'nicum n. Derivation obscure. Called leopard's bane, this genus of perennial herbs was considered useful in destroying and warding off wild beasts. *D. pardalianches*, meaning to strangle leopards, is reputedly poisonous. COMPOSITAE.

Dorothean'thus m. In honour of Dorothea Schwantes, mother of the German specialist on succulent plants, Dr. Martin Heinrich Schwantes (1881–1960), archaeologist and professor at Kiel. AIZOACEAE.

Dorste'nia f. Named for Theodor Dorsten (1492–1552), German botanist, professor of medicine at Marburg. MORACEAE.

Doryan'thes f. Spear lily. From Gr. *dŏry*, a spear; *anthos*, a flower. The flower stem of this Australian desert plant is very long, sometimes as much as 20 feet. AMARYLLIDACEAE (DORYANTHACEAE).

Dory'cnium n. The ancient Greek name *doryknion* for a *Convolvulus*, later transferred to these plants of the Mediterranean region. LEGUMINOSAE.

Doryo'pteris f. From G. *dŏry*, a spear; *pteris*, a fern; from the form of the fronds. ADIANTACEAE.

doshongen'sis, -is, -e Of the Doshong Pass, south-eastern Tibet.

Dougla'sia f. Named for David Douglas (1798–1834), Scottish collector for the Horticultural Society of London, who introduced many Pacific Coast plants of great merit into European gardens. He came to Oregon in 1825 for a two-year botanical exploration, returning in 1830. He was killed in Hawaii; by a strange accident he fell into a wild-cattle pit where he was gored to death by a bull already trapped. PRIMULACEAE.

dougla'sii In honour of David Douglas.

Dovya'lis f. Derivation obscure. Sometimes incorrectly spelled *Doryalis. D. caffra*, the Kei apple, with orange-like fruits, is much used in South Africa as a hedge plant. FLACOURTIACEAE.

Downin'gia f. Named in honour of Andrew Jackson Downing (1815–1852), American landscape gardener and pomologist. He was the founder of a great tradition of American landscape design and his book *A Treatise on the Theory and Practice of Landscape Design* has run through many editions, with re-editing at least up to 1921. His ideas were naturalistic and had great influence on subsequent practice. CAMPANULACEAE.

Doxan'tha f. Cat's claw. From Gr. *dŏxa*, glory; *anthos*, a flower. BIGNONIACEAE.

Dra'ba f. Whitlow-grass or nailwort. Gr. *drabē*, for a cruciferous plant, probably *Cardaria draba*. This spring-flowering tufted herb was supposed to have value in poulticing whitlows. CRUCIFERAE.

drabifo'lius, -a, -um Having leaves like *Draba* or whitlow-grass.

Dracae'na f. The usual derivation of the name is from Gr. *drakaina*, a female dragon. The colour dragon's blood can be obtained from *D. draco*, the dragon tree. At least one authority (Pulteney), however, suggests that the genus was named by Clusius in honour of Sir Francis Drake, whom he met in 1581 and from whom he received some of his plant discoveries from the New World. He would naturally have regarded Drake as worth commemorating. The oldest dragon tree in the Canary Islands is now considered to be about 400 years old, according to Karl Mägdefrau.

A number of the plants popularly described and sold as *Dracaena* are in fact *Cordyline*. AGAVACEAE.

dracaenoi'des Resembling *Dracaena*.

dra'co Specific name meaning a dragon.

Dracoce'phalum n. Dragonhead. From Gr. *draco*, a dragon; *cĕphalē*, a head; from the shape of the flower. LABIATAE.

dracoce'phalus, -a, -um Dragon-headed.

Drac'ula f. Diminutive of *draco*, dragon; no association with Bram Stoker's *Dracula*. ORCHIDACEAE.

dracunculoi'des Resembling *Artemisia dracunculus* or tarragon.

Dracun'culus m. Dragon arum. Latin name for another plant, not this evil-smelling *Arum*-like plant with large blackish purple spathe. ARACEAE.

dracun'culus Latin word meaning a small dragon; specific epithet for *Artemisia dracunculus* or tarragon.

drepanophyl'lus, -a, -um With sickle-shaped leaves.

drepanostachyus, -a, -um With a sickle-shaped, i.e. curved, spicate inflorescence.

Dri'mia f. Gr. *drimys*, acrid, pungent. LILIACEAE (HYACINTHACEAE).

Drimio'psis f. From *Drimia*, name of a related genus; *opsis*, resemblance. LILIACEAE (HYACINTHACEAE).

Dri'mys f. Gr. *drimys*, acrid, pungent; from the taste of the bark of these evergreen trees and shrubs. WINTERACEAE.

Drosan'themum n. From Gr. *drŏsŏs*, dew; *anthos*, flower; with reference to the papillae on the flowers. AIZOACEAE.

Dro'sera f. Sundew. From Gr. *drŏsĕrŏs*, dewy; in allusion to the gland-tipped hairs on the leaves which give a dewy appearance to these carnivorous plants. DROSERACEAE.

dru'cei In honour of George Claridge Druce (1850–1932), Oxford pharmacist and celebrated amateur botanist who was a keen student of British plants and the author of several floras.

drummon'dii In honour of one or other of two brothers who collected for Messrs Veitch, James Drummond (*c.* 1784–1863), curator of Cork botanic garden 1809–1829, who collected in Western Australia, and Thomas Drummond (*c.* 1790–1831), who collected in Canada, Texas and elsewhere in North America.

drupa'ceus, -a, -um; dru'pifer, drupi'fera, -um Bearing fleshy fruits or drupes such as peach, cherry, and plum.

Dryan'dra f. Named for Jonas Dryander (1748–1810), Swedish botanist, librarian (1782–1810) to Sir Joseph Banks, part-author of *Hortus Kewensis* (1789), published under W. Aiton's name, and author of the *Catalogus Bibliothecae Historico-Naturalis J. Banks* in 5 volumes. PROTEACEAE.

Dry'as f. Mountain avens. From Gr. *dryas*, a wood nymph or dryad, to whom the oak was sacred. The leaves of one species, *D. octopetala*, resemble oak leaves. ROSACEAE.

drynarioi'des Resembling *Drynaria*, a fern with fronds which look like oak leaves.

dryophyl'lus, -a, -um Oak-leaved.

Dryo'pteris f. American shield fern. Wood fern. From Gr. *dryŏptĕris*, from *drys*, oak; *pteris*, a fern. ASPLENIACEAE.

Dry'pis f. Gr. *drypis*, a kind of thorny shrub mentioned by Theophrastus. CARYOPHYLLACEAE.

du'bius, -a, -um Doubtful, in the sense of not conforming to pattern.

Duboi'sia f. In honour of Charles Du Bois (1656–1740), merchant, treasurer to the Honourable East India Company, who amassed a big herbarium now at Oxford. SOLANACEAE.

Duches'nea f. Indian or mock strawberry. Trailing herbs named in honour of

Antoine Nicolas Duchesne (1747–1827), a French horticulturist, who published in 1766 a remarkable *Histoire naturelle des Fraisiers* dealing with the history and kinds of strawberry, to which this plant is related. ROSACEAE.

Dudley'a f. In honour of William Russel Dudley (1849–1911), first professor of botany at Stanford University, California. CRASSULACEAE.

dulcama'ra Latin for bittersweet. The specific epithet for climbing nightshade (*Solanum dulcamara*).

dul'cis, -is, -e Sweet.

dumeto'rum Of hedges or of bushy places.

dumo'sus, -a, -um Bushy, shrubby.

dunen'sis Of sand dunes.

du'plex; duplica'tus, -a, -um Double, duplicate.

dura'bilis, -is, -e Durable, lasting.

dura'cinus, -a, -um With hard fruit or berries.

Duran'ta f. Tropical American woody plants named for Castore Durante (*c.* 1529–1590), Papal physician and botanist in Rome. VERBENACEAE.

Du'rio m. From the Malayan name, *durian*. The heavy fruit, larger than a coconut, has a spiky outside rind which encloses an evil-smelling pith in which the pulp-covered seeds are embedded. This pulp (the aril) is the edible part, creamy white, delicious and sweet, but taints the breath like garlic. BOMBACEAE.

durius'culus, -a, -um Somewhat hard.

dur'ra Native Egyptian name for a fodder sorghum, *Sorghum vulgare* var. *durra*, fed to camels and other beasts.

du'rus, -a, -um Hard. Thus, *durior*, harder.

Duva'lia Dwarf succulent herbs named for Henri Auguste Duval (1777–1814), French medical man who published in 1809 a pamphlet on succulent plants in the Alençon garden containing descriptions of the genera *Haworthia* and *Gasteria*; this is reprinted in *The Cactus Journal*, 7, no. 4 (June 1939). ASCLEPIADACEAE.

Dy'ckia f. Succulent plants named for Prince Joseph Salm-Reifferscheid-Dyck (1773–1861), author of outstanding books on succulents. BROMELIACEAE.

dy'eri In honour of Sir William Turner Thiselton-Dyer (1843–1928), Director of the Royal Botanic Gardens, Kew, from 1885 to 1905, distinguished not only as a botanist and an administrator much disliked by many of his staff but also as a dogmatic classical scholar interested in the interpretation of ancient Greek plant-names.

dys- Greek prefix meaning 'bad, ill, hard, with difficulty', i.e. 'destroying the *good* sense of a word or increasing the *bad*' and used in the formation of many generic names.

Dyschoris'te f. Gr. *dys*, with difficulty; *choristos*, separated; referring to the valves of the capsule which adhere tenaciously. ACANTHACEAE.

dysente'ricus, -a, -um Relating to dysentery; used for plants serving as remedies for dysentery.

Dysosma f. Gr. *dys*, bad; *ŏsmē*, smell. Genus often included in *Podophyllum*. BERBERIDACEAE.

Dyso'xylum n. Gr. *dys*, bad; *xylon*, wood. The wood of some species of these trees has an unpleasant smell. MELIACEAE.

E

ebena'ceus, -a, -um Resembling ebony (*Diospyros* sp., etc.)

E'benus f. Gr. *ĕbĕnŏs*, apparently of Egyptian origin for another plant, not this genus of shrubs and sub-shrubs. LEGUMINOSAE.

e'benus, -a, -um Ebony-black.

Eberlan'zia f. In honour of Father Eberlanz of Lüderitz Bay, South-west Africa, collector of succulent plants *c.* 1926. AIZOACEAE.

eboracen'sis, -e Of York, England.

ebractea'tus, -a, -um Without bracts.

e'bulus Latin and Greek name of the dwarf elder or danewort, *Sambucus ebulus*.

ebur'neus, -a, -um Ivory-white.

Ecbal'lium n. Gr. *ekballein*, to cast out. The seeds of this trailing plant, called squirting cucumber, are violently ejected in a squirt of fluid when ripe. CUCURBITACEAE.

Eccremocar'pus m. Glory flower. From Gr. *ĕkkrĕmēs*, hanging; *karpos*, fruit. The fruit is a slender, hanging pod. BIGNONIACEAE.

Echeve'ria f. Succulent plants named for Athanasio Echeverria Godoy, botanical artist who accompanied the expedition under Sessé and Mociño to New Spain (Mexico) (1787–1797), and thus must have done many of the 1,200 drawings of Mexican plants copied within 10 days in 1816 by the citizens of Geneva for A. P. de Candolle. CRASSULACEAE.

Echidno'psis f. Gr. *echidna*, a viper; *opsis*, like; in allusion to the slender snake-like, somewhat prostrate stems. ASCLEPIADACEAE.

Echina'cea f. Cornflower (in the United States), purple cone flower. From Gr. *ĕchīnos*, a hedgehog; in allusion to the prickly scales of the receptacle. The thick black roots have a pungent flavour and are edible. COMPOSITAE.

echina'tus, -a, -um Covered with prickles like a hedgehog.

Echinocac'tus m. Gr. *ĕchīnos*, a hedgehog (*Erinaceus europaeus*); from being like hedgehogs in appearance. CACTACEAE.

echinocar'pus, -a, -um With prickly fruit.

Echinoce'reus m. Gr. *ĕchīnos*, a hedgehog. The spiny fruit differentiates this genus from *Cereus*. CACTACEAE.

Echino'chloa f. Gr. *ĕchīnos*, a hedgehog; *chlŏē*, grass. GRAMINEAE.

Echinocys'tis f. Prickly cucumber. Gr. *ĕchīnos*, a hedgehog; *kystis*, a bladder; from the prickly fruit. CUCURBITACEAE.

Echino'dorus m. Gr. *ĕchīnos*, a hedgehog; *dŏrŏs*, bag; referring to the spiny achenes. ALISMATACEAE.

Echinofossulocac'tus m. From L. *ĕchīnos*, a hedgehog; *fossula*, a little ditch; *cactus*; with reference to the spines and furrows of these cacti. CACTACEAE.

Echinomas'tus m. Gr. *ĕchīnos*, a hedgehog; *mastos*, a breast. The tubercles of this cactus are spiny. CACTACEAE.

Echino'panax m. Gr. *ĕchīnos*, a hedgehog; *Panax*, the generic name for American ginseng. The stems are densely covered with spines. ARALIACEAE.

Echi'nops m. Globe thistle. From Gr. *ĕchīnŏs*, a hedgehog; *ōps*, appearance; in allusion to the handsome spiny, globe-shaped flower heads of metallic blue. COMPOSITAE.

Echino'psis Sea-urchin cactus. From Gr. *ĕchīnŏs*, a sea-urchin (also hedgehog); *ŏpsis*, resemblance; from the resemblance of these round, spiny cacti to a sea-urchin. CACTACEAE.

echinose'palus, -a, -um With prickly sepals.

echinosper'mus, -a, -um With spiny seeds.

echioi'des Resembling viper's bugloss or *Echium*.

Echi'tes f. Gr. *ĕchis*, a viper. This shrub is poisonous and of twining habit. APOCYNACEAE.

Echi'um n. Viper's bugloss. The Greek name *ĕchīŏn*. This genus of rough herbs and shrubs was supposed to discourage serpents, while the root drunk with wine was regarded as good for snake-bites. BORAGINACEAE.

eclec'teus, -a, -um Picked out; selected.

ecornu'tus, -a, -um Without horns.

Edgeworthia f. Ornamental shrubs named in honour of Michael Pakenham Edgeworth (1812–1881) of the East India Company's Service, a keen amateur botanist who collected many new plants in India. THYMELEACEAE.

Edithcolea f. In honour of Miss Edith Cole who collected the type material in Somalia. ASCLEPIADACEAE.

Edraian'thus m. Gr. *hedraios*, sitting; *anthos*, flower, with reference to the sessile flowers clustered at the top of the flower stalk. CAMPANULACEAE.

edu'lis, -is, -e Edible.

effu'sus, -a, -um Loosely spreading, straggling, spread out.

Ehre'tia f. Named in honour of Georg Dionysius Ehret (1708–1770), celebrated German botanical artist, who settled in England and was the brother-in-law of Philip Miller, the English author of the great *Gardeners Dictionary*. BORAGINACEAE.

Eichhor'nia f. Water-hyacinth. Named for J. A. Fr. Eichhorn (1779–1856), Prussian minister of education. PONTEDERIACEAE.

eich'leri Of Wilhelm Eichler, a German who found *Tulipa eichleri* in the Caucasus, or August Wilhelm Eichler (1839–1887), German professor of botany, successively at Graz, Kiel and Berlin.

elaeagnifo'lius, -a, -um With leaves resembling *Elaeagnus*.

Elaeag'nus f. Gr. *elaiagnos*, often said to be from Gr. *elaia*, the olive tree; *agnos*, the Greek name for *Vitex agnus-castus* or chaste-tree; but, as it was applied by Theophrastus to a willow (*Salix*) growing on floating islands of Lake Copais, and was probably *hĕleagnos*, it is most likely from *hĕlōdēs*, marshy; *agnŏs*, lamb, or *hagnŏs*, pure, hence white; in allusion to the white fluffy masses of fruits on a willow as on a poplar. The use of the name *Elaeagnus* in its modern generic sense began apparently in the 16th century with Rauwolf and Camerarius. ELEAGNACEAE.

Ela'eis f. African oil palm. From Gr. *ĕlaia*, the olive tree. Oil obtained from the fruit of these feather palms, is of great commercial importance in West Africa and other tropical areas and is used in making margarine etc. PALMAE.

Elaeocar'pus m. Gr. *ĕlaia*, olive; *karpos*, fruit; from the appearance of the round fruit. ELAEOCARPACEAE.

Elaeoden'dron n. Gr. *ĕlaia*, olive; *dĕndrŏn*, a tree. The fruit is like an olive and has an oily seed. CELASTRACEAE.

Elaphoglos'sum n. From Gr. *elaphos*, a stag; *glossa*, a tongue; in allusion to the shape of the fronds in some species. The elephant-ear fern is *E. crinitum*, with broadly oblong fronds. ASPLENIACEAE.

elas'ticus, -a, -um Elastic; producing an elastic substance.

elater'ium Specific epithet of *Ecballium elaterium*, squirting cucumber. Both words have the implication of driving with force—an allusion to the way in which the seeds are thrown out.

Elati'ne f. Ancient Greek plant-name *ĕlatinē*. ELATINACEAE.

ela'tior Taller.

Elatoste'ma n. From Gr. *ĕlatŏs*, driving, striking; used by the authors J. R. and G. Foster in the sense of 'elastic'; *stēma*, penis, stamen; in allusion to the stamens springing up. URTICACEAE.

ela'tus, -a, -um Tall.

e'legans; elegan'tulus, -a, -um Elegant.

elegantis'simus, -a, -um Very elegant.

elephan'tidens Large-toothed.

elephan'tipes Like an elephant's foot.

elephan'tum Of the elephants; monstrous, big.

Eletta'ria f. Cardamon. From *elettari*, the vernacular name in Malabar, south-western India. The cardamon of commerce is the spicy seed of *E. cardamomum*. ZINGIBERACEAE.

Eleu'sine f. From the city of Eleusis in Greece where the temple of Ceres stood. This genus of coarse grasses includes various cereals such as African millet. GRAMINEAE.

elisabe'thae Of Elisabeth.

ellacombia'nus, -a, -um After Canon Henry Nicholson Ellacombe (1822–1916) of Bitton, Gloucestershire, England, celebrated gardener and author of *Plant-lore and Garden-craft of Shakespeare*.

Elliot'tia f. A handsome shrub named after Stephen Elliott (1771–1830), botanist at Charleston, South Carolina, U.S.A. ERICACEAE.

ellipsoida'lis, -is, -e Elliptic in outline but solid.

ellip'ticus, -a, -um Elliptic; longer than wide, broadest at the middle, with curved sides.

Elo'dea f. Waterweed or ditchmoss. From Gr. *hĕlŏdēs*, marshy, bred in marshes. HYDROCHARITACEAE.

elonga'tus, -a, -um Lengthened; elongated.

Elshol'tzia f. Aromatic herbs and sub-shrubs named in honour of Johann Sigismund Elsholtz (1623–1688), Prussian horticulturist and physician, author of the once popular *Vom Garten-Baw*. LABIATAE.

elwesii In honour of Henry John Elwes (1846–1922), English naturalist, sportsman, traveller, dendrologist and horticulturist, author of *Monograph of the Genus Lilium* (1877–1880).

E'lymus m. Wild rye. From the Greek name *ĕlymos*, a cereal. GRAMINEAE.

emargina'tus, -a, -um With a shallow notch at the end as though a piece had been removed.

Emboth'rium n. Gr. *en*, in; *bŏthrĭŏn*, a little pit; in allusion to the anthers being in pits of the calyx. PROTEACEAE.

eme'rus Latinized form used by Cesalpino in 1583 of the Italian vernacular name *emero* for *Coronilla emerus*.

eme'ticus, -a, -um Causing vomiting; emetic.

Emi'lia f. Tassel flower. Probably named for some individual but just who is unknown. COMPOSITAE.

e'minens Eminent; prominent.

Emmenan'the f. California golden bells. Gr. *ĕmmĕnēs*, lasting; *anthos*, a flower. The corolla lasts for a long time. HYDROPHYLLACEAE.

Emmeno'pterys f. Gr. *ĕmmĕnēs*, lasting; *ptĕryx*, a wing. One of the calyx lobes becomes large and like a wing. RUBIACEAE.

emo'di Of the Himalaya, the Emodi Montes in classical geography.

empetrifo'lius, -a, -um With leaves like crowberry or *Empetrum*.

Em'petrum n. Crowberry. Gr. *ĕmpĕtrŏn*, from *en*, on; *pĕtros*, a rock. EMPETRACEAE.

Ence'lia f. Derivation obscure. COMPOSITAE.

Encephalar'tos m. Gr. *ĕn*, in ; *cĕphalē*, a head; *artos*, bread. The inner parts of the top of the trunks of these primitive palmlike trees are farinaceous and edible. ZAMIACEAE.

endres'sii In honour of Philip Anton Christoph Endress (1806–1831), German plant collector who collected in Germany, France and Spain.

Endy'mion m. Gr. *Endymīōn*, in Greek mythology a beautiful youth beloved by Diana; the accepted name for this genus is now *Hyacinthoides*. LILIACEAE.

engelman'nii Specific epithet of Engelmann's spruce and oak (*Picea* and *Quercus engelmannii*) as well as of other less well-known plants. Named for Georg Engelmann (1809–1884), German-born St. Louis physician and botanist. He did some exploring but his greater contribution lay in his botanical publications on *Cactaceae*, North American conifers, oaks, *Cuscuta*, etc.

eng'leri In honour of H. G. Adolf Engler (1844–1930), extremely industrious German botanist, monographer of *Saxifraga* and many tropical families, successively professor of botany at Kiel, Breslau and Berlin, initiator of important works such as *Pflanzenfamilien*, *Pflanzenreich* and *Vegetation der Erde*.

Enkian'thus m. Gr. *enkyos*, pregnant; *anthos*, a flower. In the type-species (*E. quinqueflorus*) of this genus the corolla has marked swellings at the base as if, to quote Loureiro, 'the flowers are gravid with other flowers'. The flower gives the appearance of being a flower within a flower. ERICACEAE.

enneacan'thus, -a, -um With nine spines.

enneaphyl'lus, -a, -um With nine leaves or leaflets.

ensa'tus, -a, -um Sword-shaped.

ensifo'lius, -a, -um With sword-shaped leaves.

ensifor'mis, -is, -e Quite straight and with a sharp point like a sword or the leaf of an iris.

Enta'da f. A southern Indian (Malabar) name used by van Rheede. The large brown seeds, often called 'sea beans', of the West Indian species *Entada gigas* float across the Atlantic Ocean and are cast up on the shores of north-west Europe. LEGUMINOSAE.

Ente'lea f. Gr. *ĕntĕlēs*, complete, perfect. The stamens of this shrub or small tree are all fertile unlike those of a related genus. TILIACEAE.

entomo'philus, -a, -um Insect-loving. Applied as a specific epithet to certain plants that by means of colour, smell, or nectar attract the insects which carry their pollen to other plants.

Eome'con f. Snow poppy. Gr. *ēōs*, dawn, the east; *mēkōn*, a poppy; a Far Eastern genus. PAPAVERACEAE.

E'pacris f. Gr. *epi*, upon; *akris*, a summit; the upper regions of hills are often the habitat of some species. EPACRIDACEAE.

Ephe'dra f. Greek name *ĕphĕdra* for the common mare's tail (*Hippuris*) which it somewhat resembles. Certain species of these low evergreen shrubs are the source of ephedrine which, as *ma-huang*, the Chinese have used in medicine for centuries. EPHEDRACEAE.

Epiden'drum n. Gr. *ĕpĭ*, upon; *dendron*, a tree. These are epiphytic or tree-perching orchids. ORCHIDACEAE.

Epiga'ea f. Ground-laurel, mayflower or trailing arbutus. From Gr. *ĕpĭ*, upon; *gaia*, the earth; a reference to the creeping habit. ERICACEAE.

epiga'eus, -a, -um Growing close to the ground or, sometimes, growing on land and not in water.

Epilo'bium n. Willow-herb. From Gr. *ĕpĭ*, upon; *lobos*, a pod. The petals surmount the podlike ovary. ONAGRACEAE.

Epime'dium n. Barrenwort. The Greek name *ĕpĭmēdĭon* for a plant quite different from these herbaceous perennials which were unknown to the Greeks. BERBERIDACEAE.

Epipac'tis f. Gr. *ĕpipaktis*, name adopted for this genus. ORCHIDACEAE.

Epiphroni'tis f. Name applied to orchid hybrids resulting from crossing *Epidendrum* and *Sophronitis*.

Epiphyl'lum n. Gr. *ĕpĭ*, upon; *phyllon*, a leaf. The flowers are borne on flattened green leaf-like stems once thought to be leaves. CACTACEAE.

epipsi'lus, -a, -um Somewhat bare, i.e. referring to sparse leaves or hair-covering.

Epis'cia f. From Gr. *ĕpiskios*, shaded; from the natural habitat of these herbaceous perennials. GESNERIACEAE.

episcopa'lis, -is, -e Of bishops; resembling a bishop's mitre.

Epithelan'tha f. Gr. *ĕpĭ*, upon; *thēlē*, a nipple; *anthos*, a flower. The flowers are borne on tubercles. CACTACEAE.

eques'tris, -is, -e Pertaining to horses or horsemen.

equi'nus, -a, -um Of horses.

equisetifo'lius, -a, -um With leaves like horsetail or *Equisetum*.

Equise'tum n. Horsetail. L. *equus*, a horse; *seta*, a bristle. Of more historical than garden interest, the genus links modern plants with some of the most ancient orders of vegetation. In the Carboniferous period, the earth was covered with immense forests of gigantic woody *Equisetum*-like plants, the fossilized remains of which comprise most of our coal measures. It was called scouring rush in the 17th-century because the presence of silica gave a fine finish in scouring pewter. EQUISETACEAE.

Eragros'tis f. Love-grass. From Gr. *erōs*, love; *agrostis*, grass. The reason for the name is unknown. GRAMINEAE.

Eran'themum n. Gr. *ēranthĕmon*, an old name transferred to this genus. ACANTHACEAE.

Eran'this f. Winter aconite. From Gr. *ēr*, spring; *anthos*, a flower. The plant is one of the earliest blooming spring flowers. RANUNCULACEAE.

Ercil'la f. Named for Alonso de Ercille (1533–1595) of Madrid. PHYTOLACCACEAE.

erec'tus, -a, -um Erect; upright.

Erema'ea f. Gr. *ĕrēmŏs*, lonely, solitary; the flowers are borne singly at the ends of shoots. MYRTACEAE.

Ereman'thus m. Gr. *ĕrēmos*, lonely solitary; *anthos*, flower; each head has only one flower. COMPOSITAE.

Ere'mia f. Gr. *ĕrēmŏs*, lonely, solitary. There is but one seed in each chamber of the fruit. ERICACEAE.

Eremocit'rus f. Australian desert kumquat. From Gr. *ĕrēmia*, desert; *Citrus*. RUTACEAE.

Eremo'phila f. From Gr.*ĕrēmia*, desert; *philos*, friend; i.e. desert-lover; Robert Brown collected the two original species in a sandy desert. MYOPORACEAE.

Eremo'stachys f. From Gr. *ĕrēmia*, desert; *Stachys*, woundwort; another genus of *Labiatae*. LABIATAE.

Eremu'rus m. Desert-candle or foxtail lily. From Gr. *ĕrēmia*, desert; *oura*, a tail; from the appearance of the flower spike of these giant desert- and steppe-inhabiting plants. LILIACEAE.

Ere'psia f. From Gr. *ĕrĕpo*, to cover with a roof, or Gr. *ĕrĕpsis*, a roof. In allusion to the stamens hidden under the staminodes. AIZOACEAE.

eri- Gr. prefix indicating woolliness (the medieval Latin epithet *erigena* means, however, 'Irish-born'); but see *Erigenia*. Thus:
eriacan'thus (woolly-spined); **erianthe'rus** (with woolly anthers);
erian'thus (with woolly flowers); **eriogy'nus** (with a woolly ovary).

E'ria f. Gr. *ĕriŏn*, wool. The flowers and pedicels are often woolly. ORCHIDACEAE.

Erian'thus m. Plume-grass. From Gr. *ĕriŏn*, wool; in allusion to the silvery woolly effect of the inflorescence. GRAMINEAE.

Eri'ca f. Heath. Latin *ĕrīcē*; Gr. *ĕrĕikē*. The Mediterranean *E. arborea* is the source of brier (or briar) for tobacco pipes turned from the roots, which are sometimes very large and solid. (French *bruyère*, heath.) ERICACEAE.

ericae'us -a, -um Relating to *Erica* or heath.

ericifo'lius, -a, -um With leaves like *Erica* or heath.

ericoi'des Resembling *Erica* or heath.

Erige'nia f. Harbinger-of-spring. From Gr. *ēr*, spring; *genia*, born; *Erigĕnĕia* being another name for Aurora, a reference to this small North American plant's early flowering. UMBELLIFERAE.

eri'genus, -a, -um Irish-born. From Old Irish *Eriu*, *Erin*, Ireland.

Eri'geron m. Fleabane. Possibly from Gr. *ēri*, early; *gĕrōn*, an old man. COMPOSITAE.

erina'ceus, -a, -um Resembling a hedgehog, hence *Erinacea* f., a spiny genus of Leguminosae.

E'rinus m. Greek name *ĕrinos*, used by Dioscorides for a plant like basil, transferred to this genus. SCROPHULARIACEAE.

Eriobot'rya f. Gr. *ěriŏn*, wool; *botrys*, a cluster of grapes; from the woolly, clustered panicles. ROSACEAE.

eriobotryoi'des Resembling *Eriobotrya*.

eriocar'pus, -a, -um Woolly-fruited. Also: **erioce'phalus** (woolly-headed); **erio'phorus** (wool-bearing); **erio'spathus** (with a woolly spathe); **erio'stachys** (with a woolly spike); **erioste'mon** (with a woolly stamen).

Erioce'phalus m. Gr. *ěriŏn*, wool; *cěphale*, a head. The heads of these scented shrubs become woolly. COMPOSITAE.

Erio'gonum n. Gr. *ěriŏn*, wool; *gŏny*, a knee. The stems of these herbs or sub-shrubs are downy at the nodes. POLYGONACEAE.

Erio'phorum n. Cottongrass. From Gr. *ěriŏn*, wool; *phŏros*, bearing; from the woolly heads. CYPERACEAE.

Eriophyl'lum n. Gr. *ěriŏn*, wool. *phyllon*, a leaf. The leaves of these perennial herbs are woolly. COMPOSITAE.

Eriop'sis f. Gr. resembling *Eria*, another genus of epithytic or tree-perching orchids. ORCHIDACEAE.

Erioste'mon m. Gr. *ěriŏn*, wool; *stēmōn*, a stamen. The stamens are woolly. RUBIACEAE.

Eritri'chium n. Gr. *ěriŏn*, wool; *thrix*, hair; from the woolly covering. BORAGINACEAE.

Erlan'gea f. Named in honour of the University of Erlangen in Bavaria, where the author, C. H. Schulz Bipontinus, studied botany in 1825–1826. COMPOSITAE.

ernes'tii In honour of Ernest Henry Wilson (1876–1930); see *Sinowilsonia*.

Ero'dium n. Heron's bill. From Gr. *erōdios*, a heron. The carpels before expulsion of seeds resemble the head and beak of a heron. GERANIACEAE.

Ero'phila f. From Gr. *ēr*, spring; *philos*, loving. CRUCIFERAE.

ero'sus, -a, -um Jagged, as if irregularly gnawed or bitten off.

erra'ticus, -a, -um Unusual; sporadic; departing from the normal of the genus.

erro'menus, -a, -um Gr. *ěrrōměnŏs*, in good health, vigorous.

erubes'cens Blushing; becoming red.

Eru'ca f. Latin name for *Eruca sativa* or rocket-salad, grown for its oil-rich seed and sometimes for use as a salad plant. CRUCIFERAE.

erucas'trum; erucoi'des Resembling *Eruca* or rocket-salad.

Eryn'gium n. Sea-holly, eryngo. The Greek name, *ēryngiŏn*, for *Eryngium campestre*. The roots of sea-holly (*E. maritimum*) are sweet and aromatic and were formerly candied. UMBELLIFERAE.

Ery'simum Blistercress. Gr. *ěrysimŏn* used by Theophrastus. CRUCIFERAE.

Erythe'a f. Named for one of the three Hesperides who lived far away in the West on the border of the ocean where the sun set, guarding the golden apples which Earth had given to Hera. PALMAE.

Erythri'na f. Coral tree. From Gr. *ĕrythrŏs*, red; an allusion to the colour of the flower. LEGUMINOSAE.

erythro- As a prefix in compound words it signifies red. Thus: **erythroca'lyx** (with a red calyx); **erythrocar'pus** (having red fruit); **erythroce'phalus** (red-headed); **erythro'podus** (having a red foot or stem); **erythro'pterus** (having red wings); **erythroso'rus** (having red spore cases).

Erythro'nium n. Dog's-tooth violet, trout-lily, adder's tongue. Gr. *ĕrythrŏniŏn*, from *ĕrythrŏs*, red, originally the name for another plant. LILIACEAE.

Erythro'xylon n. From Gr. *ĕrythrŏs*, red; *xylŏn*, wood. Some species have red wood. *E. coca*, the only source of cocaine, is a bush cultivated in the tropics for that product. In Bolivia and Peru, where it is native, the inhabitants chew the leaves both for their euphoric effect and to sustain themselves in feats of endurance. ERYTHROXYLACEAE.

Escallo'nia f. Shrubs and small trees named for Señor Escallon, an 18th-century Spanish traveller in South America. GROSSULARIACEAE.

Eschschol'zia f. California poppy. Named in honour of Johann Friedrich Eschscholtz (1793–1831) of Tartu (Dorpat), Estonia, who accompanied Otto von Kotzebue on his first expedition round the world (1815–1818). PAPAVERACEAE.

Escoba'ria f. Named for two Mexicans, Romulo and Numa Escobar. CACTACEAE.

Escon'tria f. Named after Don Blas Escontria, a Mexican of distinction. CACTACEAE.

esculen'tus, -a, -um Edible, good to eat.

Espo'stoa f. In honour of Nicolas E. Esposto, botanist at Lima, Peru. CACTACEAE.

esserteauia'nus -a, -um. In honour of Dr. Esserteau, bacteriologist in China. *c*. 1908–1910.

estria'tus, -a, -um Without stripes.

etrus'cus, -a, -um Of Tuscany, the classical Etruria, in Italy.

etubero'sus, -a, -um Without tubers.

Euan'the f. From Gr. *euanthēs*, blooming, well-flowered; with reference to the showy flowers. ORCHIDACEAE.

eucalyptoi'des Resembling *Eucalyptus*.

Eucalyp'tus f. Gum tree, blue gum. From Gr. *eu*, well; *kalypto*, to cover, as with a lid; an allusion to the united calyx-lobes and petals forming a lid or cap which is shed when the flower opens. Among the largest trees in the world (highest recorded, 326 feet) they have very small and even minute seeds. MYRTACEAE.

Eu'charis f. Amazon lily. Greek word meaning pleasing, charming; from the fragrant beauty of the flowers. AMARYLLIDACEAE.

Euchlae'na f. Gr. *eu*, good; *chlaina*, cloak, covering; referring to the stigma of these grasses. GRAMINEAE.

euchlo'rus, -a, -um Fresh green, from Gr. *eu*, good; *chlōrŏs*, green.

Eucni'de f. Gr. *eu*, good; *knidē*, stinging nettle. LOASACEAE.

Eu'comis f. From Gr. *eu*, good; *kome*, hair; implying a beautiful head, from the tufted leaves crowning the flower spike. LILIACEAE.

Eucom'mia f. Gr. *eu*, good; *kommi*, gum. Its single species (*E. ulmoides*) is the only hardy rubber-producing tree. When the leaves are torn gently across, threads of rubber remain strong enough to support the torn portion. EUCOMMIACEAE.

Eucry'phia f. Gr. *eu*, well; *kryphios*, covered. The sepals of these white-flowered trees and shrubs, cohering at the tips, form a cap. EUCRYPHIACEAE.

eudo'xus, -a, -um Of good repute.

Euge'nia f. Clove tree, rose apple. Named in honour of Prince Eugene of Savoy (1663–1736). Born in Paris but spurned by the French court, he became a very successful Austrian general, defeating first the Turks, then the French, pursuing war with rare energy, imagination, and resolution. Wounded thirteen times in fifty years in the field, he did not spend two years together without fighting, almost invariably with success. Among his best-known campaigns were those he fought as an ally of the Duke of Marlborough. MYRTACEAE.

eugenioi'des Resembling *Eugenia*.

Eulo'phia f. From Gr. *eu*, good; *lŏphŏs*, crest; referring to the crest on the lip. ORCHIDACEAE.

Euo'nymus f. Spindle-tree. The ancient Greek name *euōnymŏn dĕndron* whence the Latin *euonymus*, meaning 'of good name', referred ironically to its being poisonous to animals. CELASTRACEAE.

eupatorioi'des Like *Eupatorium*.

Eupato'rium n. Joe-Pye-weed, thoroughwort, boneset. The Greek name for these herbaceous and shrubby plants, commemorating Mithridates VI Eupator (132–63 B.C.), King of Pontus, enemy of Rome in Asia Minor. The English names indicate a traditional use in medicine. COMPOSITAE.

Euphor'bia f. Spurge. Classically supposed to have been named for Euphorbus, physician to the King of Mauretania. EUPHORBIACEAE.

euphorbioi'des Resembling spurge or *Euphorbia*.

Euphra'sia f. Eye bright. Gr. *euphrasia*, delight, mirth. SCROPHULARIACEAE.

Eupritchar'dia f. Gr. *eu*, good; *Pritchardia*, a related genus of ornamental Pacific palms, named for William Thomas Pritchard, author of *Polynesian Reminiscences* (1866). PALMAE.

Eupte'lea f. Gr. *eu*, good; *ptĕlĕa*, an elm; in reference to the edible fruit. EUPTELEACEAE.

europae'us, -a, -um European.

Euro'tia f. Gr. *euros*, mould; in reference to the grey-white downy or mouldy appearance. CHENOPODIACEAE.

Eu'rya f. Gr. *eury*, broad; possibly referring to broad petals. THEACEAE.

Eury'ale f. Fox nut. Named after one of the three Gorgons, the monstrous daughters of the sea god, who had venomous snakes for hair and of whom only Medusa was mortal. The allusion here is to the very thorny, prickly character of this large and handsome water-lily. NYMPHAEACEAE.

Eury'ops m. Derivation uncertain. A large African genus. COMPOSITAE.

Eu'scaphis f. Gr. *eu*, good; *scaphis*, a vessel; from the character of the seed pod of this ornamental woody plant. STAPHYLEACEAE.

Eu'stoma n. Prairie gentian. From Gr. *eu*, good; *stŏma*, a mouth, or (less literally) a pretty face; in allusion to the showy flowers. GENTIANACEAE.

Eu'strephus m. Gr. *eu*, well; *strepho*, to twine; from the character of these Australian climbers. SMILACACEAE.

Euta'xia f. Greek word meaning modesty; from the nature of these Australian flowering shrubs. LEGUMINOSAE.

Euter'pe f. Named for one of the nine Muses, goddesses of the liberal arts. Euterpe had charge of music. The genus includes the cabbage palm (*E. oleracea*), the terminal bud of which is eaten fresh as a salad or pickled. PALMAE.

evec'tus, -a, -um Extended, springing forwards.

ever'tus, -a, -um Overturned.

Evo'dia f. Gr. *euōdia*, a sweet scent. Now included in *Tetradium*. RUTACEAE.

Evol'vulus m. L. *evolvo*, to untwist or unravel. These plants of the convolvulus family do not twine. CONVOLVULACEAE.

Evonymus see *Euonymus*.

ewer'sii In honour of Joseph Philipp Gustav Ewers (1781–1830), German statesman in Russian employ who promoted Ledebour's botanical exploration in Siberia.

E'xacum n. Variant of the Gallic name *exacon* for *centaurium* transferred to this genus. GENTIANACEAE.

exalta'tus, -a, -um Very tall; lofty.

exara'tus, -a ,-um Engraved; furrowed.

exaspera'tus, -a, -um Roughened.

excava'tus, -a, -um Hollowed out.

excel'lens Excelling; excellent; distinguished.

excel'sior Taller.

excel'sus, -a, -um Tall.

exci'sus, -a, -um Cut away; cut out.

excortica'tus, -a, -um Without bark or cortex; stripped of bark.

exi'guus, -a, -um Very little; meagre; poor.

exi'mius, -a, -um Out of the ordinary; distinguished.
exitio'sus, -a, -um Pernicious; destructive.
Exochor'da f. Pearl-bush. From Gr. *exo*, outside; *chorde*, a cord; referring to a technical characteristic of the ovary; there are fibres outside the placenta. ROSACEAE.
exole'tus, -a, -um Mature; dying away.
exonien'sis, -is, -e Of Exeter, Devon, England.
exo'ticus, -a, -um From a foreign country.
expan'sus, -a, -um Expanded.
explo'dens Bursting suddenly; exploding.
exscul'ptus, -a, -um Dug out.
exser'tus, -a, -um Protruding.
exsur'gens Rising up.
exten'sus, -a, -um Extended.
exu'dans Exuding.
eystet'tensis. Epithet of a *Narcissus* grown in the richly stocked garden of the prince-bishop Johann Conrad von Gemmingen (*c.*1561–1612) at Eichstätt, south Germany.

F

faasse'nii Of J. H. Faassen, Dutch nurseryman, of Tegelen, Netherlands.
Fa'ba f. Latin for broad bean (*Vicia faba*); more appreciated in Europe than in the United States where the climate is generally too hot and dry. LEGUMINOSAE.
faba'ceus, -a, -um Resembling the broad bean.
Fabia'na f. False heath. Named in honour of Archbishop Francisco Fabian y Fuero (1719–1801) of Valencia, Spain, a promoter of botanical study. SOLANACEAE.
fagifo'lius, -a, -um Beech-leaved.
Fagopy'rum n. Buckwheat. From *Fagus*, beech; *pȳrŏs*, wheat. POLYGONACEAE.
Fa'gus f. Beech. The Latin name. FAGACEAE.
falca'tus, -a, -um; falcifor'mis, -is, -e Sickle-shaped; falcate.
falcifo'lius, -a, -um With sickle-shaped leaves.
falcinel'lus, -a, -um Resembling a small sickle.
falco'neri In honour of Hugh Falconer (1808–1865), Scottish doctor, geologist and botanist in India from 1830 to 1855.
fal'lax Deceptive; false.
Fallopia f. In honour of Gabriele Fallopi (1523–1562), Italian anatomist, professor of anatomy at Pisa and Padua after whom the Fallopian tube was named. Often included in *Polygonum*. POLYGONACEAE.

Fallu'gia f. In honour of Abbot V. Fallugi of Vallombrosa, *c.* 1840. ROSACEAE.
farge'sii In honour of Paul Guillaume Farges (1844–1912), French missionary and naturalist in Central China.
farina'ceus, -a, -um Yielding starch; mealy, like flour.
farino'sus, -a, -um Mealy; powdery.
farleyen'sis, -is,-e Of Farley Hill Gardens in Barbados, West Indies.
farnesia'nus, -a, -um Of the gardens of the Farnese Palace, Rome.
farre'rae In honour of Mrs Farrer, wife of an Honourable East India Company captain.
farre'ri Specific epithet of *Gentiana farreri* and other plants. In honour of Reginald John Farrer (1880–1920), English plant hunter and botanist, who specialized in Alpines; he collected in the European Alps, Upper Burma and the mountain regions of Kansu (Gansu) whence he introduced *G. farreri* among other good plants. He wrote vividly several books on his travels and on rock-garden plants, possibly the best known being *The English Rock Garden* (1919).
fascia'tus, -a, -um Bound together.
fascicula'ris, -is, -e; fascicula'tus, -a, -um Clustered or grouped together in bundles.
fascina'tor Bewitcher, enchanter.
fastigia'tus, -a, -um Having branches close together and erect, often forming a column; fastigiate.
fastuo'sus, -a, -um Proud.
Fatshe'dera f. Name of intergeneric hybrid of *Fatsia* and *Hedera*.
Fat'sia Latinized adaptation of an obsolete or misrendered Japanese name, *Fatsi*, for *Fatsia japonica*. ARALIACEAE.
fa'tuus, -a, -um Simple or foolish but, in this relationship, insipid or, merely, not good.
Fauca'ria f. L. *faux*, a gullet or throat. The paired leaves with tooth-like edges of these succulents resemble the wide open mouth of a carnivore. AIZOACEAE.
febri'fugus, -a, -um Fever-dispelling.
Fe'dia f. Meaning of the name unknown. VALERIANACEAE.
Feijo'a f. A small tree or shrub with pineapple- or guava-flavoured fruit. Named for Don de Silva Feijoa, 19th-century Brazilian botanist. MYRTACEAE.
Feli'cia f. Blue daisy. After Felix, a German official at Regensburg, who died in 1846. COMPOSITAE.
fe'mineus, -a, -um Female, feminine.
Fendle'ra f. In honour of August Fendler (1813–1883), German plant-collector in North and Central America. HYDRANGEACEAE.
fenestra'lis, -is, -e Pierced with openings resembling windows.
Fenestra'ria f. From L. *fenestra*, a window; in allusion to the transparent area at the top of the leaves. AIZOACEAE.

fen'nicus, -a, -um Finnish, of Finland.
fe'rax Fruitful.
ferdinandi-coburgi. Epithet of species of *Haberlea* and *Saxifraga* named for Ferdinand (1861-1948) of Saxe-Coburg-Gotha, German prince, from 1887 to 1918 King of Bulgaria; see also *borisii*.
Ferocac'tus m. From L. *ferox*, ferocious, savage; in allusion to the horrible spines with which this cactus is armed. CACTACEAE.
Fero'nia f. Wood-apple. Named in honour of the Roman nymph who presided over woods and groves. An earlier name is *Limonia*. RUTACEAE.
Feroniel'la f. Diminutive of *Feronia* to which it is closely related. RUTACEAE.
fe'rox Ferocious; very thorny.
Ferra'ria f. In honour of Giovanni Battista Ferrari (1584–1655) of Siena, Italian botanist, author of *Flora* (1633) and *Hesperides* (1646). IRIDACEAE.
fer'reus, -a, -um Pertaining to iron or, sometimes, iron-hard.
ferrugi'neus, -a, -um Rust-coloured; rusty.
fer'tilis, -is, -e Fruitful; producing numerous seeds.
Fe'rula f. Giant fennel. The classical Latin name. The word also means the rod used to chastise schoolboys and slaves for minor offences. The giant fennel has tall sticklike stems with a slow-burning pith which enabled Prometheus to bring fire to mankind from heaven. The true fennel, a flavouring and a vegetable, belongs to the genus *Foeniculum*, *q.v.* UMBELLIFERAE.
ferulifo'lius, -a, -um With leaves resembling giant fennel or *Ferula*.
fe'rus, -a, -um Wild.
festa'lis, -is, -e; festi'vus, -a, -um Festive; gay; bright.
Festu'ca f. Fescue. Latin word meaning a grass stalk or straw. GRAMINEAE.
fibrillo'sus, -a, -um; fibro'sus, -a, -um Fibrous; composed of fibres.
Fica'ria f. Medieval plant-name probably from *Ficus*, fig, with reference to the tubers somewhat resembling little figs. RANUNCULACEAE.
ficifo'lius, -a, -um With leaves like a fig.
ficoi'des; ficoi'deus, -a, -um Resembling a fig or *Ficus*.
fic'tus, -a, -um False; feigned; hence *fictolacteum*, for a species of *Rhododendron* confused with *R. lacteum*.
Fi'cus f. Latin name for *Ficus carica*, the edible fig. The genus includes some quite extraordinary species such as the banyan (*F. benghalensis*) of India which, starting from a single main trunk, sends down aerial roots which themselves become trunks, thus extending the original tree over areas which may be very large. *F. elastica* is possibly the most familiar in temperate climates. Naturally quite a large tree and once an important source of rubber, it is now grown as the common household rubber plant; most commercial rubber is obtained from the genus *Hevea* and from synthetics.
Fila'go f. Medieval Latin plant-name from *filum*, thread. COMPOSITAE.

filamento'sus, -a, -um; fila'rius, -a, -um Furnished with filaments or threads.

fili- As a prefix, signifying threadlike. Thus:
filicau'lis (with a threadlike stem); **fi'lifer** (thread-bearing); **filifor'mis** (threadlike); **fi'lipes** (with threadlike stalks).

filicifo'lius, -a, -um With leaves resembling fern.

filici'nus, -a, -um; filicoi'des Resembling fern. From L. *filix*, a fern.

Filipen'dula f. Meadow-sweet, dropwort. L. *filum*, a thread; *pendulus*, hanging. The root tubers in some species hang together with threads. ROSACEAE.

filipen'dulus, -a, -um Resembling meadow-sweet or *Filipendula*.

fimbriat'ulus, -a, -um With a small fringe.

fimbria'tus, -a, -um Fringed.

Firmia'na f. Named for Karl Josef von Firmian (1716–1782), Governor of Lombardy, when that was a province of the Austrian Empire, and a patron of the Padua botanic garden. STERCULIACEAE.

fir'mus, -a, -um Strong.

fis'silis, -is, -e; fissura'tus, -a, -um; fis'sus, -a, -um Cleft; split.

fistulo'sus, -a, -um Hollow, like a pipe.

Fitto'nia f. Ornamental-leaved perennials named in honour of the sisters Elizabeth and Sarah Mary Fitton (d. 1866), authors of *Conversations on Botany* (1817). ACANTHACEAE.

Fitzroy'a f. Named for Captain (later Vice-Admiral) Robert Fitzroy (1805–1865), Royal Navy, who commanded the five-year surveying expedition of H.M.S. *Beagle*. Charles Darwin was aboard as official naturalist and, on the voyage, laid the foundation for his life's work on the origin of species. Fitzroy became a celebrated meteorologist and instituted a system of storm warnings that led to daily weather reports. CUPRESSACEAE.

flabella'tus, -a, -um Like an open fan.

flabel'lifer, flabelli'fera, -um Fan-bearing.

flabellifor'mis, -is, -e Fan-shaped.

fla'ccidus, -a, -um Feeble; weak; soft.

Flacour'tia f. Shrubs and small trees named for Etienne de Flacourt (1607–1660), director of the French East India Company. FLACOURTIACEAE.

flagella'ris, -is, -e; flagellifor'mis, -is, -e Whiplike; having long, thin, supple shoots like whips.

flagel'lum A whip or flail.

flam'meus, -a, -um Flame-coloured or flamelike.

flam'mula A small flame. Application obscure but possibly alluding to a burning taste. The pre-Linnaean name for *Clematis flammula*.

flavanthe'rus, -a, -um With yellow anthers.

fla'vens Yellow.

flave'olus, -a, -um Yellowish.

flaves'cens Becoming yellow, yellowish.
fla'vidus, -a, -um Yellowish.
flavis'simus, -a, -um Deepest yellow.
flavoru'fus, -a, -um Yellow-red.
fla'vus, -a, -um Pure yellow. Also:
flavi'comus (yellow-haired); **flavispi'nus** (yellow-spined).
Flemin'gia f. Shrubs and sub-shrubs named in honour of John Fleming (1747–1829), President of the Medical Board of Bengal, Hon. East India Company. This genus is also called *Moghania*. LEGUMINOSAE
flet'cheri In honour of Harold Roy Fletcher (1907–1978), director of the Royal Horticultural Society's Gardens, Wisley, then Regius Keeper of Royal Botanic Garden, Edinburgh.
flexicau'lis, -is, -e Having a pliant stem.
fle'xilis, -is, -e Pliant; limber.
flexuo'sus, -a, -um Tortuous; zigzag.
floc'ciger, flocci'gera, -um Bearing flocks of wool.
flocco'sus, -a, -um Woolly.
flocculo'sus, -a, -um Somewhat woolly.
flo're-al'bo With white flowers.
florenti'nus, -a, -um Of Florence, Italy.
flo're-ple'no With double flowers. This phrase is now usually replaced by the epithets *duplex*, *pleniflorus* or *plenus*.
floribun'dus, -a, -um; flo'ridus, -a, -um; flo'rifer, -a, -um Free-flowering; producing abundant flowers.
florida'nus, -a, -um Of Florida.
florin'dae Specific epithet of a *Primula* discovered by Kingdon-Ward in Tibet and named after his first wife Florinda (*neé* Norman-Thompson).
-florus, -a, -um In Latin compounds means -flowered, e.g. *multiflorus*, many-flowered; *viridiflorus*, green-flowered.
flos Latin word meaning a flower, as in:
flos-cu'culi cuckoo flower (*Lychnis*)
flos-jo'vis Jove's flower (*Lychnis*)
flu'itans Floating.
fluminen'sis, -is, -e Of Rio de Janeiro (Flumen Januarii), Brazil.
fluvia'lis, -is, -e; fluvia'tilis, -is, -e Growing in a river or running water.
foe' minus, -a, -um Female, feminine.
foenicula'ceus, -a, -um; foenicula'tus, -a, -um Resembling fennel (*Foeniculum*).
Foeni'culum n. Fennel. the Latin name for this traditional salad and potherb which, in Italian, is *finocchio*. UMBELLIFERAE.
foe'tidus, -a, -um Bad-smelling. Also *foetidissimus*, very bad-smelling.

folia'ceus, -a, -um Leaflike.

folia'tus, -a, -um Provided with leaves.

foliolo'tus, -a, -um; foliolo'sus, -a, -um Furnished with leaflets.

folio'sus, -a, -um Full of leaves; leafy.

-folius, -a, -um In Latin compounds means -leaved, e.g. *ilicifolius*, holly-leaved; *latifolius*, broad-leaved.

follicula'ris, -is, -e Bearing follicles. These are dry, one-chambered fruits splitting along only one suture, e.g. peony, milkweed.

Fontane'sia f. Named for René Louiche Desfontaines (1750–1833), author of *Flora Atlantica*; see *Desfontainia*. OLEACEAE.

fonta'nus, -a, -um; fontina'lis, -is, -e Pertaining to springs or fountains; growing in fast-running water.

Forestie'ra f. Shrubs named for Charles Le Forestier, 18th-century French physician and naturalist. OLEACEAE.

formosa'nus, -a, -um Of the island of Formosa (Taiwan).

formo'sus, -a, -um Handsome; beautiful. Also *formosissimus*, very handsome.

forres'tii In honour of George Forrest (1873–1932), Scottish plant-collector who between 1904 and 1932 made immense collections of seeds and specimens in western China.

forskoh'lei In honour of Pehr Forsskål (1732–1763), Swedish botanist and traveller who collected specimens in Egypt and Arabia on a Danish expedition on which he died.

Forsy'thia f. Golden-bell. Named in honour of William Forsyth (1737–1804), Scottish superintendent of the Royal Gardens of Kensington Palace and author, among other works, of *A Treatise on the Culture and Management of Fruit Trees* which in its day was probably the most widely read work on the subject. From 1789 until his death Forsyth was a controversial figure in horticultural circles as a result of his claims for the concoction known as 'Forsyth's plaister'. This mixture, made up of lime, dung, wood-ashes, soapsuds, sand, urine, and so on, was not very different from others of the day designed to heal wounds in the bark of trees. Forsyth claimed that his 'plaister' was capable 'of curing defects in growing trees' and would restore oak trees 'where nothing remained but the bark'. Thomas A. Knight, later President of the Horticultural Society of London for many years, took active exception to statements made either by or on behalf of Forsyth, such as that holes in trees could be filled with the 'plaister' and the trees themselves 'brought to such a degree of soundness that no one can know the new wood from the old'; in due course the 'plaister' was thoroughly discredited. OLEACEAE.

fortu'nei Specific epithet after Robert Fortune (see *Fortunella*).

Fortunel'la f. Kumquat. Evergreen shrubs named for Robert Fortune (1812–

1880), Scottish horticulturist and collector in China. With the exception of *Rhododendron fortunei*, which he found wild, and one or two other less notable plants, most of his finds were in Chinese gardens and nurseries. In his day the difficulties of travel were such that he had small chance of going further inland. With a brief interval as curator of the Chelsea Physic Garden, he was in China practically continuously from 1843 to 1861. He learned Chinese.

In 1848, he was engaged by the Hon. East India Company to bring tea plants to India. Despite the justifiable secretiveness of the Chinese, he was able to investigate the growing and manufacture of tea and eventually exported the plants, Chinese workers, and implements which established the industry, first in India and then in Ceylon. It has since, of course, grown to enormous commercial importance.

Fortune wrote several readable books on his adventures and, having started as a poor garden apprentice, he died reasonably well off as a result of his plant sales and of his flourishing business in curios. RUTACEAE.

Fothergil'la f. Dwarf alder. Named for Dr. John Fothergill (1712–1780), Quaker physician of Stratford, Essex, England, who specialized in growing American plants. HAMAMELIDACEAE.

Fouquie'ra f. Candlewood. Small trees or shrubs named for Pierre Edouard Fouquier, 19th-century French physician. FOURQUIERIACEAE.

fourcroi'des Resembling the genus *Furcraea*.

fournie'ri In honour of Eugène Pierre Nicolas Fournier (1834–1884), French botanist.

foveola'tus, -a, -um Slightly pitted.

Fraga'ria f. Strawberry. From L. *fraga*, strawberry; presumably from *fragrans*, fragrant, in allusion to the perfume of the fruit. ROSACEAE.

fragariflo'rus, -a, -um With flowers like those of *Fragaria*, or else red as crushed strawberry.

fragarioi'des Resembling strawberry or *Fragaria*.

fra'gilis, -is, -e Brittle; easily broken; or, sometimes, wilting quickly.

fra'grans Fragrant.

fragrantis'simus, -a, -um Very fragrant.

Frai'lea f. In honour of Manuel Fraile (b. 1850), Spanish gardener for many years in charge of the cactus collection of the U.S. Department of Agriculture, Washington. CACTACEAE.

franchetii; franchetianus, -a, -um In honour of Adrien René Franchet (1834–1900), French botanist who worked in the Paris Museum d'Histoire naturelle on the plants of Japan and China.

Franco'a f. Named in honour of Francisco Franco, 16th-century physician of Valencia, Spain. SAXIFRAGACEAE.

Franke'nia f. Named for Johan Frankenius (1590–1661), professor of anatomy and botany at Uppsala, Sweden. FRANKENIACEAE.

Frankli'nia f. Named in honour of Benjamin Franklin (1706–1790), American printer, scientist, philosopher and statesman, this remarkable tree was discovered in Georgia in 1764 and exterminated in the wild by 1803 through excessive collecting by nurserymen in its only known locality. THEACEAE.

Fra'sera f. Named for John Fraser (1750–1811), Scottish collector of North American plants who also kept a nursery in Chelsea, England. He made several collecting expeditions, to Newfoundland and as far south as the Carolinas. His close friend, Thomas Walter of South Carolina, named this genus and *Magnolia fraseri* in his honour. GENTIANACEAE.

fra'seri Of John Fraser (see *Frasera*).

fraxi'neus, -a, -um Resembling ash or *Fraxinus*.

fraxinifo'lius, -a, -um With leaves resembling ash or *Fraxinus*.

Fra'xinus f. Ash. The classical Latin name. OLEACEAE.

Free'sia f. Named in honour of a German physician, Friedrich Heinrich Theodor Freese (d. 1876), of Kiel, a friend of Ecklon who named the genus. IRIDACEAE.

Fremon'tia f.; **Fremontoden'dron** n. Flannel-bush. Discovered by and named in honour of Major-General John Charles Fremont (1813–1890) who made four hazardous journeys exploring the Far West of the United States between 1842 and 1848. He brought back many notable trees and shrubs to be studied and named by Torrey, Gray, and Darlington and in doing so made great contributions to both botany and horticulture. See *Darlingtonia, Torreya*. STERCULIACEAE.

fresno'ensis, -is, -e Of Fresno county, California. Spanish *fresno*, ash-tree; referring here to *Fraxinus oregona*.

Freycine'tia f. Climbing shrubs named for Admiral Louis Claude De Saulses de Freycinet (1779–1842), French navigator. He made various exploring voyages to Australasia and the Pacific, returning with important collections. PANDANACEAE.

fri'gidus, -a, -um Growing in cold regions.

Fritilla'ria f. Fritillary. From L. *fritillus*, a dicebox; from the spotted markings on the flowers of *F. meleagris* suggestive of a dice-board. LILIACEAE.

Froeli'chia f. Named for Joseph Aloys von Froelich (1766–1841), German physician and botanist, author of a work on gentians. AMARANTHACEAE.

frondo'sus, -a, -um Leafy.

fruc'tifer, fructi'fera, -um Fruit-bearing; fruitful.

frumenta'ceus, -a, -um Pertaining to grain; grain-bearing.

frutes'cens; fru'ticans; frutico'sus, -a, -um Shrubby; bushy. From L. *frutex*, a shrub.

frutici'cola A dweller in bushy places.

fruticulos'us, -a, -um Shrubby and dwarf.

fuca'tus, -a, -um Painted; dyed.

Fu'chsia f. Named for Leonhart Fuchs (1501–1566), German physician and herbalist, professor at Tübingen, who published in 1542 and 1543 a herbal with unusually beautiful woodcuts of plants but never saw a *Fuchsia*. ONAGRACEAE.

fuchsioi'des Resembling *Fuchsia*.

fu'gax Withering or falling off quickly; fleeting.

ful'gens; ful'gidus, -a, -um Shining; glistening.

fuligino'sus, -a, -um Dirty-brown; sooty.

fullo'num Pertaining to fullers. Specific epithet of the fuller's teasel (*Dipsacus fullonum*) used in the manufacture of woollen cloth to raise the nap.

fulves'cens Becoming tawny, yellow-brown.

fulvi- Latin prefix meaning tawny. Thus *fulvicaulis*, tawny-stemmed; *fulvinervis*, tawny-nerved.

ful'vus, -a, -um Tawny-orange; fulvous. Thus, *fulvidus*, slightly tawny.

Fuma'ria f. Fumitory. Medieval name from L. *fumus terrae*, smoke of the earth. FUMARIACEAE.

fumariifo'lius, -a, -um With leaves resembling fumitory or *Fumaria*.

fu'midus, -a, -um Smoky; dull and greyed in colour.

fumo'sus, -a, -um Smoky.

fu'nebris, -is, -e Funereal; belonging to graveyards.

fungo'sus, -a, -um Relating to or resembling fungus; spongy.

funicula'tus, -a, -um Resembling a thin cord or rope.

Fun'kia Plantain lily. Now named *Hosta*.

fur'cans; furca'tus, -a, -um Forked; furcate.

Furcra'ea f. Named for Antoine François Fourcroy (1755–1809), French chemist and naturalist, who played an important part in the French Revolution. AGAVACEAE.

furfura'ceus, -a, -um Mealy; scurfy.

fu'riens Exciting to madness.

furseo'rum In honour of Rear Admiral J. Paul W. Furse (1904–1978), botanical artist specially interested in bulbous and alpine plants, and Mrs. (Polly) Furse who made extensive plant collections in Turkey, Iran and Afghanistan after his retirement from the Royal Navy.

fusca'tus, -a, -um Brownish.

fuscifo'lius, -a, -um With dusky-brown leaves.

fusco-ru'ber, -rubra, -rubrum Brownish-red.

fus'cus, -a, -um Brown; dusky.

fusifor'mis, -is, -e Spindle-shaped, that is thickest in the middle and drawn out at each end.

fu'tilis, -is, -e Useless.

G

Ga'gea f. Named for Sir Thomas Gage (1761–1820) of Hengrave Hall, Suffolk, who botanized in Ireland and Portugal. The greengage plum, the English name for the Reine Claude, was named for his grandfather who introduced it from France. LILIACEAE.

Gaillar'dia f. Blanket-flower. Named in 1786 by Fougeroux de Bondaroy for a French magistrate and patron of botany, Gaillard de Charentonneau. COMPOSITAE.

galacifo'lius, -a, -um With leaves like *Galax*.

galacti'nus, -a, -um Milky.

Galacti'tes f. Gr. *gala*, milk. The veins of the leaves are milky-white. COMPOSITAE.

Galan'thus m. Snowdrop. From Gr. *gala*, milk; *anthos*, flower; in allusion to the colour of the flowers. AMARYLLIDACEAE.

galan'thus, -a, -um With milky-white flowers.

Ga'lax f. Wand plant. Gr. *gala*, milk; possibly an allusion to the white flowers. DIAPENSACEAE.

ga'le English vernacular name cognate with German 'Gagel' adopted as specific epithet for *Myrica gale* by Linnaeus.

Galean'dra f. Gr. *galea*, a helmet; *andros*, a stamen. The cap over the anther is helmet-shaped. ORCHIDACEAE.

gelea'tus, -a, -um; galericula'tus, -a, -um Helmet-shaped.

Gale'ga f. Goatsrue. From Gr. *gala*, milk, it being thought that feeding on these perennials would improve the milk flow. LEGUMINOSAE.

galegifo'lius With leaves resembling goatsrue or *Galega*.

Galeop'sis f. Gr. *galēŏpsis*, from *galē*, weasel; *ŏpsis*, appearance. LABIATAE.

Galinso'ga f. Weeds, ironically and punningly named 'Gallant Soldier' in England, commemorating Mariano Martinez Galinsoga, 18th-century Spanish doctor in Madrid, whose botanical accomplishments match the smallness of their flowers. COMPOSITAE.

galioi'des Resembling *Galium* or bedstraw.

Ga'lium n. Bedstraw, cleavers. From Gr. *gala*, milk. Yellow bedstraw (*G. verum*) can be used in cheese-making to curdle the milk. RUBIACEAE.

gal'licus, -a, -um Of France, French.

galpin'ii Named for Ernest Galpin, South African banker and diligent collector of South African plants.

Galto'nia f. Named in honour of Sir Francis Galton (1822–1911), explorer, anthropologist and geneticist who travelled widely in Africa where he used a theodolite to measure at a distance the dimensions of Hottentot women. He established fingerprinting as a method of human identification, investigated the inheritance of talent, and wrote a distinguished book on meteorology. LILIACEAE (HYACINTHACEAE).

Gamo'lepis f. Gr. *gameo*, to marry; *lĕpis*, a scale. In these South African herbs and shrubs the bracts of the urn-shaped involucre are united. This is now included in *Steirodiscus*. COMPOSITAE.

gandaven'sis Relating to Ghent, Belgium.

gange'ticus, -a, -um Of the region of the Ganges, India.

Garci'nia f. Mangosteen. Named in honour of Laurent Garcin (1683–1751), French botanist, who travelled widely in India. The mangosteen (*G. mangostana*) is possibly the most delectable of all East Indian fruits. Native in Malaya, it does not do as well elsewhere. GUTTIFERAE.

Garde'nia f. Named for Dr. Alexander Garden (1730–1791), correspondent of Linnaeus and a native of Aberdeenshire, Scotland and for many years a physician in Charleston, South Carolina. *G. jasminoides* is the species commonly offered by florists. It is often called Cape jasmine, but it is not a jasmine and it originated in China. RUBIACEAE.

garga'nicus, -a, -um Of Monte Gargano in southern Italy.

Gar'rya f. Silk-tassel bush, fever bush. Named for Nicholas Garry, Secretary of the Hudson's Bay Company who, during the decade 1820–1830, assisted David Douglas in his explorations of the Pacific North-west. GARRYACEAE.

Gaste'ria f. Gr. *gaster*, belly. The asymmetrically swollen base of the flower distinguishes this genus from allied genera. LILIACEAE (ALOACEAE).

Gastrochi'lus m. From Gr. *gaster*, belly; *cheilos*, a lip. The lip of the flower on these herbs is swollen. ORCHIDACEAE.

Gaulnet'tya f. Name for intergeneric hybrids between *Gaultheria* and *Pernettya*, antedating × *Gaulthettya*. ERICACEAE.

Gaulthe'ria f. Named for Jean François Gaultier (*c.* 1708–1756), also called Gaulthier, French physician and botanist of Quebec. Perhaps the best-known member of the family is the wintergreen or checker-berry (*G. procumbens*) of the north-eastern woods of America. ERICACEAE.

Ga'ura f. Gr. *gauros*, superb; in allusion to the striking flowers of these herbs. ONAGRACEAE.

Gaylussa'cia f. Huckleberry. Named in honour of Joseph Louis Gay-Lussac (1778–1850), celebrated French chemist who is best remembered for his law concerning the volume of gases. His work laid the foundation of the food-canning industry. He was famous for his balloon ascents made in the interests of scientific investigation. In 1804 he ascended to 12,000 feet and then to 22,000 feet. ERICACEAE.

Gaza'nia f. Treasure flower. Named in honour of Theodore of Gaza (1398–1478), who translated the botanical works of Theophrastus from Greek into Latin. COMPOSITAE.

Geitonople'sium n. Gr. *geiton*, a neighbour; *plesion*, near; from a close relationship to another Australian genus, *Eustrephus*. SMILACACEAE.

ge'lidus, -a, -um From icy-cold regions.

Gelse'mium n. Latinized version of *gelsomino*, the Italian name for jasmine. *G. sempervirens* is called false jasmine. LOGANIACEAE.

gemina'tus, -a, -um Paired.

geminiflo'rus, -a, -um Having twin or several flowers. Also *geminispinus*, having twin or many spines.

gemma'tus, -a, -um Jewelled.

gem'mifer, gemmi'fera, -um Bearing buds.

genera'lis, -is, -e Prevailing; normal.

genestieria'nus, -a, -um In honour of A. Genestier (b. 1858), French missionary in China.

geneven'sis, -is, -e Of Geneva, Switzerland, the home of many distinguished botanists.

genicula'tus, -a, -um Bent sharply like a knee.

Geni'pa f. Latinized version of the Guyana vernacular name of one species, the Genipap tree, the juicy fruit of which is used to make a drink locally called genipapo. RUBIACEAE.

Genis'ta f. Broom. The Latin name from which the Plantagenet kings and queens of England took their name (*planta genista*). Dyer's greenweed (*G. tinctoria*) is now well established as an adventive wild flower in north-eastern United States; it was first introduced for the yellow dye it yields, about 1630. LEGUMINOSAE.

Genistel'la f. Diminutive of *Genista*. LEGUMINOSAE.

genistifo'lius, -a, -um With leaves like broom or *Genista*.

Gentia'na f. Gentian. Named for King Gentius of Illyria, *c*. 500 B.C., who was reputed to have discovered the medicinal virtues of the root of the yellow gentian or bitterwort (*G. lutea*) from which a tonic bitters is still made. GENTIANACEAE.

Gentianella f. Diminutive of *Gentiana*. GENTIANACEAE.

geocar'pus, -a, -um With fruit ripening in the earth.

Geogenan'thus m. Seersucker plant. From Gr. *gĕ*, earth; *gĕnĕa*, race, birthplace; *anthos*, flower; in allusion to the flowers arising almost at ground level. COMMELINACEAE.

geoi'des *Geum*-like.

geomet'ricus, -a, -um; geomet'rizans With markings arranged in a formal pattern.

Geo'noma f. Gr. *gĕōnŏmŏs*, colonist. These feather palms send out buds from the apex of the stems which eventually grow into trees themselves. PALMAE.

geonomifor'mis, -is, -e In the form of *Geonoma*.

geor'gei In honour of George Forrest (1873–1932); see *forrestii*.

georgia'nus, -a, -um Of the state of Georgia, in the United States.

geor'gicus, -a, -um Of Georgia, Caucasus.

geraniifo'lius, -a, -um Geranium-leaved.

geranioi'des Resembling *Geranium.*

Gera'nium n. Cranesbill, herb-Robert. *Gĕranion* is the classical Greek name, from *gĕranos*, a crane; in allusion to the long beak of the carpels; see *Erodium, Pelargonium.* GERANIACEAE.

Gerar'dia f. Named for John Gerard (1545–1612), barber-surgeon (at the end of his life the master of the Barber-Surgeon's Company), garden superintendent to Lord Burghley, Minister to Queen Elizabeth I. He grew many exotic plants in his Holborn garden but he is best known for his *Herball*, first published in 1597 with a much amended edition by Thomas Johnson, the so-called 'immaculate' one cited as *Ger. emac.* by old authors, in 1633. The first edition is very rare.

The *Herball* is to some extent an adaptation and translation from Latin of the writings of Rembert Dodoens with many entertaining personal additions by Gerard and later by his editor Thomas Johnson. The illustrations were mostly impressions from the wood blocks employed by Tabernaemontanus in his book published in Frankfurt am Main seven years previously. The name *Gerardia* has now been abandoned in favour of *Agalinis* (SCROPHULARIACEAE) and *Stenandrium* (ACANTHACEAE).

gerardia'nus, -a, um. Named for one or other of three brothers, Alexander Gerard (1792–1839) and Patrick Gerard (1795–1835), both Indian Army officers, and James Gilbert Gerard (1794–1828), an army surgeon in India, who collected plants in the Himalaya.

Gerbe'ra f. Transvaal daisy. Named for Traugott Gerber, a German naturalist who travelled in Russia and who died in 1743. COMPOSITAE.

germa'nicus, -a, -um Of Germany; German.

Gesne'ria f. Named for Conrad von Gessner (1516–1565) of Zurich, the most celebrated naturalist of his day, the father of bibliography, and one of the most learned men of all time. Although he died from plague when only forty-nine, he published the vast *Historiae Animalium* (1551–1587), a critical review of zoological knowledge from the time of Aristotle. His early death prevented him from publishing a similar botanical study for which he prepared many beautiful illustrations. His accomplishments were extraordinary since he had to contend with great poverty in his youth and continuous ill health. He is credited with having brought the tulip into repute in Western Europe after seeing a specimen growing in a garden in Augsburg in 1559. See *Tulipa.* GESNERIACEAE.

Ge'um n. Avens, herb-bennet. The classical Latin name. ROSACEAE.

Gevu'ina f. Chilean hazel. Native Chilean name. PROTEACEAE.

gibbero'sus, -a, -um Hunchbacked; humped on one side.

gibbiflo'rus, -a, -um Having flowers with a swelling or hump on one side.

gibbo'sus, -a, -um; gib'bus, -a, -um Swollen on one side.

gibbsiae In honour of Lilian Suzette Gibbs (1870–1925), English botanist and intrepid plant collector in East Africa, Borneo and Australia.

gib'bsii In honour of the Hon. Vicary Gibbs (1853–1932), English merchant banker and genealogist, with a fine garden at Aldenham; see *aldenhamensis.*

gibralta'ricus, -a, -um Of Gibraltar.

gigante'us, -a, -um Unusually tall or large.

gi'gas Giant. The Gigantes, huge men of incredible strength and terrifying aspect, were the sons of Terra, who stormed the heavens but were killed by the lightning of Jove and the blows of Hercules.

Gi'lia f. Named for Filippo Luigi Gilii (1756–1821), astronomer in Rome, author with Caspar Xuarez of *Osservazioni fitologiche* (1789–1792). POLEMONIACEAE.

Giliber'tia f. Named after Jean Emmanuel Gilibert (1741–1814), French physician and botanist, professor in Vilna, then in Lyons. ARALIACEAE

gilliesii In honour of John Gillies (1792–1834), Scottish naval surgeon, who collected plants in Chile and Argentina where he acquired much land.

Gille'nia f. Named for Arnold Gille (Gillenius), 17th-century German botanist, who had a garden at Cassel; this genus is now called *Porteranthus.* ROSACEAE.

gilvus, -a, -um Dull yellow.

Gin'gko f. Maidenhair tree. The puzzling name *Gingko* is a misrendering of the Japanese name *gin* (silver), *kyō* (apricot), used in the 17th century in Japan but now obsolete. This genus, of which many fossil forms exist to indicate a former wide distribution, now consists of one species, only wild in a small area of central China, but extensively cultivated over much of China and Japan, whence introduced into Europe in the 18th century. The male and female flowers are on separate trees. The female trees are undesirable because of the foul-smelling fruit but the kernel is edible. GINGKOACEAE.

Gin'seng See *Panax*.

giraldia'nus, -a, -um; giral'dii In honour of Giuseppe Giraldi (1848–1901), Italian missionary in China who collected plants in Shensi (Shaanxi) province, 1890 to 1895.

glabel'lus, -a, -um; glabra'tus, -a, -um; glabre'scens; glabrius'culus, -a, -um Rather or somewhat glabrous.

gla'ber, gla'bra, -um Without hairs; glabrous. Also *glaberrimus,* completely glabrous.

glacia'lis, -is, -e From icy-cold regions, especially the neighbourhood of glaciers.

gladia'tus, -a, -um; gladiiformis, -is, -e Swordlike.

Gladi'olus m. Latin for a small sword; in allusion to the shape of the leaves.

The plants are also sometimes called sword-lilies but people generally use the Latin plural gladioli.

glandifor'mis, -is, -e In the form of a gland or shaped like one. Also: **glandu'lifer** (gland-bearing); **glanduliflo'rus** (with glandular flowers); **glandulo'sus** (glandular).

glaphy'rus, -a, -um Smoothed; polished.

glauces'cens Having some bloom; somewhat blue or sea-green.

Glauci'dium n. From *Glaucium* in allusion to the form of the few-petalled poppy-like flowers. This botanically isolated Japanese genus has been placed in RANUNCULACEAE, *Glaucidiaceae*, and *Paeoniaceae*.

glaucifo'lius, -a, -um Having grey-green (glaucous) leaves.

glauciifo'lius, -a, -um Having leaves like *Glaucium*.

Glau'cium n. Horned poppy; sea poppy. From Gr. *glaukos*, greyish-green; referring to the colour of the leaves. PAPAVERACEAE.

glaucocar'pus, -a, -um With glaucous fruits.

glaucoi'des Appearing to be coated with bloom.

glaucopep'lus, -a, -um With a greyish or bluish covering.

glaucophy'llus, -a, -um Having grey-or bluish-green leaves; having bloom on the leaves.

glau'cus, -a, -um Having bloom, the fine, whitish, powdery coating which occurs on certain leaves (e.g. cabbage) and fruits (e.g. plum).

Glaux f. Sea milkwort. Gr. name used by Dioscorides for another plant. PRIMULACEAE.

Glecho'ma f. From Gr. *glēchōn*, a kind of mint. LABIATAE.

Gledit'sia (also, incorrectly, **Gleditschia**) f. Honey locust. Ornamental trees named for Johann Gottlieb Gleditsch (1714–1786), director of the Botanical Garden, Berlin. LEGUMINOSAE.

Gleiche'nia f. Named in honour of Friedrich Wilhelm von Gleichen (1717–1783), German botanist interested in microscopical studies. GLEICHENIACEAE.

Glirici'dia f. L. *glis*, a dormouse; *caedo*, kill; the name is a Latinization of the Colombian vernacular name *mata-raton* and refers to the use of the seeds and bark to poison rats and mice. LEGUMINOSAE.

glischrus, -a, -um Gluey; sticky.

Glob'ba f. Name based on an Indonesian plant name, *galoba*. ZINGIBERACEAE.

glo'bifers, globi'fera, -um Bearing globe-shaped or spherical clusters.

globo'sus, -a, -um Round; spherical.

Globula'ria f. Globe daisy. L. *globulus*, a small round ball; from the rounded flower heads of these herbs and shrubs. GLOBULARIACEAE.

globula'ris, -is, -e Pertaining to a small sphere or ball.

globu'lifer, globuli'fera, -um Bearing clusters in the form of small globes or spheres.

globulo'sus, -a, -um Small and globular.

glomera'tus, -a, -um Clustered into more or less rounded heads. Also *glomerulifer*, having small rounded heads or clusters.

glomerula'tus, -a, -um In small clusters or glomerules.

glomeruliflo'rus, -a, -um Having flowers in glomerules. A glomerule is an inflorescence consisting of a tightly clustered cyme (a group or cluster of flowers opening from the centre outwards as in sweet cherry).

Glorio'sa f. Glory-lily. From L. *gloriosus*, glorious. Gloriosa daisies are tetraploid hybrids of *Rudbeckia q.v.* LILIACEAE.

glorio'sus, -a, -um Superb; glorious.

glosso'phorus, -a, -um Tongue-bearing.

Glottiphyl'lum n. Gr. *glōtta*, a tongue; *phyllon*, a leaf. The leaves of these succulent plants are very thick, soft, and fleshy, and shaped like tongues. They are brittle and easily snapped. AIZOACEAE.

Gloxi'nia f. What is usually called *Gloxinia* by florists is, in fact, *Sinningia*. The true *Gloxinia*, much less common, is named for Benjamin Peter Gloxin, a doctor at Colmar, who published *Observationes botanicae* at Strasbourg in 1785. GESNERIACEAE.

gloxinioi'des Resembling *Gloxinia*.

gluma'ceus, -a, -um Having glumes, the chaffy bracts enclosing the flowers of grasses or sedges.

glutino'sus, -a, -um Sticky; gluey; glutinous.

Glyce'ria f. Manna grass. From Gr. *glykys*, sweet; with reference to the edible sweet tasting grains of *G. fluitans*. GRAMINEAE.

Glyci'ne f. Soybean. From Gr. *glykys*, sweet; in allusion to the sweetness of the roots and leaves of some species, none of which begins to compare in importance with the soybean (*G. max*). Quick to mature, drought-resistant and easy to grow, this plant has been in cultivation for at least three thousand years in eastern Asia. It yields oil, meal, cheese curds, and cake as well as milk. It is imported by Western countries for use in soap, paint, synthetic-rubber manufacture, etc., as well as for food, both for human beings and cattle. LEGUMINOSAE.

glycinioi'des Resembling soybean or *Glycine*.

Glycos'mis f. Gr. *glykys*, sweet; *osme*, smell. Both the flowers and the leaves are fragrant. RUTACEAE.

Glycyrrhi'za f. Liquorice. From Gr. *glykys*, sweet; *rhiza*, a root. The root of *G. glabra* provides the liquorice of commerce. LEGUMINOSAE.

glyptostroboi'des Resembling the genus *Glyptostrobus*.

Glypto'strobus m. From Gr. *glypto*, to carve; *strŏbilŏs*, a cone; an allusion to the depressions and pits on the cone scales. TAXODIACEAE.

Gmeli'na f. Named for Johann Gottlieb Gmelin (1709–1755), German traveller and naturalist, who made extensive travels in Siberia and Kamchatka in 1733–1743, then returned to Tübingen, his native town, where he was a professor from 1749–1755. VERBENACEAE.

Gnapha'lium n. Cudweed. From Gr. *gnaphalĭŏn*, a downy plant with soft white leaves used for stuffing cushions. COMPOSITAE.

gnaphaloi'des Resembling cudweed or *Gnaphalium.*

Gni'dia f. From Latin geographical name referring to Gnidus (Gr. *Knidos*) in Caria, Asia Minor, where a kind of Daphne grew. THYMELAEACEAE.

Gode'tia f. Showy annuals named for Charles H. Godet (1797–1879), Swiss botanist, author of a *Flore du Jura* (1858–1869). The genus is now included in *Clarkia*. ONAGRACEAE.

Goe'thea f. In honour of the German poet, dramatist and scientist Johann Wolfgang von Goethe (1749–1832), whose most important contribution to botany, *Die Metamorphose der Pflanzen* (1790), stresses the equivalence of leafy and floral organs. MALVACEAE.

Goldfus'sia f. In honour of Georg August Goldfuss (1782–1848), professor of zoology and mineralogy in Bonn, Germany.

Gome'sa f. Epiphytic orchids named for Bernardino Gomez, Portuguese naval surgeon, who published a book on the plants of Brazil in 1803. ORCHIDACEAE.

Gomphocar'pus m. Milkweed. Gr. *gŏmphŏs*, bolt, nail; *karpos*, fruit. Sometimes included in *Asclepias*. ASCLEPIADACEAE.

Gompholo'bium n. Gr. *gŏmphŏs*, bolt; *lŏbŏs*, a pod; from the shape of the pod. LEGUMINOSAE.

Gomphre'na f. Globe amaranth. Latin name for a kind of amaranth, possibly this one; usually grown as an 'everlasting'. AMARANTHACEAE.

Gon'gora f. Named for Don Antonio Cabellero y Gongora (1740–1818), Spanish Viceroy of New Granada, now Colombia, and patron of José Celestino Mutis (1732–1818), naturalist and plant explorer, who made Bogota a renowned centre of scholarship and research. ORCHIDACEAE.

gongylo'des Swollen; roundish.

gonio'calyx With prominently angled calyx.

Goode'nia f. In honour of Samuel Goodenough (1743–1827), Bishop of Carlisle from 1808 onwards, a keen amateur botanist, authority on *Carex* and a founder of the Linnean Society of London. GOODENIACEAE.

Goo'dia f. Named for Peter Good (d. 1803), Kew gardener who accompanied Robert Brown on Flinders' voyage round Australia. LEGUMINOSAE.

Goodye'ra f. Rattlesnake plantain. Named in honour of John Goodyer (1592–1664) of Mapledurham, Oxfordshire, who assisted Johnson in his edition of Gerard's *Herball*, and translated Dioscorides's *Materia medica* into English. ORCHIDACEAE.

Gordo'nia f. Named for James Gordon (d. 1781), correspondent of Linnaeus and nurseryman of Mile End, London. *Franklinia* has been placed in this genus. THEACEAE.

Gorma'nia f. Plants often included in *Sedum* named for M. W. Gorman of Portland, Oregon, collector of plants of the Pacific North-west. CRASSULACEAE.

gossy'pinus, -a, -um Cottony; resembling cotton or *Gossypium*.

Gossy'pium n. Cotton. L. *gossypion*. The English name derives ultimately from Arabic. Cotton has been cultivated since prehistoric times in the Old World and was found already in use in the New when the Spaniards arrived. MALVACEAE.

Gourli'ea f. Named for Robert Gourlie (d. 1832), Scottish botanist in Chile and Argentina. This genus is sometimes included in *Geoffroea*. LEGUMINOSAE.

Grabow'skia f. Named for Henri Emanuel Grabowski (1792–1842), German botanist at Breslau. SOLANACEAE.

gracilen'tus, -a, -um Slender; thin.

gra'cilis, -is, -e Graceful; slender. Also:
graciliflo'rus (with slender or graceful flowers); **graci'lior** (more graceful); **graci'lipes** (with a slender stalk); **gracilisty'lus** (with a slender style); **gracil'limus** (most graceful).

graebneria'nus, -a, -um In honour of Karl Otto, Peter Paul Graebner (1871–1933), German botanist, keeper of Berlin-Dahlem herbarium.

grae'cus, -a, -um Greek; Grecian.

grami'neus, -a, -um Resembling grass; grassy.

graminifo'lius, -a, -um With leaves resembling grass.

Grammatophyl'lum n. From Gr. *grammata*, letters; *phyllon*, a leaf; in allusion to the markings on the flowers. ORCHIDACEAE.

grammope'talus, -a, -um With striped petals.

granaden'sis, -is, -e; granaten'sis, -is, -e Relating to Granada, Spain or to Colombia, South America.

gran'dis, -is, -e Big; showy. Also:
gran'diceps (large-headed); **grandicus'pis** (having big points); **grandidenta'tus** (having big teeth); **grandiflo'rus** (large-flowered); **grandifo'lius** (with big leaves); **grandifor'mis** (on a large scale); **grandipuncta'tus** (having large spots).

grani'ticus, -a, -um Growing on or in crevices of granite or other hard quartz rocks.

granula'tus, -a, -um With grain-like tubercles or knobs.

granulo'sus, -a, -um Composed of minute grains or appearing as though covered with them.

Graptope'talum n. Gr. *graptos*, painted, marked with letters, to write; *petalon*, a petal; from the variegated markings on the petals. CRASSULACEAE.

Graptophyl'lum n. Gr. *graptos*, painted, marked with letters, to write; *phyllon*, a leaf; from the variegated markings on the leaves. ACANTHACEAE.

gratianopolita'nus,-a, -um Of Grenoble, Dauphiné, France.

Gra'tiola f. L. *gratia*, agreeableness or pleasantness; in allusion to the medical uses of these herbs. SCROPHULARIACEAE.

gratiss'imus, -a, -um Very pleasing or agreeable.

gra'tus, -a, -um Pleasing.

grave'olens Heavily scented.

gray'i In honour of Asa Gray (1810–1888), leading American botanist, professor at Harvard University, Cambridge, Mass., author of many botanical works, notably *Gray's Manual of Botany* (1848; 8th ed. 1950).

Greeno'via f. Named in honour of the English geologist George Bellas Greenough (1778–1855). CRASSULACEAE.

Grei'gia f. Named for Major-General Samuel Alexeivich Greig (1827–1887), President of the Russian Horticultural Society. BROMELIACEAE.

Grevil'lea f. Named for Charles Francis Greville (1749–1809), a founder of the Horticultural Society of London and a Vice-President of the Royal Society. PROTEACEAE.

Gr'ewia f. Named for Nehemiah Grew (1641–1712), English physician, a pioneer microscopist, author of a celebrated *Anatomy of Plants* (1682). TILIACEAE.

Grey'ia f. Named for Sir George Grey (1812–1898), explorer and colonial administrator, governor of South Australia, then of Cape Colony, South Africa lastly of New Zealand. GREYIACEAE.

griersonia'nus, -a, -um In honour of R.C. Grierson, customs officer at Tengyueh, Yunnan, China, friend of George Forrest.

Griffi'nia f. In honour of William Griffin (d. 1827), London nurseryman noted for his cultivation of bulbs. AMARYLLIDACEAE.

griffithia'nus, -a, -um; griffi'thii In honour of William Griffith (1810–1845), British botanist and medical man who collected in India and Afghanistan but died before he could publish his extensive and detailed observations.

Grinde'lia f. Gumplant. Named for David Hieronymus Grindel (1766–1836), German professor of botany at Riga. COMPOSITAE.

griseba'chii In honour of August Heinrich Rudolf Grisebach (1814–1879), German botanist, professor at Göttingen, noted for his work on plants of the Balkan Peninsula (Rumelia), the West Indies and Argentina, on *Gentianaceae* and plant geography.

Griseli'nia f. Named for Francesco Griselini (1717–1783), Venetian naturalist. CORNACEAE.

gris'eus, -a, -um Grey.

groenlan'dicus, -a, -um Of Greenland.

grono'vii Specific epithet in honour of Johan Fredrik Gronovius (1690–1762) of Leyden, a keen naturalist and supporter of Linnaeus during his stay in Holland; he belonged to the celebrated Gronovius family of classical scholars and was the author of a *Flora Virginica* (1739–1743).

gro'sii Named for Enrique Gros (1864–1949), Spanish laboratory assistant and plant collector in mainland Spain, Balearic Isles and Morocco.

grosse-serra'tus, -a, -um With large saw-like teeth.

Grossula'ria f. Gooseberry. Latinization of the French *groseille*. GROSSULARIACEAE.

gros'sus, -a, -um Very large.

grui'nus, -a, -um Resembling a crane.

Guaia'cum n. From the South American vernacular word *guaiac*, the name for lignum vitae (*G. officinale*), meaning wood of life, so named because of its high repute in medicine. Guaiacum resin is used for a number of acute and chronic conditions, as well as in certain blood tests. The wood of these trees is hard, dense, and virtually unsplittable. Lignum vitae was used in the lock-gate hinges on the Erie Canal where they lasted for a century. ZYGOPHYLLACEAE.

guestha'licus,-a, -um Of Westphalia, Germany.

guianen'sis, -is, -e Of Guiana, northern South America.

Guizo'tia f. Named for François Pierre Guillaume Guizot (1787–1874), French historian and statesman. COMPOSITAE.

gum'mifer, gummi'fera, -um Producing gum.

gummo'sus, -a, -um Gummy.

Gunne'ra f. Named in honour of Johan Ernst Gunnerus (1718-1773), Norwegian bishop (at Trondheim) and botanist, the author of *Flora Norvegica* (1766–1772). Of all foliage plants one of the most handsome and impressive is *G. manicata* with leaves up to six feet across. GUNNERACEAE.

gunnerifo'lius With leaves like *Gunnera*.

gunnii In honour of Ronald Campbell Gunn (1808–1881), magistrate, landowner and botanist in Tasmania.

gussonii In honour of Giovanni Gussone (1787–1866), Italian botanist and physician in Naples, who published important works on the flora of Sicily.

gutta'tus, -a, -um Spotted; speckled with small dots.

Guzma'nia f. Named for an 18th-century Spanish naturalist, Anastasio Guzman. BROMELIACEAE.

Gymnade'nia f. Gr. *gymnos*, naked; *adēn*, gland; the viscidia of the stamens of these orchids are not enclosed in a pouch. ORCHIDACEAE.

Gymnocaly'cium n. Gr. *gymnos*, naked; *kalyx*, a bud; because the flower buds of these succulent plants are naked. CACTACEAE.

gymnocar'pus, -a, -um Having naked fruit.

Gymno'cladus f. Gr. *gymnos*, naked; *klados*, a branch. The branches are bare of leaves for many months in winter. One species, the Kentucky coffee-tree (*G. dioica*), has seeds which the pioneers used as a coffee substitute. LEGUMINOSAE.

Gymnogram'ma f. Gr. *gymnos*, naked; *gramma*, a line; in allusion to the naked sori of these ferns being in lines. ADIANTACEAE.

Gymno'pteris f. Gr. *gymnos*, naked; *pteris*, fern; in allusion to the naked sori of these ferns. ADIANTACEAE.

Gynandris f. Gr. *gynē*, female; *andros*, male; *Iris*; in allusion to the union of the stamens with the pistil.

Gyne'rium n. Gr. *gynē*, female; *erion*, wool; a reference to the hairy spikelets of the female plants of these grasses. GRAMINEAE.

Gynu'ra f. Gr. *gynē*, female; *oura*, a tail; from the long rough stigma of these foliage plants. COMPOSITAE.

Gypso'phila f. Baby's breath. Gr. *gypsos*, gypsum; *philos*, loving. Some species of these small-flowered plants are lime-loving. CARYOPHYLLACEAE.

gy'rans Going around in circles.

H

haagea'nus, -a, -um In honour of J. N. Haage (1826–1878), seed grower of Erfurt, Germany.

Habena'ria f. Fringed orchis, rein orchis. From L. *habēna*, a rein, strap. In some species the spur is long and shaped like a strap; the divisions of the lip are also narrow and strap-like. ORCHIDACEAE.

Haber'lea f. Named for Carl Constantin Haberle (1764–1832), professor of botany at Budapest. GESNERIACEAE.

Habli'tzia f. Named for Carl Ludwig Hablizl (1752–1821), naturalist and traveller in the Caucasus, Iran and the Crimea, of which he was vice-governor. CHENOPODIACEAE.

Habran'thus m. From Gr. *habros*, graceful, beautiful, pretty; *anthos*, flower. AMARYLLIDACEAE.

habro'trichus, -a, -um With beautiful or soft hairs or bristles.

Hacque'tia f. Named for Balthasar Hacquet (1740–1815), Austrian writer on Alpine plants. UMBELLIFERAE.

hadria'ticus, -a, -um Of the shores of the Adriatic.

haema'leus, -a, -um Blood-red.

Haeman'thus m. Red Cape tulip, blood-lily. From Gr. *haima*, blood; *anthos*, a flower. LILIACEAE.

haeman'thus, -a, -um With blood-red flowers. Also:

haema'stomus (with blood-red mouth); **haematoca'lyx** (with blood-red calyx); **haematocar'pus** (with blood-red fruits); **haematochi'lus** (with blood-red lip); **haemato'des** (blood-red).

Haematoxy'lum n. Logwood. From Gr. *haima*, blood; *xylon*, wood. Since the 16th century it has been a very valuable export from Central America and the West Indies for the sake of the brilliant red dye produced from the heart wood. *H. campechianum* is known by several other names, such as blood-tree and campeachy-wood. LEGUMINOSAE.

Ha'kea f. Named for Baron Christian Ludwig von Hake (1745–1818), German patron of botany, councillor in Hanover. PROTEACEAE

hakeoi'des Resembling *Hakea*.

hakusan'ensis, -is, -e Of Mount Haku-san, central Japan.

Hale'sia f. Silver bell. Named in honour of the Reverend Stephen Hales (1677–1761), curate of Teddington, near London, physiologist, chemist, and inventor whose wide-ranging curiosity resulted in important and far-reaching experiments in plant physiology as well as, possibly, the first measurement of blood pressure. It was performed one Sunday on a white mare, thrown and tied to a gate, whose blood, transferred through a goose's windpipe (an efficient medium before the invention of rubber tubing), rose in a glass tube to a height of 8 feet 3 inches.

He also invented an effective method of ventilating prisons to control jail fever and ships to prevent dry rot; a 'sea gauge' for taking soundings; and a method of fumigating wheat with brimstone to preserve it from weevils. STYRACEAE.

halimifo'lius, -a, -um With leaves like *Atriplex halimus* (Gr. *halimŏn*), the tree purslane with silvery ovate and obovate leaves used in seaside planting in Britain.

Halimio'cistus m. Hybrids between *Cistus* and *Halimium*. CISTACEAE.

Halimoden'dron n. Salt tree. From Gr. *halimos*, maritime; *dĕndrŏn*, a tree. A salt-tolerant shrub from Central Asia. LEGUMINOSAE.

Hali'mium n. From Gr. *halimion*, a maritime plant. CISTACEAE.

hal'leri In honour of the many-sided Albrecht von Haller (1708–1777), Swiss anatomist, botanist, bibliographer, administrator and poet, author of the first textbook of physiology and numerous other works.

hallia'nus, -a, -um; ha'llii Plants so named may commemorate Herman Christin van Hall (1801–1874), Dutch botanist, professor in Groningen, if Indonesian; Elihu Hall (1822–1882), American surveyor, or Harvey Monroe Hall (1874–1932), professor of botany, University of California, pioneer in experimental taxonomy, if North American; George Rogers Hall (1820–1899), American medical man, then trader in the China Sea, who introduced many plants from Japan into the United States, if Japanese.

halli'eri In honour of Hans Hallier (1868–1932), German botanist.

halo'philus, -a, -um Salt-loving.

Hamame'lis f. Witch-hazel. The Greek name for a plant with a pear-shaped fruit, possibly the medlar. The bark and twigs of *H. virginiana* supply the witch-hazel of pharmacy. The twigs are also a favourite choice of dowsers or water-diviners. HAMAMELIDACEAE.

Hamatocac'tus n. L. *hamatus*, hooked, plus *cactus*; referring to the long central spine of each areole which is usually hooked at its tip. CACTACEAE.

hama'tus, -a, -um; ham-o-'sus, -a, -um Hooked.

Hame'lia f. Named in honour of Henri Louis du Hamel du Monceau (1700–1781), celebrated French writer on trees and shrubs. RUBIACEAE.

hancea'nus, -a, -um In honour of Henry Fletcher Hance (1827–1856), British consul and collector of plants in China.

hanco'ckii In honour of William Hancock (1847–1914), Irish customs official in Chinese Imperial Maritime Customs, who collected in China and Taiwan.

handelii, handelianus, -a, -um In honour of Heinrich von Handel-Mazzetti (1882–1940), Austrian botanist and plant collector in Mesopotamia (1910) and China (1914–1918), monographer of *Taraxacum*, etc.

Haplopap'pus m. Gr. *haplos*, single; *pappos*, pappus. COMPOSITAE.

Harbou'ria f. Named for J. P. Harbour, collector of Rocky Mountain plants. UMBELLIFERAE.

Hardenber'gia f. Australian sarsaparilla. Named in honour of Franziska, Countess von Hardenberg; she was the sister of Austrian baron Carl A. A. von Hügel (1795–1870), an enthusiastic patron of horticulture, who travelled widely in the Philippines and elsewhere. LEGUMINOSAE.

Harpephyl'lum n. Gr. *harpē*, sickle; *phyllon*, leaf. ANACARDIACEAE.

harpophyl'lus, -a, -um With leaves shaped like sickles.

Harri'sia f. Named for William Harris (1860–1920), Irish, Kew-trained Superintendent of the Public Gardens, Jamaica, whose collecting throughout the island vastly increased knowledge of its flora. CACTACEAE.

harrovia'nus, -a, -um In honour of George Harrow, manager of Messrs Veitch's Coombe Wood nursery where many Chinese plants introduced by E. H. Wilson were first raised between 1900 and 1905.

harryanus, -a, -um In honour of Sir James Harry Veitch (1840–1924), horticulturist who sent E. H. Wilson out to China as well as other collectors elsewhere for the firm of James Veitch and Sons.

harrysmi'thii In honour of Karl August Harald (Harry) Smith (1889–1971), Swedish botanist at Uppsala who made extensive collections in China in 1921–1922, 1924–1925 and 1934.

Hartwe'gia f. Epiphytic orchids named for Carl Theodor Hartweg (1812–1871),

German gardener who collected for the Horticultural Society of London in Mexico, 1836–1837, and California, 1846–1847. ORCHIDACEAE.

harveyan'us In honour of William Harvey (1811–1866), Irish botanist, who collected in South Africa and elsewhere, chiefly renowned as an algologist.

hasta'tus, -a, -um Spear-shaped. Also:
has'tifer (bearing a spear); **hastila'bius** (with a spear-shaped lip);
hasti'lis (of a spear); **hastula'tus** (somewhat spear-shaped).

Hatio'ra f. Anagram of *Hariota*, a genus of cacti which had to be renamed on account of the earlier use of the name *Hariota* for another genus. Originally named in honour of Thomas Hariot (1560–1621), mathematician and cartographer on the 1585 expedition to Virginia and author of *A briefe and true Report of the Newfound Land of Virginia* (1588). CACTACEAE.

Hawor'thia f. Named for Adrian Hardy Haworth (1768–1833), English authority on succulent plants and *Lepidoptera*, the author of a monograph on *Mesembryanthemum* (1794), etc. ALOEACEAE.

He'be f. Named in honour of the goddess of youth, cup-bearer to the gods. *Hebe* differs from *Veronica* primarily in the form of the capsule. SCROPHULARIACEAE.

hebecar'pus, -a, -um With fruit covered in down; from Gr. *hēbē*, youth, down of puberty; *karpos*, fruit.

Hebenstrei'tia f. Named for Johann Ernst Hebenstreit (1702–1757), professor of medicine at Leipzig. GLOBULARIACEAE.

hebephyl'lus, -a, -um Downy-leaved.

Hech'tia f. Named for J. H. G. Hecht (d. 1837), a counsellor to the King of Prussia. BROMELIACEAE.

hecto'rii In honour of Sir James Hector (1834–1907), New Zealand geologist and botanist of Scottish origin, who took a very active part in the geological survey of New Zealand.

Hedeo'ma f. American pennyroyal. Variant of *hedyosmos*, mint, from Gr. *hēdys*, sweet; *ŏsmē*, scent; referring to the aromatic foliage. LABIATAE.

He'dera f. Ivy. The Latin name. Ivy was held sacred to Dionysos (Bacchus), the god of wine, and was intimately connected with his revels. ARALIACEAE.

hedera'ceus, -a, -um Resembling ivy.

hederifo'lius Ivy-leaved.

Hedy'chium n. Ginger-lily. From Gr. *hēdys*, sweet; *chion*, snow. The flower of one species is white and fragrant. ZINGIBERACEAE.

Hedyotis From Gr. *hēdys*, sweet; *ŏus, ōtŏs* ear; significance unknown. RUBIACEAE.

Hedy'sarum n. Greek name *hēdysarŏn*, from *hēdys*, sweet. This genus of ornamental herbs or sub-shrubs includes the handsome and fragrant French-honeysuckle (*H. coronarium*). LEGUMINOSAE.

Hedy'scepe f. Umbrella palm. From Gr. *hēdys*, sweet; *scĕpē*, a covering; from the manner in which the flowers of this feather palm are produced in a dense cluster from the leaf crown. PALMAE.

Hei'mia f. Shrubs named for Dr. Heim (d. 1834), of Berlin. LYTHRACEAE.

heldrei'chii In honour of Theodor H. H. von Heldreich (1822–1902), German botanist who settled in Greece in 1844 and became its indefatigable botanical explorer, discovering some 700 Greek species new to science.

he'lenae In honour of Helen, daughter of the German botanist B. A. E. Koehne (1848–1918); see *Conradinae*.

Hele'nium n. Sneezeweed. From the Greek name *hĕlĕniŏn*, for another plant named for Helen of Troy. COMPOSITAE.

Heliam'phora f. Gr. *hĕlos*, marshy ground; *amphora*, a jar. From the habitat and leaf-shape of these pitcher plants. SARRACENIACEAE.

Helianthel'la f. Diminutive of *Helianthus* to which it is closely related.

Helian'themum n. Sunrose. From Gr. *hēlios*, the sun; *anthemon*, a flower. CISTACEAE.

helianthoi'des Resembling *Helianthus*.

Helian'thus m. Sunflower. From Gr. *hēlios*, the sun; *anthos*, a flower. *H. tuberosus* is the 'Jerusalem artichoke'. COMPOSITAE.

Helichry'sum m. Everlasting. Immortelle. From Gr. *hēlios*, the sun; *chrysos*, golden. COMPOSITAE.

Helicodi'ceros m. Dragon's mouth. Twist-arum. From Gr. *hĕlix*, anything spirally twisted; *dis*, twice; *kĕras*, a horn. The basal divisions of the leaves are twisted and stand erect like horns. ARACEAE.

Helico'nia f. After Mount Hĕlĭcōn in Greece where it was supposed that the Muses lived, so named from its relationship to *Musa*. HELICONIACEAE.

Helioce'reus m. Gr. *hēlios*, the sun; *cereus*, cactus; because of their desert habitat. CACTACEAE.

helio'lepis With sun-like (i.e. bright or glittering) scales.

Helio'phila f. Gr. *hēlios*, the sun; *philein*, to love; from the habitat of these South African plants. CRUCIFERAE.

Heliop'sis f. Gr. *hēlios*, the sun; *ŏpsis*, resembling; in allusion to the rayed yellow flower heads. COMPOSITAE.

Heliotro'pium n. Heliotrope, cherry-pie. Gr. *hēliotrōpiŏn*, from *hēlios*, the sun; *trŏpē*, a turning; in allusion to an old disproved idea that the flower heads turned with the sun. The leaves and flowers of many plants do this and are known as heliotropic. BORAGINACEAE.

Helip'terum n. Everlasting. From Gr. *hēlios*, the sun; *pteron*, a wing; a technical reference to the plumed pappus. COMPOSITAE.

hel'ix Greek for anything of spiral shape, hence applied to twining plants.

Helle'borus m. Gr. *hĕllĕbŏros*, name of the Lent hellebore (*H. orientalis*). RANUNCULACEAE.

helle'nicus, -a, -um Of Greece; Grecian.

helo'des Of bogs.
helodox'a Glory of the marsh.
Helo'nias f. Stud flower, swamp pink. From Gr. *hĕlos*, a marsh; the natural habitat of these plants. LILIACEAE (MELANTHIACEAE).
Helonio'psis f. *Helonias; opsis*, resemblance. LILIACEAE (MELANTHIACEAE).
helve'ticus, -a, -um Of Switzerland; Swiss.
hel'volus, -a, -um Pale brownish yellow.
hel'vus, -a, -um Dim-coloured, light bay.
Helwin'gia f. Named after Georg Andreas Helwing (1666–1748), German pastor who wrote on *Pulsatilla* and the flora of Prussia. CORNACEAE.
Helxi'ne f. Baby's tears. A creeping herb with a classical name formerly applied to a related plant (pellitory or *Parietaria*); see *Soleirolia*.
Hemerocal'lis f. Day-lily. Gr. *hēmera*, day; *kallos*, beauty. Each flower lasts but one day. LILIACEAE (HEMEROCALLIDACEAE).
Hemian'dra f. Gr. *hēmĭ*, half; *aner, andros*, male. This genus, which has only one anther cell per stamen, was discovered by Robert Brown, who visited Australia with Flinders, from 1801 to 1805. LABIATAE.
hemidar'tus, -a, -um Half-flayed, i.e. with patchy hair-covering.
Hemigra'phis f. Gr. *hēmĭ*, half; *graphis*, brush; i.e. for one half of a brush, referring to the dense hair-covering on the filaments of the outer stamens. ACANTHACEAE.
Hemioni'tis f. Gr. *hēmiŏnītis*, from *hēmiŏnŏs*, a mule; because these ferns were once regarded as promoting sterility and hence worn by women as a charm against pregnancy. ADIANTACEAE.
Hemipte'lea f. Gr. *hēmĭ*, half; *ptĕlĕa*, elm; in allusion to the fruits, which have a wing only on the upper half, whereas in the elm (*Ulmus*) the fruit is encircled by a wing. ULMACEAE.
hemisphae'ricus, -a, -um Hemispherical; in the shape of half a ball.
hemitricho'tus, -a, -um Hairy on one side but not the other.
hemsle'yi; hemsleya'nus, -a, -um In honour of William Botting Hemsley (1843–1924), from 1899 to 1908, Keeper of the Kew Herbarium, a leading authority on the plants of Central America and China.
hen'rici In honour of Prince Henri d'Orléans (1867–1901), French explorer, who collected in western China.
hen'ryi In honour of Augustine Henry (1857–1930), Irish medical man, plant-collector and dendrologist, who made vast collections in central China and Formosa and later became a professor of forestry.
Hepa'tica f. Liverwort. From Gr. *hēpar*, the liver. This is a double allusion to the colour and shape of the leaves, whence the belief, according to the Doctrine of Signatures, that they would be good for complaints of the liver. RANUNCULACEAE.

hepaticifo'lius, -a, -um With leaves like *Hepatica.*

hepta- In compound words signifying seven. Thus *heptaphyllus*, seven-leaved.

heracleifo'lius, -a, -um With leaves like *Heracleum* or cow-parsnip.

Heracle'um n. Cow-parsnip. Variant of Gr. *hērakleia* or *panakes hērakleiŏn* in honour of Hercules (Gr. *Hērakles*). UMBELLIFERAE.

herba'ceus, -a, -um Not woody; herbaceous.

Herman'nia f. In honour of Paul Hermann (1646–1695), German botanist, professor of botany in Leiden, Holland, who spent some years in Ceylon. STERCULIACEAE.

Hermodac'tylus m. Variant of old Gr. name for a *Colchicum*, from *Hĕrmēs*, Mercury; *daktylos*, finger; referring to the tubers. IRIDACEAE.

Hernia'ria f. Herniary, rupturewort, burstwort. From *hernia*, a rupture; for which the plant was once supposed to be effective. CARYOPHYLLACEAE.

Hespera'loe f. Gr. *hĕspĕrŏs*, western. The plants are New World but in habit like the Old World aloes. AGAVACEAE.

Hesperan'tha f. Gr. *hĕspĕra*, evening; *anthos*, flower. The flowers open late in the day. IRIDACEAE.

Hes'peris f. Sweet rocket, dame's violet. From Gr. *hĕspĕra*, the evening. The plants have a marked fragrance at sunset and later. CRUCIFERAE.

hes'perus, -a, -um Of the west, western.

heter-, hetero- In Greek compound words signifying various, diverse. Thus: **heteracan'thus** (diversely spined); **heteran'thus** (diversely flowered); **heterocar'pus** (diversely fruited); **heterochro'mus** (with various colours); **hete'rodon** (diversely toothed); **heterodo'xus** (differing from the type of the genus); **heteroglos'sus** (diversely tongued); **hetero'lepis** (diversely scaled); **heteromor'phus** (differing in form); **heterope'talus** (diversely petalled); **heterophyl'lus** (diversely leaved); **hetero'podus** (diversely stalked).

Heteranthe'ra f. From Gr. *hĕtĕros*, varying, different; *anthera*, anther; in allusion to one anther being larger and different in shape from the other two. PONTEDERIACEAE.

Heterocen'tron n. Gr. *hĕtĕros*, different; *kĕntrŏn*, a spur, sharp point. The longer anthers have two bristle-like appendages, the shorter anthers merely two basal swellings. MELASTOMATACEAE.

Heterome'les f. Toyon, Christmas-berry. From Gr. *hĕtĕros*, different; *mēlŏn*, apple. ROSACEAE.

Heterospa'the f. Gr. *hĕtĕros*, differing. The spathes of these palms are of unequal length. PALMAE.

Heu'chera f. Alum-root. In honour of Johann Heinrich von Heucher (1677–1747), professor of medicine at Wittenberg, Germany. Pronunciation can be either *hew-ke-ra* or *hoy-sher-ra*. SAXIFRAGACEAE.

Heucherel'la f. Name for intergeneric hybrids between *Heuchera* and *Tiarella*.

Heur'nia See *Huernia*.

Heve'a f. Latinized form of the Guyanan name *hevé* for *H. guianensis*. The related Para rubber tree (*H. brasiliensis*) is the most important source of natural rubber. After the discovery of the secret of vulcanization (Hancock in England, 1834; Goodyear in the United States, 1839), the demand for rubber soon exceeded the Amazon supplies. The Director of Kew, Sir Joseph Hooker, determined in the 1870's to obtain seed. Difficulties in transporting it in fertile condition were solved by H. A. Wickham who in 1870 chartered a special steamship to carry his prepared baskets of 70,000 seeds. Hothouses at Kew were summarily emptied and within two weeks there were over 2,000 seedlings, most of which went to Ceylon. Thus started the great rubber-growing industry in south-east Asia which, thanks largely to Ridley's work on tapping the latex economically, by 1910 eclipsed the rubber-collecting industry of the Amazon basin, largely based on the barbarous exploitation of native people. EUPHORBIACEAE.

hexa- In compound words signifying six. Thus:
hexagono'pterus (six-angled wings); **hexago'nus** (six-angled);
hexan'drus (with six stamens); **hexape'talus** (six-petalled); **hexaphyl'lus** (six-leaved).

Hexaglot'tis f. From Gr. *hex*, six; *glotta*, tongue; the six narrow divisions of the stigma suggest the tongues of small birds. IRIDACEAE.

hi'ans Gaping.

Hibber'tia f. Named in honour of George Hibbert (1757–1837) who had a private botanic garden at Clapham, near London, England and employed a collector, James Niven, to collect plants in South Africa. DILLENIACEAE.

hiberna'lis, -is, -e Pertaining to winter.

hiber'nicus, -a, -um Irish; cf. *erigenus*.

hiber'nus, -a, -um Winter-flowering or winter-green.

hibiscifo'lius, -a, -um With leaves resembling *Hibiscus*.

Hibis'cus m. Rose-mallow. The Greek name for mallow. *H. rosa-sinensis*, generally known simply as *Hibiscus*, is a magnificent shrub widely grown in frost-free areas and in greenhouses. In Jamaica it has been known as 'shoe flower' or 'shoe black' because the mucilaginous petals can be used to polish shoes, 'a use which seems little consistent with their elegance and beauty', as a Jamaican writer said. MALVACEAE.

Hicke'nia f. In honour of Cristobál Maria Hicken (1875–1933), professor of botany at Buenos Aires. CACTACEAE.

Hidalgo'a f. Climbing dahlia. In honour of a Mexican, Miguel Hidalgo (*fl.* 1824). COMPOSITAE.

hiema'lis, -is, -e Of the winter; winter-flowering; cf. *hibernus*.

Hiera'cium n. Hawkweed. The classical name *hiĕrakion* for various yellow-flowered Mediterranean composites, supposedly from *hierax*, falcon. COMPOSITAE.

Hiero'chloë f. Gr. *hiĕrŏs*, sacred; *chlŏē*, grass. In some parts of northern Europe these fragrant grasses are strewn before church doors on saints' days. GRAMINEAE.

hierochun'ticus, -a, -um Specific epithet of rose-of-Jericho or resurrection plant (*Anastatica*). From the classical name of the town of Jericho, *Hiĕrikŏus* or *Hiĕrichō*.

hierogly'phicus, -a, -um With markings suggestive of hieroglyphs.

hilli'eri In honour of Messrs Hillier and Sons, nurserymen specializing in trees and shrubs at Winchester, Hampshire, England, notably Sir Harold G. Hillier (1905–1985).

himala'icus, -a, -um; himalayen'sis, -is, -e Of the Himalaya; cf. *emodi*.

Hind'sia f. In honour of Richard Brinsley Hinds (d. before 1861), naval surgeon in Royal Navy, naturalist on voyage of H.M.S. *Sulphur*, 1836–1842. RUBIACEAE.

Hippeas'trum n. Evidently from Gr. *hippŏs*, borse; *hippeus*, rider; apparently suggested by the epithet *equestre*, belonging to a horseman, of an included species; as stated in *Bot. Mag.* 9: t. 305 (1795), 'the spatha is composed of two leaves, which standing up at a certain period of the plant's flowering like ears give to the whole flower a fancied resemblance to a horse's head'. These bulbous plants are popularly known as *Amaryllis*. AMARYLLIDACEAE.

Hippo'broma f. From Gr. *hippŏs*, horse; *brŏmŏs*, rage, fury; in allusion to its poisonous nature, driving horses mad. CAMPANULACEAE.

hippocas'tanum Latin name for the horse-chestnut, for which see *Aesculus*. The Turks used the chestnut-like seeds as a medicine for horses.

Hippocre'pis f. Horseshoe vetch. Gr. *hippŏs*, a horse; *krēpis*, a shoe; from the shape of the pod of this trailing herb. LEGUMINOSAE.

hippoma'nicus, -a, -um From Gr. *hippŏmanēs*, used for herbs which drive horses mad or are poisonous to them or of which they are madly fond.

Hippo'phaë f. Sea-buckthorn. The classical Greek name for another plant, probably prickly spurge. ELEAGNACEAE.

hippophaeoi'des Resembling *Hippophaë*.

hirci'nus, -a, -um Smelling of goat or goatlike.

hirsu'tus, -a, -um Hairy. Thus:
hirsutiss'imus (very hairy); **hirsu'tulus** (somewhat hairy); **hirtel'lus** (rather hairy); **hirtiflo'rus** (hairy-flowered); **hir'tipes** (hairy-footed); **hir'tus** (hairy).

hispa'nicus, -a, -um Spanish.

his'pidus, -a, -um Bristly.

hjelmqvi'stii Named for Hakon Hjelmqvist (b. 1904), Danish botanist.

hodgso'nii Named for Bryan Houghton Hodgson (1800–1894), British resident in Nepal, amateur naturalist especially interested in Himalayan birds and mammals.

Hoffman'nia f. Named in honour of George Franz Hoffmann (1761–1826), professor of botany at Göttingen, 1792–1804, and Moscow, 1804–1826. RUBIACEAE.

Hohe'ria f. Latinized version of the Maori name *houhere*. MALVACEAE.

Holboel'lia f. In honour of Frederik Ludvig Holbøll (1765–1829), curator of the Copenhagen botanic garden. LARDIZABALACEAE.

Hol'cus m. Greek name for a kind of grain. GRAMINEAE.

hollan'dicus, -a, -um Of Holland, Netherlands.

Holmskiol'dia f. Chinese hat. Named in honour of Theodor Holmskiold (1732–1794). Danish physician and professor of botany, natural history and medicine at Sorø academy, Denmark. Name best pronounced as 'Holm-shol-dia'. VERBENACEAE.

holo- In compound words signifying completely. Thus:
holocar'pus (whole-fruited, that is not lobed or split); **holochry'sus** (completely golden); **hololeu'cus** (wholly white); **holope'talus** (with entire (undivided) petals); **holoseri'ceus** (covered all over with silky hairs).

Holodis'cus m. Gr. *hŏlŏs*, entire; *diskos*, a disc; in allusion to the disc of the flower. ROSACEAE.

holos'tea Specific epithet for *Stellaria*, starwort. From the Greek *hŏlŏstĕŏn* (entire bone), jokingly applied to a weak chickweed-like plant.

holttum'ii In honour of Richard Eric Holttum (1895–1990), English botanist for many years director of Singapore botanic garden, specialist on tropical ferns, bamboos, gingers and orchids.

Homalan'thus m. Gr. *hŏmalos*, flat; *anthos*, a flower; from the appearance of the flower. EUPHORBIACEAE.

Homaloce'phala f. Gr. *hŏmalos*, flat; *kĕphalē*, a head; referring to the flat top of the flower. CACTACEAE.

Homalome'na f. Name coined by Rumphius and adopted by Schott, from Gr. *hŏmalos*, flat; *mēnē*, moon; as a translation of a Malayan vernacular name, which, according to Backer, was misrendered. ARACEAE.

Home'ria f. Not in honour of the Greek poet Homer but from Gr. *hŏmēreo*, to meet; in allusion to the joining of the filaments of the stamens into a tube. IRIDACEAE.

homo'lepis Having structurally similar scales.

hondoen'sis, -is, -e Of Hondo, Japan.

hookeri In honour of Sir William Jackson Hooker (1785–1865), unsuccessful brewer and extremely successful professor of botany, from 1841 to 1865 Director of the Royal Botanic Gardens, Kew; or his son Sir Joseph

Dalton Hooker (1817–1911), much-travelled botanist and author of many important works, from 1865 to 1885 Director of the Royal Botanic Gardens, Kew.

Hor'deum n. Barley. Latin name for barley (*H. vulgare*). Squirrel-tail grass is *H. jubatum*. GRAMINEAE.

horizonta'lis, -is, -e Flat to the ground; horizontal.

Hormi'num n. Greek name for sage. LABIATAE.

hormo'phorus, -a, -um Bearing a necklace; from Gr. *hŏrmŏs*, chain, necklace, collar, wreath.

hor'ridus, -a, -um Very prickly.

horten'sis, -is,-e; hor'torum; hortula'nus, -a, -um Of or pertaining to gardens.

hortula'norum Of gardeners.

Hosa'ckia f. Named in honour of David Hosack (1769–1835), professor of botany and materia medica at Columbia College. In 1801 he purchased land, part of which is now occupied by the Rockefeller Centre, to form the Elgin Botanic Garden maintained at his own cost. LEGUMINOSAE.

Hos'ta f. Plantain lily. Named for Nicholaus Tomas Host (1761–1834), physician to the Emperor of Austria. LILIACEAE (HOSTACEAE).

Hotto'nia f. Water-violet. In honour of Peter Hotton (1648–1709), Dutch physician and botanist, professor at Leiden. PRIMULACEAE.

Houlle'tia f. Epiphytic orchids. Named for M. Houllet (1811–1890), head gardener at the Jardin des Plantes, Paris. ORCHIDACEAE.

Housto'nia f. Bluet. Named for William Houstoun (1695–1733), Scottish surgeon, who collected plants in Central America and the West Indies, making many drawings and notes which are now in the British Museum (Natural History), London. This genus is often included in *Hedyotis*. RUBIACEAE.

Houttuy'nia f. Named for Martin (or Maarten) Houttuyn (1720–1794), Dutch naturalist and physician. SAURURACEAE.

Ho'vea f. Named in honour of Anton Pantaleon Hove, 18th-century Polish botanist and collector for the Kew botanic garden and Sir Joseph Banks in West Africa, India and the Crimea. LEGUMINOSAE.

Hove'nia f. Japanese raisin tree. Named for David ten Hove (1724–1787), Dutch senator, who helped to finance Thunberg's collecting expedition in 1772–1776 to South Africa, Java and Japan. RHAMNACEAE.

Ho'wea f. Of Lord Howe Island, east of Australia, where alone this genus of palms is found. PALMAE.

howel'lii In honour of Thomas Howell (1842–1912), pioneer investigator and industrious collector of the flora of Oregon and Washington, western North America. He himself printed his *Flora of Northwest America* (1897–1903).

Hoy'a f. Wax-flower. Named for Thomas Hoy (*fl.* 1788–1809), 18th-century head gardener to the Duke of Northumberland at Syon House, Isleworth, by the Thames opposite the Royal Botanic Gardens, Kew. ASCLEPIADACEAE.

Hudso'nia f. Beach heather. Named for William Hudson (1730–1793), London apothecary and Praefectus of the Chelsea Physic Garden, author of *Flora Anglica*. CISTACEAE.

Huer'nia f. Succulent dwarf perennials named in honour of Justin Heurnius (1587–1652), Dutch missionary and the first collector of Cape of Good Hope plants. The genus was named by Robert Brown who originated the misspelling. ASCLEPIADACEAE.

hugo'nis Specific epithet of *Rosa hugonis* introduced by John Aloysius Scallon (1851–1928), Irish missionary in China 1886–1928 who on entering the Franciscan order took the religious name of Father Hugh (Hugo).

Huma'ta f. L. *humatus*, covered with earth, from L. *humus*, soil; apparently in allusion to the terrestrial rather than epiphytic habit of these ferns. DAVALLIACEAE.

Hu'mea f. Named for Lady Amelia Hume (1751–1809), wife of Sir Abraham Hume of Wormleybury, Hertfordshire, and a pupil of J. E. Smith. *Humea elegans* is a synonym of *Calomeria amaranthoides*, published a month or so earlier.

humifu'sus, -a, -um Sprawling on the ground.

hu'milis, -is, -e Low-growing; more dwarfish than most of its kindred.

humulifo'lius, -a, -um Hop-leaved.

Hu'mulus m. Hop. Medieval name apparently Latinized from a Low German or Slav name for the hop. There are two species widely grown commercially— *H. americanus* and *H. lupulus*, the European hop (*lupulus*, meaning a small wolf, an allusion to the plant's habit of smothering the trees over which it grows). CANNABACEAE.

hunga'ricus, -a, -um Hungarian.

Hunneman'nia f. Named for John Hunneman (d. 1839), English bookseller, agent for sale of herbarium specimens and introducer of plants. PAPAVERACEAE.

hunnewellia'nus, -a, -um For the Hunnewell family of Wellesley, Massachusetts, founders of the Hunnewell Arboretum.

Huntley'a f. In honour of the Rev. J. T. Huntley (*fl.* 1837), an English clergyman and orchid enthusiast. ORCHIDACEAE.

hupehen'sis, -is, -e Of Hupeh (Hubei) province, China.

Hu'ra f. Sand-box tree, monkey-dinnerbell. Latin version of the South American name. EUPHORBIACEAE.

Hutchin'sia f. Low herbs named in honour of Ellen Hutchins (1785–1815) of Bantry, Ireland, accomplished in cryptogamic (flowerless plant) botany, who made many contributions to *English Botany*. CRUCIFERAE.

hyacin'thinus, -a, -um; hyacin'thus, -a, -um Dark purplish-blue; hyacinthine; resembling a hyacinth.

hyacinthiodo'rus, -a, -um Scented like *Hyacinthus odorus*.

Hyacin'thus m. Greek name for hyacinth, a word derived from an earlier non-Greek language (Thraco-pelasgian) and originally the name of a god associated with the rebirth of vegetation like Adonis. In Greek mythology Hyakinthos became a youth of great beauty, who was accidentally killed by Apollo when teaching him to throw the discus. From his blood sprang a plant named in his memory which is thought not to have been our hyacinth but a gladiolus, probably *Gladiolus italicus*, the lower petals of which have markings which might be taken to represent the mournful exclamation 'AI AI' (alas, alas). LILIACEAE (HYACINTHACEAE).

hya'linus, -a, -um Transparent or nearly so.

hy'bridus, -a, -um Mixed; hybrid.

Hydrange'a f. Gr. *hydōr*, water; *angos*, a jar. The fruit of these shrubs is cup-shaped. HYDRANGEACEAE.

hydrangeoi'des Resembling *Hydrangea*.

Hydras'tis f. Golden seal, orange root. Gr. *hydōr*, water. Name suggested by the leaf of *Hydrophyllum canadense*, water-leaf, with which *Hydrastis* was confused. RANUNCULACEAE.

Hydriaste'le f. Gr. *hydōr*, water; *stēlē*, a column. In the wild the trees are often near water. PALMAE.

Hydril'la f. Said to be probably an ill-formed diminutive of *hydra*, water serpent. A widely distributed aquatic plant. HYDROCHARITACEAE.

Hydro'charis f. Gr. *hydōr*, water; *charis*, grace. HYDROCHARITACEAE.

Hydro'cleys f. Water poppy. Gr. *hydōr*, water; *kleis*, a key. There is no clear reason for the allusion. LIMNOCHARITACEAE.

Hydroco'tyle f. Water pennywort. Gr. *hydōr*, water; *kŏtylē*, a small cup; from the form of the leaves in *H. vulgaris*. UMBELLIFERAE.

Hydro'lea f. Gr. *hydōr*, water; L. *olea* (Gr. *elaia*), olive; from its habitat near water and some resemblance of the leaves of *H. spinosa* to those of the olive. HYDROPHYLLACEAE.

Hydromy'stria f. From Gr. *hydōr*, water; *mystriŏn*, a little spoon; in allusion to the leaves. HYDROCHARITACEAE.

Hydrophyl'lum n. Water-leaf. Gr. *hydōr*, water; *phyllon*, leaf. The stems and leaves of the original species are very watery.

Hydros'me f. Gr. *hydra*, water snake; *osmē*, smell; evidently in allusion to the foul smell emitted from the anal stink glands of the water-loving ringed snake (*Natrix natrix*) when captured and the foul smell of these aroids when in flower. ARACEAE.

hyema'lis, -is, -e Of winter; flowering in winter.

hylae'us, -a, -um Belonging to woods.

Hyloce'reus m. Gr. *hyle*, a wood; *cereus*, cactus. These epiphytes prefer some shade. CACTACEAE.

Hylome'con f. Gr. *hyle*, a wood; *mēkōn*, poppy. PAPAVERACEAE.

hylo'philus, -a, -um Loving woods.

hymen- In compound words signifying membranous. Thus:
hymenan'thus (with flowers bearing a membrane); **hymeno'des** (resembling a membrane); **hymenorrhi'zus** (membranous rooted); **hymenose'palus** (with membranous sepals).

Hymenae'a f. After Hymen, the Greek god of marriage, in allusion to the twin leaflets. LEGUMINOSAE.

Hymenanthe'ra f. Gr. *hymēn*, a membrane; the anthers are terminated by a membrane. VIOLACEAE.

Hymenocal'lis f. Spider lily. Gr. *hymēn*, a membrane; *kallos*, beauty; in allusion to the membrane uniting the stamens. AMARYLLIDACEAE.

Hymenophyl'lum From Gr. *hymēn*, a membrane; *phyllum*, leaf; in allusion to the membranous fronds of these ferns. HYMENOPHYLLACEAE.

Hymenospo'rum n. Gr. *hymēn*, a membrane; *spora*, a seed. The seeds have a membranous wing. PITTOSPORACEAE.

Hyophor'be f. Pignut palm. Gr. *hys*, a pig; *phorbe*, food. The fleshy fruits are eaten by pigs. PALMAE.

Hyoscy'amus m. Henbane. The Greek name meaning pigbean (from *hys*, pig; *kyamos*, bean), the epithet 'pig' being used here purely in a derogatory sense, as this plant is highly poisonous and yields hyoscyamine which has medical applications as a narcotic and hypnotic. SOLANACEAE.

hyperbo'reus, -a, -um Far-northern.

hypericifo'lius, -a, -um With leaves like *Hypericum*.

hypericoi'des Resembling *Hypericum*.

Hyperi'cum n. St. John's wort. The Greek name *hypereikon* is supposed to be from either *ereike*, heath or *hyper*, above; *eikon*, picture; the plant was supposed to keep evil spirits at bay, in revenge for which the Devil pierced the leaves with a needle. The flowers of some species were placed above images to ward off evil at the ancient midsummer festival Walpurgisnacht, which later became the feast of St. John (24 June), when they are in flower. GUTTIFERAE.

Hyphae'ne f. Gr. *hyphaino*, to entwine; in allusion to the fibres of the fruit on these palms. PALMAE.

hypnoi'des Mosslike.

hypo- In compound words signifying under. Thus:
hypoglau'cus (glaucous beneath); **hypoke'rina** (waxy below); **hypoleu'cus** (white beneath); **hypophae'us** (dusky below); **hypophyl'lus** (under the leaf).

Hypocalym'ma n. Gr. *hypo*, under; *kalymma*, a veil. The calyx falls like a veil or cape. MYRTACEAE.

Hypochoe'ris f. Cat's ear. A name used by Theophrastus for this or a related genus. COMPOSITAE.

hypochondri'acus, -a, -um Of melancholy appearance; with sombre-coloured flowers.

Hypocyr'ta f. From Gr. *hypo*, under; *kyrtos*, bulging, swelling; in allusion to the swollen lower side of the corolla. GESNERIACEAE.

hypoga'eus, -a, -um Underground; developing underground.

Hypo'lepis f. Gr. *hypo*, under; *lĕpis*, a scale; from the position of the sori on these ferns. DENNSTAEDTIACEAE.

hypoma'darus, -a, -um Epithet of a *Malvastrum*; from Gk. *hypo-*, under, below; *madaros*, bald; in allusion to the carpels being hairy only at the tip.

Hypo'xis f. Star grass, yellow star grass. An old Greek plant name from Gr. *hypo*, under; *oxys*, sharp; transferred to these plants to which it has no particular relevance. HYPOXIDACEAE.

Hypse'la f. From Gr. *hypsēlos*, high; apparently in allusion to the habitat of the type-species high in the Andes. CAMPANULACEAE.

hyrca'nus, -a, -um Of the Caspian area, ancient Hyrcania.

hyssopifo'lius, -a, -um With leaves like hyssop.

Hysso'pus m. Hyssop. The classical name for this sweet herb adapted from a Semitic plant name *ēzōb*. The plant to which it is now applied is almost certainly not the hyssop of the Bible, for *Hyssopus* is not indigenous in Palestine. The biblical hyssop is generally thought to be *Origanum maru*. LABIATAE.

hys'trix Bristly; porcupine-like.

I

ian'thinus, -a, -um Violet-blue.

iber'icus, -a, -um Of Iberia, i.e. Spain and Portugal, or Georgia, Caucasus.

iberi'deus, -a, -um Resembling *Iberis*.

iberidifo'lius, -a, -um *Iberis*-leaved.

Ibe'ris f. Candytuft. Gr. *ibēris*, indicating a plant from Iberia. CRUCIFERAE.

Ibicel'la f. Diminutive of *Ibex*, in allusion to the fruit with two long slender curved horns. MARTYNIACEAE.

ibo'ta Japanese name of *Ligustrum ibota*.

ica'co Spanish name for the cocaplum (*Chrysobalanus icaco*).

ichan'gensis, -is, -e From Ichang (Ychang), Hunan, South China.

icosan'drus, -a, -um Having twenty stamens (*icos-* in compound words signifies twenty).

idae'us, -a, -um Of Mount Ida, Crete, less often Asia Minor.

ida-mai'a Specific epithet of *Brevoortia* (*q.v.*) after Ida May Burke, daughter of a California stage-coach driver, for whom the plant was named in 1867.

Ide'sia f. Named for Eberhard (or Evert) Ides, German or Dutch explorer of northern Asia in the service of the Russian Tzar Peter the Great between 1691 and 1695. FLACOURTIACEAE.

ido'neus, -a, -um Suitable; worthy.

ignes'cens; ig'neus, -a, -um Fiery red.

igno'tus, -a, -um Unknown hitherto.

ika'riae Of the island of Nikaria (Ikaria), Aegean Sea.

I'lex f. Holly. From the Latin name for the holm oak (*Quercus ilex*). AQUIFOLIACEAE.

ilicifo'lius, -a, -um *Ilex*- or holly-leaved.

illecebro'sus, -a, -um Enticing, charming.

Illi'cium n. L. *illicium*, allurement, inducement; from the enticing aromatic scent. ILLICIACEAE.

illini'tus, -a, -um Smeared; smirched.

illustra'tus, -a, -um Pictured.

illus'tris, -is, -e Brilliant, lustrous.

illy'ricus, -a, -um Of Illyria, or Illyricum, region corresponding roughly to what was western Jugoslavia and comprising Liburnia and Dalmatia.

imbecil'lis, -is, -e; imbecil'lus, -a, -um Weak, feeble.

imber'bis, -is, -e Without a beard.

imbrica'tus, -a, -um Overlapping in regular order like tiles, e.g. scales.

immacula'tus, -a, -um Spotless.

immer'sus, -a, -um Immersed, growing under water.

Impa'tiens f. Touch-me-not, balsam. L. *impatiens*, impatient; in allusion to the violent seed discharge from the ripe pods. BALSAMINACEAE.

impa'tiens Impatient; with seed capsules dehiscing as soon as ripe.

impedi'tus, -a, -um Hindered, difficult of passage, tangled.

Impera'ta f. After Ferrante Imperato (1550–1625), an apothecary of Naples. Another name for *Miscanthus*, *q.v.* GRAMINEAE.

impera'tor L. for leader, chief, emperor, used as the epithet for an outstanding species.

imperatri'cis Of the empress, i.e. of the Empress Joséphine (1763–1814) who, as the wife of Napoleon, established an outstanding garden at Malmaison, with rare plants brought from Belgium, Holland and England for its adornment, and employed the celebrated flower-painter P. J. Redouté as her botanical artist. The genera *Josephinia* and *Lapageria* (based on her earlier name Joséphine de la Pagerie) were named in her honour; see p. 184.

imperia'lis, -is, -e Imperial; showy.

imple'xus, -a, -um Tangled.

impres'sus, -a, -um Sunken or impressed, as veins may be.

impu'dicus, -a, -um Shameless; lewd.

inaequa'lis, -is, -e Unequal; in compound words, *inaequi-*, e.g. *inaequidens*, with unequal teeth.

inaper'tus, -a, -um Not open; closed.
inca'nus, -a, -um Hoary; quite grey.
incarna'tus, -a, -um Flesh-coloured.
Incarvil'lea f. Named after Pierre d'Incarville (1706–1757), French missionary in China and botanical correspondent of the great botanist Bernard de Jussieu; he introduced various north Chinese plants, including *Lycium barbarum*, into European gardens. BIGNONIACEAE.
incer'tus, -a, -um Doubtful; uncertain.
incisifo'lius, -a, -um With cut leaves.
inci'sus, -a, -um Incised; deeply and irregularly cut.
inclina'tus, -a, -um Bent down.
incompara'bilis, -is, -e Incomparable.
incomp'tus, -a, -um Unadorned.
inconspi'cuus, -a, -um Inconspicuous.
incrassa'tus, -a, -um Thickened.
incurva'tus, -a, -um; incur'vus, -a, -um Bent inward, as in the rays of some chrysanthemums.
in'dicus, -a, -um Literally of India but also applied to plants originating throughout the East Indies and from as far away as China. *Rosa indica*, for instance, came from near Canton. It seems as if any plant that came home in an Indiaman might be given this specific epithet without further ado—to the confusion of future generations.
Indigo'fera f. Indigo. From *indigo*; plus L. *fero*, to bear. A widely distributed genus in tropical and subtropical countries, from which the dye indigo is obtained. LEGUMINOSAE.
indivi'sus, -a, -um Undivided.
indura'tus, -a, -um Hard, hardened.
ine'brians Intoxicating.
iner'mis, -is, -e Unarmed; without prickles.
infaus'tus, -a,-um Unfortunate.
infecto'rius, -a, -um Dyed; coloured.
infes'tus, -a, -um Dangerous; troublesome.
infla'tus, -a, -um Swollen; inflated.
infle'xus, -a, -um Bent inwards.
infortuna'tus, -a, -um Unfortunate, hence applied to poisonous plants.
infrac'tus, -a, -um Curving inward.
infundibulifor'mis, -is, -e Funnel- or trumpet-shaped.
In'ga f. The Latinized West Indian name. LEGUMINOSAE.
in'gens Enormous.
ingramii In honour of Douglas C. Ingram (1882–1929), American forester, killed in a forest fire in Washington, U.S.A., or Collingwood Ingram

(1880–1981), English traveller, ornithologist and horticulturist, author of *Birds of the Riviera* (1926), *Isles of the Seven Seas* (1936), *Ornamental Cherries* (1948) and *Garden of Memories* (1970).

innomina'tus, -a, -um Unnamed.

inno'xius, -a, -um Harmless, e.g. without prickles.

inodo'rus, -a, -um Unscented.

inopi'nus, -a, -um Unexpected.

inorna'tus, -a, -um Without ornament; unadorned.

in'quinans Stained; flecked.

inscrip'tus, -a, -um Inscribed; marked with what might seem to be letters.

insig'nis, -is, -e Distinguished; remarkable.

insi'pidus, -a, -um Tasteless; insipid.

insiti'tius, -a, -um Grafted.

insub'ricus, -a, -um Relating to southern Switzerland (Ticino) and adjacent northern Italy.

insula'nus, -a, -um; insula'ris, -is, -e Pertaining to islands; growing on an island.

intac'tus, -a, -um Untouched; intact.

intamina'tus, -a, -um Unsullied.

in'teger, inte'gra, -um Entire, undivided, without toothing.

integer'rimus, -a, -um Completely entire.

integrifo'lius, -a, -um With entire or uncut leaves.

interme'dius, -a, -um Intermediate in colour, form, or habit.

interrup'tus, -a, -um Interrupted; not continuous, as with scattered leaves or flowers.

intertex'tus, -a, -um Intertwined.

intor'tus, -a, -um Twisted.

intrica'tus, -a, -um Tangled.

intror'sus, -a, -um Turned inward toward the axis.

intumes'cens Swollen.

intyba'ceus, -a, -um Chicory-like.

in'tybus Specific epithet for chicory; from the Latin name *intubus* for wild chicory.

I'nula f. The Latin name. *I. helenium* is elecampane. COMPOSITAE.

inunda'tus, -a, -um Flooded, hence growing in places liable to flooding.

inu'tilis, -is, -e Useless.

inver'sus, -a, -um Turned over.

invi'sus, -a, -um Unseen; not visible.

involucra'tus, -a, -um Provided with an involucre, a ring of bracts surrounding several flowers.

involu'tus, -a, -um Rolled inward, as leaves in a bud.

invol'vens Wrapped around, covered.

Iochro'ma n. Gr. *ĭŏn*, violet; *chrōma*, colour; from the colour of the flowers of the type species. SOLANACEAE.

io'des Violet-like, from Gr. *ĭŏn*, violet or rust-like, rust-coloured, from Gr. *ĭŏs*, rust.

ioen'sis, -is, -e From Iowa, U.S.A.

ionan'drus, -a, -um With violet-coloured stamens.

ionan'thus, -a, -um With violet-coloured flowers.

Ionopsid'ium n. Gr. *ĭŏn*, violet; *opsis*, like. These annuals somewhat resemble tufted violets.

iono'pterus, -a, -um Violet-winged.

Iphei'on n. Derivation obscure. *I. uniflorum*, the spring starflower often called *Brodiaea* (or *Triteleia*) *uniflora*, is now put in *Tristagma*. ALLIACEAE.

Ipomoe'a f. Morning-glory. From Gr. *ips*, worm; *hŏmoios*, similar to. CONVOLVULACEAE.

Iresi'ne f. Apparently a contraction of Gr. *ĕirĕsīōnē*, a branch wound round with wool, from Gr. *ĕiros*, *ĕirĕos*, wool. AMARANTHACEAE.

Iriar'tea f. In honour of Bernardo de Yriarte, 18th-century Spanish patron of learning. PALMAE.

i'ricus, -a, -um Irish; cf. *erigenus*, *hibernicus*.

irides'cens Iridescent.

iridiflor'us, -a, -um Iris-flowered. Also:

iridifo'lius, -a, -um (iris-leaved); **iridoi'des** (resembling iris).

I'ris f. Named for the Greek goddess of the rainbow. IRIDACEAE.

irra'sus, -a, -um Unshaven, hence hairy or rough.

irregula'ris, -is, -e With parts of dissimilar size.

irri'guus, -a, -um Watered, soaked.

ir'ritans Causing discomfort.

irrora'tus, -a, -um Bedewed, dew-sprinkled, hence minutely spotted.

isabelli'nus, -a, -um Yellowish, tawny yellow.

isan'drus, -a, -um With equal stamens.

I'satis f. Woad. The classical Greek name for this dye plant. *I. tinctoria* provided the blue dye used by ancient Britons to stain their bodies. Up to 1930 it was cultivated on a small scale in Britain as one of the best of the blue vegetable dyes. CRUCIFERAE.

Iser'tia f. Named for Paul Erdmann Isert (1757–1789), a German surgeon. RUBIACEAE.

islan'dicus, -a, -um Of Iceland.

Isolo'ma n. Gr. *isos*, equal; *lōma*, a border. The name *Isoloma* of J. Smith for ferns has priority over *Isoloma* of Decaisne which is now *Kohleria*. DENNSTAEDTIACEAE.

isophyl'lus, -a, -um With equal-sized leaves.

Isople'xis f. Gr. *isos*, equal; *plĕkō*, to plait, devise, contrive. Application to these shrubs related to foxgloves is not clear, but apparently it refers to the almost equal upper and lower lips of the corolla being 'equally contrived'. SCROPHULARIACEAE.

Isopo'gon m. Gr. *isos*, equal; *pōgōn*, a beard. The fruit is equally hairy all round in contrast to that of *Petrophila*. PROTEACEAE.

Isopy'rum n. Gr. *isos*, equal; *pyros*, wheat. The ancient Greek name of a *Fumaria*, probably transferred to these low-growing perennial herbs on account of similar foliage. RANUNCULACEAE.

Iso'toma f. Gr. *isos*, equal; *tŏma*, a section. The segments of the corolla are equal. CAMPANULACEAE.

Isotre'ma n. Gr. *isos*, equal; *trēma*, hole; in allusion to the almost regular calyx-limb around the open mouth of the flower, which distinguishes this group from *Aristolochia* proper. ARISTOLOCHIACEAE.

istri'acus, -a, -um Of Istria, northern Adriatic peninsula.

ita'licus, -a, -um Of Italy, Italian.

I'tea f. Greek name for the willow. GROSSULARIACEAE.

iteophyl'lus Willow-leaved.

I'xia f. African corn lily. Gr. *ixia*, bird-lime; from the sticky sap of these South African plants. IRIDACEAE.

ixioi'des *Ixia*-like.

Ixioli'rion n. *Ixia* plus *leirion*, a lily; in allusion to the similarity of appearance to *Ixia*. AMARYLLIDACEAE.

ixocar'pus, -a, -um With sticky or glutinous fruit.

Ixo'ra f. Portuguese rendering of Sanskrit *Isvara*, lord, referring to the god Siva. RUBIACEAE.

J

jabotapita Brazilian plant name.

Jacaran'da f. Latinized Brazilian name. BIGNONIACEAE.

jack'ii Specific epithet honouring John George Jack (1861–1949), Canadian dendrologist at the Arnold Arboretum, Boston, Massachusetts.

jackma'nii Named after George Jackman (1801–1867) and his son George Jackman junior (1837–1887), nurserymen at Woking, Surrey, England, who raised many *Clematis* hybrids.

jacobae'us, -a, -um Specific epithet applied to various plants, in honour of St. James (Jacobus), whose saint's day is 25 July, or referring to the island of St. Iago, Cape Verde Islands.

Jacobi'nia f. Possibly named for the town of Jacobina near Bahia, Brazil. ACANTHACEAE.

Jacquemon'tia f. Named for Victor Jacquemont (1801–1832), French naturalist and traveller in the West Indies and in India, where he died. CONVOLVULACEAE.

jacqui'nii In honour of Nicolaus Joseph von Jacquin (1727–1817), Austrian botanist of Dutch origin, who travelled extensively in the Caribbean region as a young man and later published many sumptuous folio botanical works when a professor at Vienna.

jagus A gardener's corruption of *gigas*.

jala'pa Of Jalapa (once Xalapan), Veracruz, Mexico. Specific epithet of marvel of Peru (*Mirabilis jalapa*) from which the drug jalap was once supposed to be derived. It comes, in fact, from *Ipomoea purga*.

jamaicen'sis, -is, -e Of Jamaica; Jamaican.

Jame'sia f. Named in honour of Edwin James (1797–1861), American botanical explorer of the Rocky Mountains area. HYDRANGEACEAE.

Jankae'a f. Named for the Hungarian botanist and cavalry officer Victor de Janka (1837–1890). GESNERIACEAE.

japo'nicus, -a, -um Japanese. 'Japonica' is a common name for Japanese quince (*Chaenomeles speciosa*) and *Kerria japonica*.

Jasi'one f. Sheep's-bit scabious. Greek name for another plant. CAMPANULACEAE.

jasmi'neus, -a, -um; jasminoi'des Resembling jasmine. Also *jasminiflorus*—jasmine-flowered.

Jasmi'num n. Jasmine (sometimes called jessamine). Late medieval Latin version of the Persian name *yāsmīn* or *yāsamīn*. OLEACEAE.

Jat'ropha f. Gr. *iatros*, a physician; *trophe*, food. *J. curcas*, the physic nut, is nutritious in small amounts but can be a drastic purgative. EUPHORBIACEAE.

java'nicus, -a, -um Of Java; Javan.

Jefferso'nia f. Twin leaf. Named in honour of Thomas Jefferson (1743–1826), President of the United States, 1801–1809. A truly great man of many parts, he was an outstanding plantsman and patron of botany with deep interest in horticulture and farming. BERBERIDACEAE

jonquil'la Variant of Spanish vernacular name *junquillo*, from *Juncus*, rush, referring to the rush-like leaves.

Jovella'na f. Named for Gaspar Melchor de Jovellanos, 18th-century patron of botany in Peru. SCROPHULARIACEAE.

Jovibarba f. From medieval Latin name *Jovis barba*, beard of Jupiter, used by Charlemagne (*c.* 742–814) for houseleek *(Sempervivum tectorium)*, which he ordered to be planted on the house of an imperial gardener, presumably in accordance with the ancient belief that it protected against lightning. The French name is *Joubarbe*. CRASSULACEAE.

Jubae'a f. Named for King Juba of Numidia who committed suicide in 46 B.C. when his ancient kingdom in North Africa was absorbed as a province by the Romans. PALMAE.

juba'tus, -a, -um Crested, hence with long awns.

jucun'dus, -a, -um Agreeable; pleasing.

juga'lis, -is, -e; jugo'sus, -a, -um Yoked.

Ju'glans f. Walnut. The Latin name. From *jovis*, of Jupiter; *glans*, an acorn. JUGLANDACEAE.

ju'liae In honour of Julia Mlokosiewicz, who discovered *Primula juliae* in the Caucasus in 1900: see *mlokosewitschii*.

julia'nae In honour of Juliana Schneider, first wife of C.K. Schneider (1876–1951).

jun'ceus, -a, -um Rushlike.

juncifolius, -a, -um. With rush-like leaves.

Jun'cus, m. Rush. The classical Latin name. JUNCACEAE.

juniperifo'lius, -a, -um Juniper-leaved.

juniperi'nus, -a, -um Juniper-like; bluish-black like juniper berries.

Juni'perus f. Latin name for the juniper. Apart from ornament, a principal use of juniper is as the source of the berries which flavour gin. CUPRESSACEAE.

Juri'nea f. In honour of André Jurine (1780–1804), son of the Geneva professor Louis Jurine, student of medicine and botany. COMPOSITAE.

Jussiae'a f. Primrose willow. Named for Bernard de Jussieu (1699–1777), French botanist, who laid the foundation for the natural system of plant classification. This genus is now included in *Ludwigia*. ONAGRACEAE.

Justi'cia f. Named for James Justice (*fl.* 1730–1763), celebrated 18th-century Scottish horticulturist. ACANTHACEAE

Juttadin'tera f. In honour of Jutta Dinter, wife of Professor K. Dinter (see *Dinteranthus*). This was once included in *Mesembryanthemum*. AIZOACEAE.

K

Kadsu'ra f. The Japanese name. SCHISANDRACEAE.

Kaempfe'ria f. Named for Engelbert Kaempfer (1651–1716), a very learned and adventurous German physician, who travelled widely throughout the East, lived for two years in Japan, and wrote a book *Amoenitates exoticae*, which includes accurate descriptions of hitherto unknown Japanese plants, and a posthumously published history of Japan. ZINGIBERACEAE.

kaki Japanese name for *Diospyros kaki*, in full, *kaki-no-ki*; in no way connected with 'khaki'.

Kalan'choe f. From corrupted and unintelligible Chinese name of one species of these succulent herbs or sub-shrubs. CRASSULACEAE.

Kal'mia f. Mountain laurel, sheep laurel, swamp laurel, calico bush. Named in honour of Pehr Kalm (1715–1779), Finnish pupil of Linnaeus, who was sent by the Swedish government in 1748 to report on the natural resources of North America. His descriptions of the domestic economy and natural history are both interesting and trustworthy. ERICACEAE.

kalmiiflo'rus, -a, -um With flowers like *Kalmia*.

Kalmio'psis f. Western American shrub named from its resemblance (*opsis*) to *Kalmia*. ERICACEAE.

Kalo'panax m. Gr. *kalos*, beautiful; *Panax*, name of a related genus. ARALIACEAE.

kamtscha'ticus, -a, -um Of Kamchatka.

kansuen'sis, -is, -e Of Kansu (Gansu), the north-western province of China whence Reginald Farrer introduced some good hardy plants.

karatavien'sis, -is, -e Of the Kara Tau (black mountain) range in Kazakhstan, Central Asia.

kashmiria'nus, -a, -um Of Kashmir.

keiskea'nus, -a, -um; keiskei In honour of Keisuke Ito (1803–1901), celebrated Japanese physician and botanist, who as a young man studied under von Siebold at Nagasaki, promoted western medicine in Japan and became professor of botany at Tokyo, honoured in his 98th year as one of the 'Twelve Heroes of Modern Japan'.

kele'ticus, -a, -um Gr. *kēlētikos*, charming.

kellog'gii In honour of Albert Kellogg (1813–1887), American physician and pioneer Californian botanist.

Kel'seya f. Named for Harlan P. Kelsey (1872–1958), nurseryman of Boxford, Massachusetts, U.S.A. ROSACEAE.

kemaonen'sis, -is, -e Of Kumaun, region of Western Himalaya, India; also spelled *kamaoensis*, *kumaoensis*.

Kenne'dia f. Named for John Kennedy (1775–1842), the original partner of Lee and Kennedy, celebrated Scottish nurseryman at Hammersmith, London. LEGUMINOSAE.

kermesi'nus, -a, -um Carmine.

Ker'nera f. Named for Johann Simon von Kerner (1755–1830), professor of botany at Stuttgart. CRUCIFERAE.

ker'neri In honour of Anton Joseph Kerner von Marilaun (1831–1898), professor of botany at Innsbruck, then at Vienna, author of a monumental *Pflanzenleben* or *Natural History of Plants*, which remains a treasure-house of botanical information, as well as a monograph of *Pulmonaria*.

Ker'ria f. Jew's mallow. Named for William Kerr (d. 1814), Kew gardener and collector, who collected in China, Java and the Philippines. ROSACEAE.

Ketelee'ria f. Named in honour of Jean Baptiste Keteleer (1813–1903), French nurseryman of Belgian origin. PINACEAE.

kewen'sis, -is, -e Of the Royal Botanic Gardens, Kew, England; mostly applied to interspecific hybrids which originated there.

khasia'nus, -a, -um Of the Khasi hills of Assam, India.

Kick'xia f. Named for Jean Kickx (1775–1831), apothecary of Brussels and writer on cryptogamic plants. SCROPHULARIACEAE.

Kige'lia f. Sausage-tree. Latinized from the Mozambique vernacular name *kigeli-keia*. BIGNONIACEAE.

Kingdonwardia f. In honour of F. Kingdon-Ward. See *wardii*. GENTIANACEAE.

Kirengesho'ma f. Japanese name for a yellow-flowered perennial, from *ki*, yellow; *renge*, lotus blossom; *shoma*, hat; *Rengeshoma* being the Japanese vernacular name for *Anemonopsis macrophylla*. HYDRANGEACEAE.

kir'kii Specific epithet in honour of Thomas Kirk (1828–1898), author of *The Forest Flora of New Zealand*, etc. Born in Coventry, England, he worked in a timber mill before migrating to New Zealand where his outstanding botanical work led to his gaining university teaching posts, and finally to his appointment as Chief Commissioner of State Forests. Also in honour of his son Harry Bower Kirk (d. 1948), professor of biology in the University of New Zealand (1903–1944), outstanding naturalist, humanist and teacher. Also used in remembrance of Sir John Kirk (1832–1922), for many years British Consul at Zanzibar, ardent botanist and father of Colonel J. W. C. Kirk (1878–1962), author of *A British Garden Flora*.

Kitaibe'la f. In honour of Paul Kitaibel (1757–1817), Hungarian botanist, professor of botany at Pest. MALVACEAE.

kiusia'nus, -a, -um Of Kyushu, the large southern island of Japan.

Klei'nia f. In honour of Jacob Theodor Klein (1685–1759), German botanist in East Prussia. COMPOSITAE.

Knau'tia f. In honour of Christoph Knaut (1638–1694), German doctor and botanist. DIPSACACEAE.

Knigh'tia f. Named for Thomas Andrew Knight (1759–1838), a president of the Horticultural Society of London and a well-known pomologist. PROTEACEAE.

Knipho'fia f. Red-hot poker. Named for Johann Hieronymus Kniphof (1704–1763), professor of medicine at Erfurt and author of a folio of nature-printed illustrations of plants (1747). ALOACEAE.

ko'bus Japanese name for a species of *Magnolia*.

Ko'chia f. Summer cypress. Named for Wilhelm Daniel Josef Koch (1771–1849), professor of botany at Erlangen. Now included in *Bassia*. CHENOPODIACEAE.

Koellike'ria f. In honour of Rudolf Albrecht Koelliker (1817–1905), Swiss anatomist, professor at Würzburg. GESNERIACEAE.

Koellikohle'ria f. Hybrids of *Koellikeria* and *Kohleria*.

Koelreute'ria f. Named for Joseph Gottlieb Koelreuter (1733–1806), professor of natural history at Karlsruhe, a pioneer experimental investigator of plant hybridization. SAPINDACEAE.

Kohle'ria f. In honour of Michael Kohler, 19th-century teacher of natural history at Zürich, Switzerland. GESNERIACEAE.

koka'nicus, -a, -um Of Khokand, Turkistan.

Kolkwit'zia f. Beauty bush. Named for Richard Kolkwitz (b. 1873), professor of botany in Berlin *c.* 1900. CAPRIFOLIACEAE.

kolomik'ta Vernacular name in the Amur region, eastern Asia, for *Actinidia kolomikta.*

kongboen'sis, -is, -e Of Kongbo, south-eastern Tibet.

korea'nus, -a, -um (also **korianus**); **koriaen'sis, -is, -e** Of Korea; Korean.

Korthal'sia f. Named for Peter Willem Korthals (1807–1892), Dutch botanist, from 1831–1836 in Java, Sumatra and Borneo. PALMAE.

Kosteletz'kya f. Named in honour of Vincenz Franz Kosteletzky (1801–1887) of Prague, writer on medical botany. MALVACEAE.

kot'schyi; kotschya'nus, -a, -um In honour of Theodor Kotschy (1813-1866), Austrian botanist, plant-collector and traveller in North Africa and western Asia.

kou'sa Japanese name for *Cornus kousa.*

kouytchen'sis, -is, -e Of Kweichow (Guizhou) province, central China.

Krame'ria f. Named for Johann Georg Heinrich Kramer, Austrian botanist and military doctor in the first half of the 18th century. KRAMERIACEAE.

kraussia'nus, -a, -um Named for Christian Friedrich von Krauss (1812–1890), German botanist and zoologist, who collected in South Africa in 1838–1840, long associated with Stuttgart natural history museum.

Kri'gia f. Dwarf dandelion. Named for David Krieg (d. 1713 at Riga), 18th-century German physician who collected in Maryland in 1698.

Kun'zea f. Named for Gustav Kunze (1793–1851), professor of botany at Leipzig, especially interested in ferns. MYRTACEAE

kur'dicus, –a, -um Referring to the mountainous region of western Asia inhabited by the Kurds; see *carduchorum.*

kur'roo Vernacular name, better rendered as *karu*, of *Gentiana kurroo* and *Picrrorhiza kurrooa*, used in the Himalaya as a bitter tonic.

L

labia'tus, -a, -um; labio'sus, -a, -um Lipped, i.e. with well-developed lip.

Labi'chea f. In honour of a French naval officer Labiche (d. 1818). LEGUMINOSAE.

la'bilis, -is, -e Slippery; unstable.

laburnifo'lius, -a, -um *Laburnum*-leaved.

Laburnocy'tisus m. Compounded name for the graft hybrid between *Cytisus* and *Laburnum*.

Labur'num n. Latin name. LEGUMINOSAE.

lac Milk. L. *lac*, milk. Thus:
lactes'cens (becoming milky); **lac'teus** (milk-white); **lacti'color** (milk-white); **lac'tifer** (producing milky sap); **lactiflo'rus** (with milk-white flowers).

la'cer, la'cera, -um; lacera'tus, -a, -um Torn or cut into fringelike segments.

Lachena'lia f. Cape cowslip. Named for Werner de la Chenal (1736–1800), professor of botany at Basel, Switzerland. LILIACEAE HYACINTHACEAE.

Lachnan'thes f. Gr. *lachne*, wool, down; *anthos*, a flower; in allusion to the woolly flowers. HAEMODORACEAE.

lachno'podus, -a, -um With downy pedicels.

lacinia'tus, -a, -um; lacinio'sus, -a, -um Slashed or torn into narrow divisions.

lac'rimans Weeping.

lac'ryma-jo'bi Job's tears. Specific epithet of a plant in the genus *Coix*. The edible seeds form a dirty-white beadlike structure. ('My face is foul with weeping, and on my eyelids is the shadow of death.' Job 16: 16.)

lact- See above under *lac*.

Lactu'ca f. Lettuce. L. *lac*, milk; from the milky sap. COMPOSITAE.

lactucifo'lius, -a, -um Lettuce-leaved.

lacuno'sus, -a, -um With deep holes or pits.

lacus'tris, -is, -e Pertaining to lakes.

lada'nifer, ladani'fera, -um Bearing ladanum (labdanum), a fragrant resinous juice used in medicine, collected from *Cistus*.

Lae'lia f. A genus of epiphytic orchids named after one of the Vestal Virgins. ORCHIDACEAE.

Laeliocattley'a f. *Laelia* plus *Cattleya*; a collective designation for bigeneric hybrids derived from these genera.

lae'tus, -a, -um Bright; vivid. Also: *laetiflorus*, bright flowered; *laetivirens*, vivid green.

laevicau'lis, -is, -e Smooth-stemmed.

laeviga'tus, -a, -um; lae'vis, -is, -e Smooth.

Lagarosi'phon m. From Gr. *lagaros*, thin, narrow; *siphōn*, tube. HYDROCHARITACEAE.

lagas'cae Named for Marianio La Gasca y Segura (1776–1839), Spanish botanist who spent many years as an exile in England and Jersey.

Lagenan'dra f. From Gr. *lāgēnos*, flask; *anēr*, *andros*, man, male organ; in allusion to the peculiar anther-cells prolonged upwards into a slender tube like the neck of a flask. ARACEAE

Lagena'ria f. Bottle-gourd. Gr. *lāgenos*, a flask. CUCURBITACEAE.

Lagerstroe'mia f. Named for the Swedish merchant Magnus von Lagerström (1691–1759) of Göteborg by his friend Linnaeus. LYTHRACEAE.

lag'geri In honour of Franz Josef Lagger (1802–1870), Swiss physician and botanist.

lagodechia'nus, -a, -um Of Lagodekhi in the Caucasus; cf. *mlokosewitschii*.

Lago'tis f. From Gr. *lagos*, hare; *ŏus*, *ōtŏs*, ear. SCROPHULARIACEAE.

Laguna'ria f. A handsome flowering tree named for Andréa Laguna (1494–1560), Spanish botanist. MALVACEAE.

Lagu'rus m. Hare's-tail grass. From Gr. *lagos*, a hare; *oura*, a tail; referring to the hairy inflorescence. GRAMINEAE.

Lalleman'tia f. In honour of Julius Leopold Avé-Lallemant (1803–1867), of Lübeck, German botanist, for a time an official of the St. Petersburg botanic garden, who travelled in South America. LABIATAE.

Lamar'ckia f. Ornamental annual grass. Named for the Chevalier Jean Baptiste de Monet Lamarck (1744–1829), distinguished French naturalist and author of the botanical part of the *Encyclopédie méthodique* and *Flore française;* his work on evolution was an important forerunner of Darwin's. GRAMINEAE.

Lamber'tia f. Named for Aylmer Bourke Lambert (1761–1842), English botanist, author of *The Genus Pinus* (1804). PROTEACEAE.

lamella'tus, -a, -um Layered, i.e. made of or possessing thin layers of tissue.

La'mium n. Dead-nettle. The Latin name. LABIATAE.

Lampran'thus m. From Gr. *lampros*, shining, glossy; *anthos*, flower. A genus of succulent plants, formerly included in *Mesembryanthemum*. AIZOACEAE.

lana'tus, -a, -um Woolly.

lanceola'tus, -a, -um Lanceolate, spear-shaped, i.e. of narrow shape with curved sides tapering to a point.

lan'ceus, -a, -um Spear-shaped. Thus: *lancifolius*, lance-leaved.

langkongensis, -is, -e From Langkong (Eryuan), Yunnan province, China.

lan'iger, lani'gera, -um; lano'sus, -a, -um; lanugino'sus, -a, -um Woolly.

lannesia'nus, -a, -um In honour of Lannes de Montebello, a Frenchman who sent plants from Japan to France about 1870.

Lanta'na f. Late Latin name for *Viburnum*, transferred to this genus. Aggressive and difficult weeds. VERBENACEAE.

Lapage'ria f. Named after Joséphine Tascher de La Pagerie, later the Empress Joséphine (1763–1814), wife of Napoleon, who as a girl lived on the La Pagerie estate in Martinique. SMILACACEAE.

Lapeirou'sia f. Named after Baron Philippe Picot de la Peyrouse (1744–1818) of Toulouse, French botanist, author of a flora of the Pyrenees. *Davallia lapeyrousii* is, however, named after Jean François de Galaup, Comte de la Pérouse (1741–1788), a French admiral and explorer. After active service in the Seven Years and American Revolutionary wars, he sailed in 1785 with two ships to continue Cook's exploration of the Pacific, and carried with him a complement of scientists. He disappeared in 1788, the wrecks of his ships being found in 1826 off the Santa Cruz Islands. IRIDACEAE.

lapha'mii In honour of Increase Allen Lapham (1811–1875), American naturalist.

Lapor'tea f. Named after an obscure Frenchman, possibly a companion of Gaudichaud on the voyage of the corvette *L'Uranie*, 1817–1820. URTICACEAE.

lappa Latin name for a bur.

lappa'ceus, -a, -um Bur-like.

lappo'nicus, -a, -um Of Lapland, northern Scandinavia.

Lardizaba'la f. Named for Miguel Lardizábal, 18th-century Spanish patron of botany. LARDIZABALACEAE.

larici'nus, -a, -um Resembling larch. Also *laricifolius*, larch-leaved.

La'rix f. Larch. The classical name for the tree highly valued for centuries for its tough, strong wood. The English name 'larch' was introduced by William Turner in 1548 as an equivalent of the German 'Lärch', but the common larch was not introduced until the next century into Britain from Central Europe. PINACEAE.

Lar'rea f. Creosote bush. Named for Juan Antonio de Larrea, Spanish patron of science, *c.* 1810. ZYGOPHYLLACEAE.

Laserpi'tium n. The classical Latin name. UMBELLIFERAE.

lasi- In compound words signifying woolly. Thus:
lasiacan'thus (woolly-spined); **lasian'drus** (woolly-stamened);
lasiocar'pus (woolly-fruited); **lasiodon'tus** (woolly-toothed);
lasioglos'sus (having a rough or hairy tongue); **lasiope'talus** (woolly-petalled); **lasio'podus** (woolly-footed).

Lasthe'nia f. Named for the Athenian girl Lasthĕnīa who, in order to attend Plato's classes, dressed as a boy. COMPOSITAE.

Las'trea f. In honour of Charles Jean Louis Delastre (*c.* 1792–1859), author of a flora of the Département de la Vienne, France. THELYPTERIDACEAE.

Lata'nia f. From the Mauritius vernacular name. PALMAE.

latebro'sus, -a, -um Pertaining to dark or shady places.

latera'lis, -is, -e At the side. In compound words *lateri-*, as in *lateripes* (with stalk on the side).

lateri'tius, -a, -um Brick-red.

Lathra'ea f. From Gr. *lathraios*, hidden; with reference to its underground parasitic existence. OROBANCHACEAE.

la'thyris Greek name of a kind of spurge (*Euphorbia*).

La'thyrus m. Sweet pea, beach pea, everlasting pea. The Greek name *lathyros* for pea or pulse. Sweet pea is *L. odoratus*. The edible garden pea comes under *Pisum* (*P. sativum*). LEGUMINOSAE.

la'tus Broad; wide. In compound words *lati-*. Thus:
latiflo'rus (broad-flowered); **latifo'lius** (broad-leaved); **lati'frons** (with broad fronds); **latilab'rus** (wide-lipped); **latilo'bus** (wide-lobed); **latimacula'tus** (with broad spots); **la'tipes** (with a broad stalk); **latispi'nus** (with wide thorns); **latisqua'mus** (with wide scales).

laudan'dus, -a, -um; lauda'tus, -a, -um Praiseworthy.

Laure'lia f. Bay-tree. Latinization of Spanish vernacular name *laurel*, from L. *laurus*, transferred to this South American plant on account of the scent of the leaves. MONIMIACEAE.

laurifo'lius, -a, -um Bay-leaved.

lauri'nus, -a, -um Resembling *Laurus*, true laurel, or bay-tree.

Lauroce'rasus f. Cherry-laurel. From L. *laurus*, laurel; *cerasus*, a cherry. ROSACEAE.

Lau'rus f. Latin name for the laurel or bay. The laurel crown was of bay leaves. Hence the word 'laureate'. The sweet bay provides the leaves used in cookery. LAURACEAE.

Lavan'dula f. Lavender. From L. *lavo*, to wash; from its use in soaps and toiletries of various kinds. LABIATAE.

lavandula'ceus, -a, -um Lavender-violet.

lavandulifo'lius, -a, -um With lavender-like leaves.

Lavate'ra f. Tree-mallow. Named for J. R. Lavater, 17th-century Swiss physician and naturalist in Zürich. MALVACEAE.

lavateroi'des Resembling *Lavatera*.

Lawson'ia f. Henna. Named by Linnaeus to honour his friend and patron, Isaac Lawson (d. *c.* 1747), Scottish army doctor who, while in Leiden, helped to pay for the publication of Linnaeus's *Systema Naturae* (1735). Despite statements by some authors, he was not the brother of John Lawson, surveyor-general of North Carolina and author of *The History of North Carolina*. LYTHRACEAE.

lawsonia'nus, -a, -um For Charles Lawson (1794–1873), author of *Pinetum Britannicum*, to whose nursery in Edinburgh *Cupressus lawsoniana* was introduced from western North America in 1854.

la'xus, -a, -um Loose or open. Thus, *laxiflorus*, *laxifolius*, loose-flowered and loose-leaved.

La'yia f. Named for George Tradescant Lay (d. *c.*1845), naturalist on the voyage of exploration made in 1825–1828 by Frederick William Beechey to the Pacific North-west and the Bering Straits. COMPOSITAE.

la'zicus, -a, -um Epithet for an iris found in Lazistan, on the shores of the Black Sea, closely related to *Iris unguicularis*. IRIDACEAE.

ledifo'lius, -a, -um With leaves resembling *Ledum*.

Le'dum n. Labrador tea, wild rosemary, etc. From Gr. *lēdon*, cistus. It has been proposed that this group should be included in *Rhododendrum*. ERICACEAE.

Le'ea f. Tropical small trees or shrubs named for James Lee (1715–1795), nurseryman with Kennedy of Hammersmith, London. LEEACEAE.

Legou'sia f. Derivation unknown. CAMPANULACEAE.

leian'thus, -a, -um Smooth-flowered. Also:
leiocar'pus (smooth-fruited); **leio'gynus** (with a smooth pistil); **leiophyl'lus** (with smooth leaves).

leichtli'nii- For Max Leichtlin (1831–1910) of Baden-Baden, who introduced many plants into cultivation, notably from the Near East.

Leiophyl'lum n. Sand-myrtle. From Gr. *leios*, smooth; *phyllon*, a leaf; from the glossy foliage. ERICACEAE.

Leitne'ria f. Corkwood. Named in memory of Edward F. Leitner (d. 1838), German physician and naturalist killed by Indians when collecting in Florida. *L. floridana*, a shrub, is the only member of the family. LEITNERIACEAE.

Lemaireoce'reus m. Named for Charles Lemaire (1801–1871), French specialist in cacti and horticultural editor. CACTACEAE.

Lem'na f. Duckweed. The Greek name for some water weed. LEMNACEAE.

lemoi'nei For Victor Lemoine (1823–1911), and his son Emile Lemoine (1862–1942) of Nancy, France, nurserymen and plant-breeders, who introduced many new lilacs, peonies and other hybrids, including the bigeneric × *Heucherella tiarelloides*.

Lens f. Lentil. The classical name for this ancient pulse. Esau sold Jacob his birthright for a red-lentil pottage. The lens (magnifying glass) was so named from being shaped like a lentil seed. LEGUMINOSAE.

lenticula'ris, -is, -e; lentifor'mis, -is, -e Shaped like a lens.

lentigino'sus, -a, -um Freckled.

len'tus, -a, -um Tough but pliant.

leo'nis Coloured or toothed like a lion.

Leono'tis f. Gr. *lēon*, a lion; *ŏus*, *ōtis*, an ear. The corolla might be imagined to look like a lion's ear. LABIATAE.

Leon'todon m. From Gr. *lēon*, lion; *ŏdŏus*, tooth; in allusion to the toothed leaves of the dandelion (*dent de lion*) to which the name was also applied. COMPOSITAE.

leontoglos'sus, -a, -um With a throat or tongue like a lion.

Leontopo'dium n. Edelweiss. The classical name *lēŏntŏpŏdiŏn*, meaning lion's foot, for another plant but transferred to this genus. COMPOSITAE.

Leonu'rus m. Motherwort or lion's-tail. From Gr. *lēon*, a lion; *oura*, a tail; with reference to the inflorescence. LABIATAE.

leonu'rus, -a, -um Like a lion's tail.

leopardi'nus, -a, -um Conspicuously spotted like a leopard.

Le'pachys f. Cone-flower. From Gr. *lepis*, scale; *pachys*, thick; from the thickened tips of the involucral scales. This name has now been discarded in favour of the earlier *Ratibida*, the meaning of which is obscure. COMPOSITAE.

Lepi'dium m. Garden cress, pepper-grass. The classical name. CRUCIFERAE.

lepidosty'lus, -a, -um With a scaly style.

lepido'tus, -a, -um Scaly; covered with small scales.

le'pidus, -a, -um Graceful; elegant.

Lepis'mium n. From Gr. *lepis*, scale. Separated from *Rhipsalis*; with small scales surrounding the areoles. CACTACEAE.

lept- In Greek compound words, signifies thin or slender. Thus: **leptan'thus** (slender-flowered); **leptocau'lis** (thin-stemmed); **lepto'cladus** (thin-branched); **lepto'lepis** (thin-scaled); **leptope'talus** (thin-petalled); **leptophyl'lus** (thin-leaved); **lep'topus** (thin-stalked); **leptorrhi'zus** (with slender rhizome); **leptose'palus** (thin-sepalled); **lepto'stachys** (slender-spiked); **leptoth'rius, -a, -um** (with thin fine leaves); **leptu'rus** (slender-tailed).

Leptoder'mis f. Gr. *lĕptos*, thin; *derma*, skin; from the membranous fused bracteoles forming a sheath or skin covering the calyx. RUBIACEAE.

Lepto'pteris f. Gr. *lĕptos*, slender; *pteris*, a fern. OSMUNDACEAE.

Lepto'pyrum n. Gr. *lĕptos*, slender; *pyros*, wheat; from the form of the fruits. RANUNCULACEAE.

Leptosper'mum n. Gr. *lĕptos*, slender; *sperma*, seed; referring to the narrow seeds. MYRTACEAE.

Lepto'tes f. Gr. *lĕptŏtēs*, delicateness. ORCHIDACEAE.

lepturoi'des Like a hare's tail.

Leschenaul'tia f. Named for Louis Théodore Leschenault de la Tour (1773–1826), French botanist and traveller. GOODENIACEAE.

Lespede'za f. Bush clover. Named by the French botanist Michaux to honour the Spanish governor of Florida, Vincente Manuel de Céspedes, *c.* 1790, whose name was misspelled by L. C. Richard when preparing for printing Michaux's *Flora Boreali-Americana* (1802). LEGUMINOSAE

Lesquerel'la. f. Named in honour of Leo Lesquereux (1805–1889), foremost authority on American fossil botany in the latter part of the 19th century. CRUCIFERAE.

Lettso'mia f. Named for John Coakley Lettsom (1744–1815), London physician, botanist and philanthropist, an original member of the Horticultural Society of London. An advocate of prison reform, he is stated to have accomplished more in twelve months than his predecessors had in thirty years: 'no proposal for the public benefit was ever offered to his consideration in vain'. He was notable for his generosity to the poor and came to the aid of Curtis, a fellow Quaker, in 1778 with £500, a very large sum then, towards the publication of the *Flora Londinensis*. CONVOLVULACEAE.

leuc- In compound words signifies white. Thus:
leucan'thus (white-flowered); **leucocau'lis** (white-stemmed); **leucoce'phalus** (white-headed); **leucochi'lus** (white-lipped); **leucoder'mis** (with white skin or bark); **leuconeu'rus** (white-nerved); **leucope'talus** (white-petalled); **leucopha'eus** (dusky-white); **leucopha'rynx** (with white throat); **leucophyl'lus** (white-leaved); **leucorrhi'zus** (with white roots or rhizome); **leuco'stachys** (white-spiked); **leuco'trichus** (white-haired); **leucoxan'thus** (whitish-yellow); **leuco'xylon** (white-wooded).

Leucaden'dron n. Cape silver tree. From Gr. *leukos*, white; *dĕndron*, a tree; in allusion to the silky white foliage. PROTEACEAE.

Leucae'na f. Gr. *leukos*, white; in allusion to the flowers. LEGUMINOSAE.

leucanthemifo'lius, -a, -um With leaves like *Leucanthemum*.

Leucanthemum n. Gr. *leukos*, white; *anthemon*, flower. The best known species of this genus often included in *Chrysanthemum* is the ox-eye daisy (*L. vulgare*, syn. *Chrysanthemum leucanthemum*). COMPOSITAE.

leucas'pis With a white shield.

Leucheria f. Gr. *leucheres*, white; some species are white and woolly. COMPOSITAE.

Leuchtenber'gia f. Apparently named in honour of Prince Maximilian E. Y. N. von Beauharnais (1817–1852), Duke of Leuchtenberg, Ober-Pfalz, Germany. CACTACEAE.

Leucoco'ryne f. From Gr. *leukos*, white; *kŏrynē*, club; in allusion to the white staminodes of the type species. LILIACEAE (ALLIACEAE).

Leuco'crinum m. Sand lily. Gr. *leukos*, white; *crinon*, a lily. LILIACEAE.

Leuco'jum n. Snowflake, St. Agnes' flower. The Greek name for various scented white-flowered plants, from *leukŏn iŏn* (white violet) later transferred to the snowflake. AMARYLLIDACEAE.

Leuco'thoe f. Ornamental shrubs named in honour of Leucothoë, one of the many loves of Apollo. ERICACEAE.

Leu'zia f. Named for a friend of de Candolle, Joseph Philippe François Deleuze (1753–1835) of Avignon and Paris. COMPOSITAE.

Levis'ticum n. Lovage. Said to be a corruption of *Ligusticum*.

Lewi'sia f. Bitterwort. Named in honour of Captain Meriwether Lewis (1774–1809), who with Captain William Clark (1770–1838) made in 1804 to 1806 the first coast-to-coast expedition of North America. PORTULACACEAE.

Leyceste'ria f. Named for William Leycester, Chief Justice of Bengal *c.* 1820. CAPRIFOLIACEAE.

leyland'ii. Named for C.J. Leyland who grew hybrids of *Cupressos macrocarpa* and *Chamaecyparis nootkatensis* in the 1890's on his property, Haggerston Hall, Northumberland, England. One of these, called *Cupressocyparis leylandii*, propagated vegetatively has become popular as a fast growing evergreen garden tree.

Li'atris f. Unknown derivation. This genus includes plants under many English names: prairie-button, snakeroot, Kansas gay-feather, blue blazing star, etc. COMPOSITAE.

li'bani; libano'ticus, -a, -um Of Mount Lebanon, Lebanon.

libe'ricus, -a, -um Of Liberia.

Liber'tia f. Named for Marie A. Libert (1782–1863), Belgian botanist who wrote on liverworts. IRIDACEAE.

Libo'cedrus f. Incense cedar. From Gr. *libos*, tear, drop; *cedrus*, cedar; presumably in allusion to the tear-like gummy exudations and suggested by the name *Dacrydium* for an allied genus. CUPRESSACEAE.

libur'nicus, -a, -um Of Croatia (Liburnia) on the shores of the Adriatic.

liby'cus, -a, -um Referring to Libya, North Africa.

lichian'gensis, -is, -e see *likiangensis.*

Licua'la f. From the native Moluccan name *leko wala*. PALMAE.

ligno'sus, -a, -um Woody.

Ligula'ria L. *ligula*, a strap; from the strap-like ray florets. COMPOSITAE.

ligula'ris, -is, -e; ligula'tus, -a, -um Straplike; provided with ligules.

ligusticifo'lius, -a, -um Lovage-leaved.

Ligus'ticum n. Lovage. From Gr. *ligustikas*, pertaining to Liguria, the Italian province in which is the city of Genoa.

ligus'ticus, -a, -um Of Liguria (see *Ligusticum*).

ligustrifo'lius, -a, -um Privet-leaved.

ligustri'nus, -a, -um Privetlike.

Ligus'trum n. Latin for privet. OLEACEAE.

likiangen'sis, -is, -e From the Lichiang (Lijiang) range, Yunnan, Western China, a favourite collecting area of George Forrest.

lila'cinus, -a, -um Lilac in colour.

lili- In compound words signifying lily. Thus:
lilia'ceus (lilylike); **liliflo'rus** (lily-flowered); **lilifo'lius** (lily-leaved).

Li'lium n. Lily. The Latin name akin to Gr. *leirion*, lily. LILIACEAE.

lilliputia'nus, -a, -um Of very small and low growth, suitable for the land of Lilliput.

limba'tus, -a, -um Bordered.

limen'sis, -is, -e Of Lima, Peru.

Limnan'thes f. Meadow foam. Gr. *limne*, a marsh; *anthos*, a flower; from the habitat. LIMNANTHACEAE.

Limno'bium n. Gr. *limne*, a marsh; *biŏs*, life. HYDROCHARITACEAE.

Limno'charis f. Gr. *limne*, a marsh; *charis*, beauty; from the habitat and beauty of the flowers of these aquatic perennials. ALISMATACEAE.

limno'philus, -a, -um Swamp-loving.

limoniifo'lius, -a, -um With leaves like sea-lavender.

Limo'nium n. Sea-lavender, sea-pink. From Gr. *leimōn*, a meadow; in allusion to the common habitat in salt meadows. PLUMBAGINACEAE.

limo'sus, -a, -um Of marshy or muddy places.

Linan'thus m. Gr. *linon*, flax; *anthos*, a flower. Now often classified under *Gilia*. POLEMONIACEAE.

Lina'ria f. Medieval name from Gr. *linon*, L. *linum*; in allusion to the flaxlike leaves of *L. vulgaris* (toadflax, butter-and-eggs).

linariifo'lius, -a, -um *Linaria*-leaved.

lindaue'anus, -a, -um; lindavia'nus, -a, -um In honour of Gustav Lindau (1866–1923), German botanist who specialized in the study of Fungi and *Acanthaceae*.

Lindelo'fia Named for Friedrich von Lindelof, of Darmstadt, patron of botany *c.* 1850. BORAGINACEAE.

linden'ii. In honour of Jean Jules Linden (1817–1898), Luxembourg botanical explorer in South America, later nurseryman in Belgium.

Linde'ra f. Feverbush, Benjamin bush, benzoin. Named after Johann Linder (1676–1723), a Swedish botanist and physician. LAURACEAE.

Lindhei'mera f. In honour of Ferdinand Jacob Lindheimer (1801–1879), German political exile who made large botanical collections in Texas. COMPOSITAE.

lindleyi In honour of John Lindley (1799–1865), one of the most industrious of British botanists, long associated with the Horticultural Society of London, first professor of botany in London University, a specialist in the study of *Orchidaceae*, author and editor of numerous comprehensive botanical and horticultural publications.

Lindsa'ea f. In honour of John Lindsay (*fl.* 1785–1803), surgeon in Jamaica. DENNSTAEDTIACEAE.

linearifo'lius, -a, -um Linear-leaved.

linea'ris, -is, -e Narrow, with sides nearly parallel.

line'atus, -a, -um With lines or stripes.

lin'gua A tongue or tonguelike.

linguifor'mis, -is, -e Tongue-shaped.

lingula'tus, -a, -um Tongue-shaped.

liniflo'rus, -a, -um Flax-flowered.

linifo'lius, -a, -um Flax-leaved.

Linna'ea f. Twin-flower. Named for and by Carl Linnaeus (1707–1778), ennobled as Carl von Linné, and described by him as 'a plant of Lapland, lowly, insignificant, disregarded, flowering but for a brief space—from Linnaeus who resembles it'. The family name Linnaeus was a Latinization of the Swedish *linn*, linden. Born in Sweden, Linnaeus trained as a physician but became a professor of botany at Uppsala. His *Systema Naturae* (1735) began a series of encyclopaedic works dealing mostly with plants but also with animals, minerals and materia medica, seeking in a very methodical and systematic manner to classify and name the whole organic world, thereby providing a basis for further and wider research.

He was a teacher without peer. His pupils, such as Pehr Kalm (see *Kalmia*), became botanic leaders in their own right. He was a prodigious writer and his published works total more than 180 titles. It was inevitable that his so-called 'sexual system' of classifying plants into major artificial groups based primarily on the number of stamens and pistils should become modified and then replaced by more natural systems based on more characters. His most permanent contribution to biology was his introduction in 1753 of a consistent binomial system of naming species by means of a single-word generic name followed by a single-word specific epithet. (See pp. 10, 326.)

His 'herborizing lectures' while he walked through lanes and fields at the head of two or three hundred students were social events of summer at Uppsala. Attended by a band of trumpets and French horns, his pupils were divided into detached companies to be called together by music when he wished to discourse on any particular subject.

He made a close study of what he called 'floral nuptials', his metaphorical descriptions being taken too literally by his opponents. The petals of flowers he regarded as 'bridal beds which the Creator has so gloriously arranged, adorned with such noble bed-curtains and perfumed with so many sweet scents that the bride-groom may celebrate his nuptials with his bride with all the greater solemnity.' The great Linnaean collections are now at Burlington House in London, having been purchased by Sir James Smith for one thousand guineas in 1784 and later bought by the Linnean Society of London.

The epitaph on Linnaeus's monument in the cathedral at Uppsala describes him simply as 'Princeps Botanicorum'.

linnaea'nus, -a, -um; linnaei In honour of Linnaeus.

Li'num n. Flax. The Latin name. Until cotton came into supply in the 18th century, flax and hemp were the most important vegetable fibres at least as far as Western man was concerned. Flax is also the source of linseed. LINACEAE.

Lipa'ris f. Twayblade (in America). From Gr. *liparŏs*, oily or smooth; in allusion to the glossy surface of the leaves. ORCHIDACEAE.

Lip'pia f. Lemon verbena. Named for Augustin Lippi (1678–1701), Italian naturalist, killed in Abyssinia. VERBENACEAE.

Liquidam'bar f. Sweetgum. From L. *liquidus*, liquid; *ambar*, amber. A fragrant resin called liquid storax is obtained from two species, *L. orientalis* and *L. styraciflua*. HAMAMELIDACEAE.

Lirioden'dron n. Tulip tree. From Gr. *leiriŏn*, a lily; *dĕndron*, a tree. MAGNOLIACEAE.

Liri'ope f. Named for a Greek woodland nymph, Līrĭŏpē, the mother of Narcissus. LILIACEAE CONVALLARIACEAE.

lissoca'rpus, -a, -um With smooth fruit.

Lissochi'lus m. From Gr. *lissos*, smooth; *cheilos*, lip. ORCHIDACEAE.

Listera f. Named for Martin Lister (*c.* 1638–1712), British zoologist, naturalist, physician to Queen Anne. This is called twayblade in Britain, but more often adder's tongue or sweethearts in the United States. ORCHIDACEAE.

litangen'sis, -is, -e From Litang, Yunnan, China.

Li'tchi f. From the Chinese name for the handsome tree producing the well-known fruits. SAPINDACEAE.

Lithocar'pus m. Tanbark or chestnut-oak. From Gr. *lithos*, a stone; *karpos*, fruit; in allusion to the hard acorns. FAGACEAE.

Lithodo'ra f. From Gr. *lithos*, a stone; *dōron*, gift; with reference to the rocky habitat of these beautiful plants formerly included in *Lithospermum*. BORAGINACEAE.

litho'philus, -a, -um Rock-loving.

Lithophrag'ma n. Californian perennial herbs. From Gr. *lithos*, a stone; *phragma*, a fence; probably an unsuccessful attempt to render *Saxifraga* in Greek. SAXIFRAGACEAE.

Li'thops f. Pebble plants. From Gr. *lithos*, a stone; *ops*, like; from the resemblance of these succulents to small stones. AIZOACEAE.

Lithosper'mum n. Gromwell. The classical name from Gr. *lithos*, stone; *sperma*, seed; in allusion to the hard nutlets. BORAGINACEAE.

lithosper'mus, -a, -um With very hard stonelike seeds.

Lithrae'a f. From the Chilean vernacular name for these dermatitis-causing plants. ANACARDIACEAE.

litien'sis, -is, -e From Litiping, Yunnan, China.

Litse'a f. From Chinese *li*, little, *tse*, plum. LAURACEAE.

Litto'nia f. In honour of Samuel Litton (1779–1847), professor of botany at Dublin. LILIACEAE (COLCHICACEAE).

littora'lis, -is, -e; littore'us, -a, -um Of the seashore.

li'vidus, -a, -um Lead-coloured; bluish-grey.

Livisto'na f. Named by Robert Brown for Patrick Murray, Baron of Livingston, near Edinburgh, who before 1680 had a well-stocked garden; his plants later became the foundation of the Edinburgh Botanic Garden. PALMAE.

Lloydia f. In honour of Edward Lloyd (1660–1709), Welsh antiquary and botanist, for many years keeper of the Ashmolean Museum, Oxford. LILIACEAE.

Loa'sa f. Origin obscure. The Chilean name for these plants with showy flowers and stinging foliage is usually *ortiga* (nettle). LOASACEAE.

loba'tus, -a, -um Lobed. Also:

lobocar'pus (with lobed fruits); **lobophyl'lus** (with lobed leaves); **lobula'tus** (with small lobes).

lob'bii In honour of one or other of two Cornish brothers who collected for Messrs Veitch, William Lobb (1809–1864) in America, Thomas Lobb (1820–1894) in eastern Asia and Indonesia.

Lobe'lia f. Named for Mathias de l'Obel (1538–1616), Flemish botanist and physician to James I of England. Cardinal flower is *L. cardinalis*. CAMPANULACEAE.

lobelioi'des Resembling *Lobelia*.

Lobi'via f. Anagram of Bolivia where these cacti are found. CACTACEAE.

Lobula'ria f. Sweet alyssum. From L. *lobulus*, small pod. CRUCIFERAE.

loch'mius, -a, -um Of thickets.

Lockhar'tia f. Named for David Lockhart (d. 1846), Kew gardener, first superintendent of the Trinidad Botanic Garden. ORCHIDACEAE.

Lodoi'cea f. Double coconut, Coco-de-mer. Has the world's largest seed. Named for Louis XV of France (1710–1774). Louis can be Latinized as Ludovicus or Lodoicus. PALMAE.

Loese'lia f. Named for Johann Loesel (1607–1657), German botanical writer, professor at Königsberg. POLEMONIACEAE.

Loga'nia f. Named by Robert Brown in honour of James Logan (1674–1751), born in Ireland, Governor of Pennsylvania and a botanical writer, contemporary and friend of John Bartram. LOGANIACEAE.

loganobac'cus Loganberry, commemorating Judge J. H. Logan of California, *c*. 1880.

Loiseleu'ria f. Named for Loiseleur, whose full name was Jean Louis Auguste Loiseleur-Deslongchamps (1774–1849), French botanist and physician. ERICACEAE.

lok'jae In honour of Hugo Lokja (1845–1887), Hungarian teacher and lichenologist, who made extensive collections of Alpine plants in the Caucasus in 1885.

lolia'ceus, -a, -um Like *Lolium*.

Lo'lium n. Rye-grass. The classical name for one species. GRAMINEAE.

Loma'ria f. From Gr. *lōma*, border, edge; in allusion to the sori on the edge of the fronds. Now included in *Blechnum*. BLECHNACEAE.

lomariifo'lius, -a, -um With leaves like the fern genus *Lomaria*.

Loma'tia f. Gr. *lōma*, border, edge; a reference to the winged edges of the seeds of these evergreen shrubs or trees. PROTEACEAE.

Lo'nas f. African daisy. The derivation unknown. COMPOSITAE..

Lonchocar'pus m. Gr. *lonche*, a lance; *karpos*, fruit; in allusion to the shape of the pods. LEGUMINOSAE.

lon'gus, -a, -um Long. In compound words signifying long. Thus: **longebractea'tus** (with long bracts); **longepeduncula'tus** (with a long peduncle); **longicau'lis** (long-stalked); **longi'comus** (long-haired);

longicus'pis (long-pointed); **longiflo'rus** (long-flowered); **longifo'lius** (long-leaved); **longihama'tus** (with long hooks); **longi'labris** (long-lipped); **longi'lobus** (with long lobes); **longimucrona'tus** (ending in a long sharp point); **lon'gipes** (long-stalked); **longipe'talus** (with long petals); **longipinna'tus** (having leaflets arranged on each side of a long common stalk; long-pinnate); **longiracemo'sus** (with long racemes); **longiros'tris** (long-beaked); **longisca'pus** (with a long scape); **longise'palus** (with long sepals); **longi'spathus** (with long spathes); **longispi'nus** (long-thorned); **longisquama'tus** (with long scales); **longi'stylus** (with a long style); Also: **longis'simus** (very long).

Lonice'ra f. Honeysuckle. Named for Adam Lonitzer (1528–1586), German botanist, the author of a herbal (*Kreuterbuch*) many times reprinted between 1557 and 1783. CAPRIFOLIACEAE.

Lope'zia f. Named for Tomás Lopez, Spanish botanist, who wrote on the plants of South America *c.* 1540. ONAGRACEAE.

lophan'thus, -a, -um Having crested flowers.

lopho'gynus, -a, -um With a crested ovary.

Lopho'phora f. Gr. *lŏphŏs*, a crest; *phoreo*, to bear; from the tufts of hairs borne on the equivalent of leaf axils. CACTACEAE.

lorifo'lius, -a, -um With strap-shaped leaves.

Lorope'talum n. Gr. *lōrŏn*, strap, thong; *petalon*, petal. The flowers have long narrow petals. HAMAMELIDACEAE.

lotifo'lius, -a, -um *Lotus*-leaved.

Lo'tus f. The classical Greek name *lōtos*, applied to a diversity of plants, e.g. clover, birdsfoot trefoil, fenugreek, melilot, etc. LEGUMINOSAE.

louisia'nus, -a, -um Of Louisiana, U.S.A.

lownde'sii In honour of Colonel George Lowndes (1899–1956) of the Royal Garhwal Rifles, who collected in Nepal in 1950.

lu'cens; lu'cidus, -a, -um Bright; shining; clear.

luc'iae In honour of Lucie Savatier, wife of P. A. L. Savatier (1830–1891).

luci'liae In honour of Lucile Boissier (1822–1849), who died while accompanying her husband, Edmond Boissier, on an expedition in Spain.

Lucu'lia f. From the Nepalese vernacular name *lukuli swa* for *Luculia gratissima*. RUBIACEAE.

ludlowii Epithet of a beautiful tree peony and some other Himalayan and Tibetan plants named in honour of Frank Ludlow (1885–1972), political officer, ornithologist and botanist, who explored south-eastern Tibet (1933–1949) in company with Major George Sheriff.

ludovicia'nus, -a, -um Of Louisiana, U.S.A.

Ludwi'gia f. In honour of Christian G. Ludwig (1709–1773), professor at Leipzig. ONAGRACEAE.

Luet'kea f. In honour of Feodor Lütke (1797–1882), Russian admiral. ROSACEAE.

Lu'ffa f. Dish-cloth gourd. Vegetable sponge. From the Arabic name. *L. aegyptiaca* produces the elongated fruit of which the dried fibrous interior serves as a bathroom sponge (loofah). CUCURBITACEAE.

Lui'sia f. In honour of Luis de Torres, 19th-century Spanish botanist. ORCHIDACEAE.

lukiangen'sis, -is, -e From Lukiang, Yunnan, China.

Luna'ria f. Honesty. Satin-flower. L. *luna*, the moon; from the flat rounded seed-vessel suggesting the full moon. CRUCIFERAE.

luna'tus, -a, -um; lunula'tus, -a, -um Shaped like a crescent moon.

Lupi'nus m. Lupin or lupine. The classical name. Supposed to be derived from *lupus*, a wolf, because of the completely erroneous belief that these plants destroyed the fertility of the soil. LEGUMINOSAE.

lupuli'nus, -a, -um Hoplike.

lu'pulus Specific epithet of hop (*Humulus lupulus*). Literally, *lupulus* means a small wolf, the vine once having been locally called willow-wolf (*lupus salictarius*) from its habit of climbing over willow trees.

lu'ridus, -a, -um Smoky yellow; sallow; wan.

lusita'nicus, -a, -um Of Portugal (Lusitania); Portuguese.

lute'olus, -a, -um Yellowish.

lutes'cens Yellowish, becoming yellow.

lutetia'nus, -a, -um Of Paris (Lutetia).

lu'teus, -a, -um Yellow from *lutum*, dyer's greenweed, the source of a yellow dye. Also in compound words *luteo-*.

luxu'rians Luxuriant.

Luzula f. From Italian *lucciola*. JUNCACEAE.

Lycas'te f. Presumably in honour of Lycaste, a daughter of Priam. ORCHIDACEAE.

lychnidifo'lius, -a, -um *Lychnis*-leaved.

Lych'nis f. Catchfly, Maltese cross, evening campion. The classical name, said to be derived from Gr. *lychnōs*, a lamp; possibly referring to the ancient use of leaves of a woolly species for wicks.

Ly'cium n. Matrimony vine or box-thorn. From the Greek name for a thorny tree, with juice and roots used medicinally, from Lycia, a south-west region of Asia Minor. SOLANACEAE.

ly'cius, -a, -um Referring to Lycia, Asia Minor.

lycoc'tonum Specific epithet for *Aconitum lycoctonum* or wolf's-bane, one of the more poisonous species of monkshood, from Gr. *lykŏs*, wolf; *ktŏnŏs*, murder.

Lycoper'sicon n. Tomato. Gr. *lykŏs*, a wolf; *persicon*, a peach; originally the name of an Egyptian plant, later transferred to this American genus. SOLANACEAE.

lycopodioi'des Resembling *Lycopodium* or club-moss.

Lycopo'dium n. Club-moss. From Gr. *lykŏs*, a wolf; *pŏdiŏn*, a foot; from some fancied resemblance to a wolf's foot. LYCOPODIACEAE.

Ly'copus m. Water hoarhound. Gr. *lykŏs*, wolf; *pŏus*, a foot; as above. LABIATAE.

Lyco'ris f. Golden spider lily. Named for a Roman beauty, the mistress of Mark Antony, famed for intrigues. AMARYLLIDACEAE.

ly'dius, -a, -um From Lydia in Asia Minor.

Lygo'dium n. Climbing-fern. From Gr. *lygōdēs*, like a willow; in allusion to the supple twisting shoots. SCHIZAEACEAE.

Lyo'nia f. Ornamental shrubs named for John Lyon (*c.* 1765–1814), a Scottish gardener and botanist who introduced many North American plants into European gardens. He was associated with Frederick Pursh in developing William Hamilton's beautiful garden at Woodlands, Philadelphia. He achieved enthusiastic notice from Captain Mayne Reid and died exploring in North Carolina. ERICACEAE.

Lyonotham'nus m. An ornamental woody plant named for W. S. Lyon (1851–1916), an American collector who sent plants to Asa Gray. The second half of the name is from Gr. *thamnos*, a shrub. ROSACEAE.

lyra'tus, -a, -um Of lyrate form, usually referring to leaves with a broad rounded tip and sinuate sides diminishing in width towards the base.

Lysichi'ton m. From Gr. *lysis*, a loosening, releasing; *chitōn*, a cloak. As the fruits ripen, the spathe is released from the spadix. The type species, *L. camtschatcensis*, with white spathe, extends from Kamchatka to Japan, the other species, *L. americanus*, with yellow spathe, from Alaska to California. ARACEAE.

Lysima'chia f. Yellow loosestrife. Gr. *lysimacheion*, said to be named after King Lysimachos of Thrace, or from Gr. *lysimachōs*, ending strife, whence the English name loosestrife. PRIMULACEAE.

lysimachioi'des *Lysimachia*-like.

Lysiono'tus m. From Gr. *lysis*, a loosening, releasing; *nōtŏs*, the back; in allusion to the capsule opening away from the dorsal (back) suture. GESNERIACEAE.

lysiste'mon With a loose stamen or stamens, i.e. free from the other stamens.

lyso'lepis With loose scales.

Ly'thrum n. Presumably from Gr. *lythron*, blood; with reference to the colour of the flowers. Purple loosestrife is *Lythrum salicaria*. LYTHRACEAE.

M

Maa'ckia f. Named for Richard Maack (1825–1886), Russian naturalist, explorer of the Ussuri region, eastern Asia, whence he introduced many plants into cultivation. LEGUMINOSAE.

Ma'ba f. Now included in *Diospyros*. EBENACEAE.

Macada'mia f. In honour of John Macadam (1827–1865), chemist, medical man and lecturer in Melbourne, Australia. PROTEACEAE.

macedo'nicus, -a, -um Of Macedonia, Balkan Peninsula.

Machaeroce'reus m. Gr. *machaira*, a dagger, short sword; *Cereus*, cactus; with reference to the spines. CACTACEAE.

macilentus, -a, -um Thin; lean.

Mackay'a f. In honour of James Townsend Mackay (c. 1775–1862), Scottish botanist and gardener, who emigrated to Ireland in 1804 and founded the botanic garden of Trinity College, Dublin in 1806. ACANTHACEAE.

mackli'niae Specific epithet of a lily discovered by Kingdon-Ward and named after his second wife, Jean, *née* Macklin, later Rasmussen.

Maclea'nia f. In honour of John Maclean, Scottish merchant at Lima, Peru, between 1832 and 1834, who sent many plants from there to Britain. ERICACEAE.

Macleay'a f. Named for Alexander Macleay (1767–1848), Colonial Secretary for New South Wales and once Secretary of the Linnean Society of London. Plume poppy is *Macleaya cordata*, syn. *Bocconia cordata*. PAPAVERACEAE.

Maclu'ra f. Osage orange. Named for William Maclure (1763–1840), American geologist. MORACEAE.

Macrade'nia f. Gr. *makros*, long; *adēn*, a gland; an allusion to the length of the stalk of the pollen masses (pollinia). ORCHIDACEAE.

macr- In compound words signifying either long or big. Thus: **macracan'thus** (with large spines); **macrade'nus** (with large glands); **macran'drus** (with large anthers); **macranthe'rus** (with large anthers); **macran'thus** (with large flowers); **macrobot'rys** (with large grapelike clusters); **macrocar'pus** (large-fruited); **macroce'phalus** (large-headed); **macrodon'tus** (with large teeth); **macro'meris** (with large parts); **macrophyl'lus** (with long or large leaves); **macro'podus** (stout-stalked); **macrorrhi'zus** (with large roots or root stocks); **macrosper'mus** (with large seeds); **macrosta'chyus** (with long or large spikes); **macru'rus** (long-tailed).

Macroza'mia f. Gr. *makros*, large; plus *Zamia*, a related genus. These Australian plants are highly poisonous to sheep. ZAMIACEAE.

macula'tus, -a, -um; macu'lifer, maculi'fera, -um; maculo'sus, -a, -um Spotted.

madagascarien'sis, -is, -e Of Madagascar; Madagascan.

Madde'nia. f. Named for Colonel Edward Madden (1805–1856), who collected in India. ROSACEAE.

maderaspaten'sis, -is, -e Of Madras, southern India.

maderensis, -is, -e. Of Madeira, Atlantic Ocean; Madeiran.

Madhu'ca f. Latinized form of Indian name. SAPOTACEAE.

Ma'dia f. Tarweed. From the native Chilean name. COMPOSITAE.

Mae'sa f. From the Arabic name *maas*. MYRSINACEAE.
magella'nicus, -a, -um Of the area of the Straits of Magellan, South America.
magellen'sis, -is, -e Of Monte Majella, southern Italy.
magni'ficus, -a, -um Great, splendid, magnificent.
Magno'lia f. Sweet bay (in U.S.A.). Named for Pierre Magnol (1638–1715), professor of botany and director of the botanic garden, Montpellier, France. MAGNOLIACEAE.
mag'nus, -a, -um Great; big.
Maher'nia f. Honey-bell. Anagram of *Hermannia*, a genus related to these herbs and evergreen sub-shrubs. STERCULIACEAE.
Mahober'beris f. Hybrids of *Berberis* and *Mahonia*.
Maho'nia f. Named for Bernard M'Mahon (1775–1816), American horticulturist and author of *The American Gardener's Calendar* (1807). BERBERIDACEAE.
Maian'themum n. False or wild lily-of-the-valley (North America), May lily (Britain). Gr. *Maios*, May; *anthemon*, blossom. LILIACEAE (CONVALLARIACEAE).
mai'rei In honour of Edouard Ernest Maire (b. 1848), Roman Catholic missionary in Yunnan, west China, who made large collections of herbarium specimens between 1905 and 1916, if a Chinese plant; in honour of René C. J. E. Maire (1878–1949), for many years professor of botany in Algeria, if a North African plant.
maja'lis, -is, -e May-flowering.
majes'ticus, -a, -um Majestic.
ma'jor; ma'jus, -a, -um Bigger; larger.
Majora'na f. Sweet marjoram. From the Arabian *marjamie*. LABIATAE.
maki'noi In honour of Tomitaro Makino (1863–1957), celebrated Japanese botanist.
malaba'ricus, -a, -um Of the Malabar coast, southern India.
Malach'ra f. Variant of Gr. *malache*, malva. MALVACEAE.
Malacocar'pus m. Gr. *malakos*, soft; *karpos*, fruit. ZYGOPHYLLACEAE.
malacoi'des Soft; mucilaginous; like mallow.
malacophy'llus, -a, -um Soft-leaved.
malacosper'mus, -a, -um With soft seeds.
Malco'mia f. Virginia-stock. Named by Robert Brown for William Malcolm (d. 1820), London nurseryman, who published a catalogue of greenhouse plants in 1778 and for his son, also William (1769–1835). CRUCIFERAE.
malifor'mis, -is, -e Apple-shaped.
Mallo'tus m. Kamila tree. Gr. *mallōtŏs*, woolly. EUPHORBIACEAE.
Ma'lope f. Name apparently of Greek origin used by Pliny for a kind of mallow. MALVACEAE.
Malpi'ghia f. Barbados-cherry. Named for Marcello Malpighi (1628–1694), famous Italian anatomist and professor at Bologna. MALPIGHIACEAE.

Ma'lus f. Gr. *mēlon*, apple; applied also to other tree-fruits with a fleshy exterior. ROSACEAE.

Mal'va f. Mallow. MALVACEAE.

malva'ceus, -a, -um Mallow-like.

Malvas'trum n. False mallow. L. *malva*, mallow; plus *-aster* indicating incomplete resemblance. MALVACEAE.

Malvavis'cus m. L. *malva*, mallow; and possibly *viscus*, glue, then referring to the pulp around the seeds. MALVACEAE.

malviflo'rus, -a, -um Mallow-flowered.

malvi'nus, -a, -um Mauve.

Mam'mea f. Mammee-apple. From the West Indian name. GUTTIFERAE.

mammifor'mis, -is, -e Nipplelike.

Mammilla'ria f. L. *mammilla*, a nipple. CACTACEAE.

mammilla'tus, -a, -um; mammilla'ris, -is, -e; mammo'sus, -a, -um Furnished with nipples or breasts.

Mandevil'la f. Climbing shrubs named for Henry John Mandeville (1773–1861). British Minister in Buenos Aires. APOCYNACEAE.

Mandra'gora f. Mandrake. The Greek name. Long known for its poisonous and narcotic properties and used in classical times to deaden pain in surgical operations, it has also been much esteemed as an aphrodisiac and used in love potions, presumably from the shape of the root. It was supposed that the plant shrieked when uprooted ('And shrieks like mandrake torn out of the earth that mortals, hearing them, run mad,' *Romeo and Juliet*), a notion probably encouraged by the herbalists and root-gatherers of antiquity in order to protect the plant from undue exploitation. In the East the plant was supposed to facilitate pregnancy. The Greek *mandragoras* is apparently a corruption of Assyrian *nam tar ira* (male drug of the plague god Namtar). SOLANACEAE.

mandshu'ricus, -a, -um Of Manchuria; Manchurian.

manesca'vii Of Manescau (d. 1875), merchant at Pau, who botanized in the Pyrenees.

Manet'tia f. Named for Saverio Manetti (1723–1785), Florentine medical man, prefect of the Florence botanic garden. RUBIACEAE.

Mangi'fera f. Mango. The Indian vernacular name *Mango* plus L. *fero*, to bear. The Bombay Alfonso, at its best, is probably the most delectable of the many varieties. ANACARDIACEAE.

manica'tus, -a, -um With long sleeves.

Ma'nihot f. Cassava or tapioca. From the Brazilian name, *manioc*. The cyanide-containing tuberous roots in their natural state are very poisonous and become edible only after steeping in water. EUPHORBIACEAE.

Manilka'ra f. Malabar (South Indian) vernacular name cited by van Rheede in 1683. SAPOTACEAE.

manipuliflo'rus, -a, -um With few-flowered clusters; from *manipulus*, a handful.

manipuren'sis, -is, -e Of Manipur State, India.

mantegazzia'nus, -a, -um, In honour of Paolo Mantegazzi (1831–1910), Italian anthropologist and traveller.

Manti'sia f. Named from a resemblance of the fantastically shaped flowers to the Mantis insect. ZINGIBERACEAE

Manu'lea f. From L. *manus*, a hand; from the fingerlike divisions of the corolla. SCROPHULARIACEAE.

manzanil'la Specific epithet of *Arctostaphylos.* From the Spanish word meaning a small apple.

Maran'ta f. Arrowroot. Named for Bartolommeo Maranti, Venetian botanist *c.* 1559. MARANTACEAE.

Marat'tia f. Evergreen ferns named for Giovanni Francesco Maratti (d. 1777), Italian botanist, professor in Rome. MARATTIACEAE.

margarita'ceus, -a, -um; margari'tus, -a, -um Pertaining to pearls, pearly; from L. *margărīta*, a pearl.

margari'tae Of Santa Margarita Island, Baja California.

margari'tifer, margariti'fera, -um Pearl-bearing.

margina'lis, -is, -e; margina'tus, -a, -um Margined.

marginel'lus, -a, -um With a narrow margin.

Margyricar'pus m. Pearl fruit. L. *margarita*, Gr. *margaritēs*, a pearl; *karpos*, fruit; from the white berries. ROSACEAE.

maria'nus, -a, -um Specific epithet of various plants with mottled leaves, notably Our Lady's or blessed thistle (*Silybum marianum*); the spots were supposed to have resulted from drops of her milk falling on the leaves. It is also applied to plants from Maryland (*Terra Mariana*).

marie'sii In honour of Charles Maries (*c.* 1851–1902) who collected for Messrs. Veitch in Japan and China, 1877–1879.

marilan'dicus, -a, -um (also **marylandicus**) Of Maryland.

mari'timus, -a, -um Pertaining to the sea; coastal.

marlo'thii In honour of Hermann W. R. Marloth (1855–1931), South African pharmacist and botanist of German origin, author of a profusely illustrated *Flora of South Africa* (1913–1932).

marmora'tus, -a, -um; marmo'reus, -a, -um Marbled; mottled.

marocca'nus, -a, -um Of Morocco; Moroccan.

Marru'bium n. Horehound. The classical Latin name for this familiar cough remedy. LABIATAE.

Marsde'nia f. Named for William Marsden (1754–1836), secretary to the Admiralty, orientalist and traveller, author of a *History of Sumatra* (1783). ASCLEPIADACEAE.

Marshal'lia f. Named for Humphry Marshall (1722–1801), a cousin of John Bartram, and author of *Arbustum Americanum* (1785). COMPOSITAE.

Marsi'lea f. Named for Count Luigi Ferdinando Marsigli (1656–1730) of Bologna, Italian botanist. MARSILEACEAE.

mar'tagon n. Origin obscure, possibly alchemical.

Martine'zia f. Named for Baltasar Jacobo Martinez, archbishop of Santa Fé de Bogota, a helper of Ruiz and Pavon. PALMAE.

Marty'nia f. In honour of John Martyn (1699–1768), London physician, professor of botany at Cambridge, England, although for many years of his professorship he neither resided nor lectured there. His son Thomas Martyn (1735–1825) succeeded him as professor. PEDALIACEAE.

mas; mas'culus, -a, -um Male; masculine; used by early writers in a metaphorical sense to distinguish robust species from more delicate ones, red-flowered species from blue-flowered, hard-wooded from soft wooded etc.

Mascarenha'sia f. Named for the Mascarene Islands (Réunion, Mauritius and Rodriguez) in the south-west Indian Ocean, which take their name from the 16th-century Portuguese navigator Mascarenhas. APOCYNACEAE.

Masdeval'lia f. Named for José Masdevall (d. 1801), Spanish botanist and physician. ORCHIDACEAE.

massilien'sis, -is, e Of Marseilles, ancient Massilia, France.

massonia'nus, -a, -um; masso'nii In honour of Francis Masson (1741–*c.* 1805), Scottish gardener and traveller in South Africa, West Indies and North America, the first plant collector sent out from Kew.

Matrica'ria f. Mayweed. Medieval name possibly from L. *matrix*, womb, because of its one-time medical use in affections of the uterus. COMPOSITAE.

matriten'sis, -is, -e Of Madrid, Spain.

matrona'lis, -is, -e Pertaining to March 1st which was the Roman festival of the matrons, hence to married ladies.

Matteu'cia f. Ostrich fern. In honour of Carlo Matteuci (1811–1868), Italian physicist and physiologist distinguished for research on animal electricity. ASPLENIACEAE.

Matthio'la f. Stock. Named for Pierandrea Mattioli (1500–1577), Italian physician and botanist, author of a celebrated commentary on Dioscorides, which exists in Latin, Italian, German, French and Czech versions, with elaborate woodcut illustrations. CRUCIFERAE.

Mauran'dia f. Named for Catalina Pancratia Maurandy, Spanish botanist, wife of the director of the Cartagena botanic garden, late 18th century. SCROPHULARIACEAE.

maurita'nicus, -a, -um Of North Africa, particularly Morocco.

Mauri'tia f. Latinized from Surinam name 'Mauritii-boom'. PALMAE.

mauritia'nus, -a, -um Relating to Mauritius, Indian Ocean.

maw'ii; mawea'nus, -a, -um Named for George Maw (1832–1912), English manufacturer of ornamental tiles and plant collector, author of *Monograph of the Genus Crocus* (1886).
Maxilla'ria f. L. *maxilla*, the jaw. The flowers of some of these orchids resemble the jaws of an insect. ORCHIDACEAE.
maxilla'ris, -is, -e Pertaining to the jaws.
Maximilia'na f. Palms named for Maximilian A. P. zu Wied-Neuwied (1782–1867), German prince who travelled and collected in Brazil, 1813–1817. PALMAE.
maximowi'czii In honour of Carl Ivanovich Maximowicz (1827–1891), pronounced 'Maksimovich', Russian botanist at St. Petersburg, author of many scholarly publications on plants of eastern Asia.
ma'ximus, -a, -um Largest.
Maya'ca f. French Guiana vernacular name. MAYACACEAE.
mays From Mexican vernacular name for maize (Indian corn).
May'tenus m. From the Chilean name *maitén* (Mapuche, *mañtŭn*) for *M. boaria*. CELASTRACEAE.
Ma'zus m. Gr. *mazos*, a teat; from the tubercles closing the mouth of the corolla of these low-growing herbs. SCROPHULARIACEAE.
mea'dia Specific epithet for shooting star (*Dodecatheon meadia*). Named for Richard Mead (1673–1754), English physician.
Mecono'psis f. Indian poppy, blue poppy. Gr. *mēkōn*, a poppy; *ŏpsis*, likeness, appearance. The type-species is the yellow-flowered Welsh poppy (*M. cambrica*). PAPAVERACEAE.
Mede'ola f. Indian cucumber root. Named for the sorceress Medea, who helped Jason to get the golden fleece, because of the plant's reputed medicinal properties. LILIACEAE.
Medica'go f. Alfalfa, lucerne. The classical name for a crop plant, from Gr. *mēdike*, medick, introduced from Media. LEGUMINOSAE.
me'dicus, -a, -um Medicinal.
Medinil'la f. Named for José de Medinilla of Pineda, Governor of the Marianna Islands, *c.* 1820. MELASTOMATACEAE.
Mediocac'tus m. L. *medius*, middle; plus *Cactus*. This genus is intermediate between two others. CACTACEAE.
mediopic'tus, -a, -um Striped or coloured down the middle.
mediterra'neus, -a, -um Mediterranean, i.e. belonging to the Mediterranean region or, on the contrary, in the middle of the land, far inland.
me'dium From Italian vernacular name *erba media*.
me'dius, -a, -um Intermediate; middle.
medulla'ris, -is, -e; medul'lus, -a, -um Pithy.
meebol'dii In honour of Alfred Karl Meebold (1863–1952), botanical collector in many countries, novelist, poet and essayist.

mega- In Greek compound words signifies big. Thus:
megaca'lyx (with large calyx); **megacan'thus** (with big spines); **megacar'pus** (big-fruited); **megalan'thus** (big-flowered); **megalophyl'lus** (with big leaves); **megapota'micus** (of the big river, e. g. the Rio Grande); **megarrhi'zus** (big-rooted); **megasper'mus** (with big seeds); **megasta'chyus** (with a big spike); **megasti'gmus** (with a big stigma).

Megacli'nium n. From Gr. *mĕga*, big, wide; *klinion*, a little bed; in allusion to the flattened inflorescence. ORCHIDACEAE.

megaseifo'lius, -a, -um With leaves like *Megasea* (large-leaved saxifrages now called *Bergenia*).

Megaskepas'ma n. Gr. *mĕga*, large; *skĕpasma*, covering; with reference to the conspicuous bracts. ACANTHACEAE.

mege'ratus, -a, -um Gr. *mĕgērătŏs*, very lovely, much contended for.

meiacan'thus, -a, -um Small-spined or few-spined.

mekongen'sis, -is, -e From a region by the Mekong River (Lancang Jiang), usually where it flows through Yunnan, China.

Melaleu'ca f. Bottlebrush. From Gr. *mĕlas*, black; *leukos*, white. Often the trees have a black trunk and white branches. MYRTACEAE.

melancho'licus, -a, -um Drooping; sad-looking.

Melan'thium n. Bunchflower. From Gr. *mĕlas*, black; *anthos*, a flower. The flower segments are persistent and become dark after flowering. The name *mĕlanthion* was originally applied to *Nigella*. LILIACEAE (MELANTHIACEAE).

Melasphae'rula f. Gr. *mĕlas*, black; *sphaerula*, a small ball; in allusion to the small black bulbs of this South African plant. IRIDACEAE.

Melas'toma n. Gr. *mĕlas*, black; *stŏma*, a mouth. The berries of these tropical shrubs and trees stain the mouth black. MELASTOMATACEAE.

mele'agris, -is, -e Spotted like the guineafowl (*Numida mĕleāgris*).

Me'lia f. Bead-tree, chinaberry. Greek name *mĕlia* for the ash tree in allusion to the similarity of the leaves. MELIACEAE.

Melian'thus m. Honeybush. From Gr. *mĕli*, honey; *anthos*, a flower. The flowers produce abundant nectar. MELIANTHACEAE.

Me'lica f. Melic grass. From Greek name *mĕlikē* for a grass. GRAMINEAE.

Melicoc'ca f. Honeyberry. From Gr. *mĕli*, honey; *kŏkkŏs*, a berry; in allusion to the sweetness of the fruit. SAPINDACEAE.

Melicy'tus m. From Gr. *mĕli*, honey; *kytŏs*, a hollow container, jar; in allusion to the hollow staminal nectaries of the flowers. VIOLACEAE.

Melilo'tus f. Sweet clove. Gr. *mĕlilōtos*, from *mĕli*, honey; *lotos*, lotus. The foliage is fragrant, bees enjoy the flowers. LEGUMINOSAE.

meliodo'rus, -a, -um Honey-scented.

Melios'ma f. Gr. *mĕli*, honey; *ŏsmē*, fragrance. The flowers are honey-scented. SABIACEAE.

Melis'sa f. Balm. From Gr. *mĕlissa*, a honeybee, also the name of a Cretan princess who first discovered how to get honey. According to the herbalist Gerard, bees 'are delighted with this herbe above all others'. LABIATAE.

meliten'sis, -is, -e Of Malta, ancient Melita.

Melit'tis f. Bastard balm. From Attic Gr. *mĕlitta*, variant of Gr. *mĕlissa*, a honeybee. LABIATAE.

mel'leus, -a, -um Pertaining to honey.

melli'tus, -a, -um Honey-sweet. Also: *mellifer*, honey-bearing; *melliodorus*, honey-scented.

Melocac'tus m. From Latin *mēlo*, a shortened form of *mēlŏpĕpo*, an apple-shaped melon, and *Cactus*, with reference to the rounded form of these cacti. CACTACEAE.

melofor'mis, -is, -e Melon-shaped.

Melo'thria f. From the Greek name *mēlōthron* for another plant, probably *Bryonia*. CUCURBITACEAE.

membrana'ceus, -a, -um Skinlike; membranous.

meniscifo'lius, -a, -um With crescent-shaped leaves.

Menisper'mum n. Moonseed. Gr. *mēnē*, the crescent moon; *sperma*, a seed; from the shape of the seed. MENISPERMACEAE.

Men'tha f. Mint. L. *mentha*, Gr. *minthē*; one of the oldest (possibly going back 4000 years) plant names still in use. LABIATAE.

Mentze'lia f. Prairie lily, blazing star. Named for Christian Mentzel (1622–1701), German botanist, personal physician to the Elector of Brandenburg. LOASACEAE.

Menyan'thes f. Bogbean, buckbean. Gr. *mēnyanthos*, water plant mentioned by Theophrastus. MENYANTHACEAE.

Menzie'sia f. Named for Archibald Menzies (1754–1842), Scottish naval surgeon and botanist who accompanied Vancouver on his voyage of Pacific exploration, 1790–1795, then settled in London. He was long remembered by Hawaiians as 'the red-faced man who cut off the limbs of men and gathered grass'. ERICACEAE.

menzie'sii Specific epithet of *Arbutus* and many other western American and Hawaiian plants. See *Menziesia*.

Mercuria'lis f. Herb-Mercury. Originally *herba mercurialis* named in honour of the messenger of the gods. EUPHORBIACEAE.

Merende'ra f. From Spanish *Quitameriendas*, ultimately from L. *merenda*, midday meal; the flowering of *Merendera* in autumn warns the shepherds in the mountains that they must leave the pastures there. LILIACEAE.

meridia'nus, -a, -um; meridiona'lis, -is, -e Of noonday; blooming at noontime.

Merrem'ia f. In honour of Blasius Merrem (1761–1824), professor of natural sciences, Marburg, Germany. CONVOLVULACEAE.

Merten'sia f. Bluebell (in the United States), Virginia cowslip. Named for Franz Carl Mertens (1764–1831), professor of botany at Bremen. BORAGINACEAE.

Mery'ta f. Gr. *merytos*, glomerate; in allusion to the male flowers being crowded together in a head. ARALIACEAE.

Mesembryan'themum n. Fig marigold. The etymological tangle of this name began in 1684 when Breyne published it as *Mĕsēmbrianthĕmum*, derived from Gr. *mĕsēmbria*, midday; *anthemon*, flower; in allusion to the fact that the only species then known bloomed at noon. When night-flowering species became known, and this name accordingly seemed inappropriate, Dillenius in 1719 ingeniously renamed the genus *Mĕsĕmbryanthĕmum*; by changing the i to y he altered the derivation, to Gr. *mĕsos*, middle; *ĕmbryon*, embryo; *anthemon*, flower, with reference to the position of the ovary. The group has now been divided into numerous smaller genera based on habit of growth and fruit-characters. AIZOACEAE.

mes'nyi In honour of William Mesny (1842–1919), a Channel Islander from Jersey who became a general in the Chinese Imperial Army and travelled widely in China, occasionally collecting plants, among them the large-flowered yellow jasmine (*Jasminum mesnyi*).

Mes'pilus f. Medlar. The Latin name for this fruit. ROSACEAE.

metal'licus, -a, -um Metallic.

Metasequo'ia f. Dawn redwood. From Gr. *mĕta*, with, after, sharing, changed in nature; *Sequoia* (which see), to which it is related and to which fossil specimens were first referred. TAXODIACEAE.

Metroside'ros f. Iron tree. From Gr. *mētra*, core, heart-wood; *sidērŏs*, iron; from the hardness of the heart-wood; the noble *M. tomentosa* discovered by Banks and Solander in 1769 is the crimson-flowered New Zealand Christmas tree or pohutukawa. MYRTACEAE.

metterni'chii In honour of Prince Clemens L.W. Metternich (1773–1859), Austrian statesman unpopular as a supporter of autocracy and despotism.

Me'um n. Baldmoney. The Greek name for a herb, possibly *Meum athamanticum*. UMBELLIFERAE.

mexica'nus,-a, -um Mexican.

meyenia'nus, -a, -um; meye'nii In honour of Franz Julius Ferdinand Meyen (1804–1840), German physician and naturalist, who collected many plants on a voyage round the world, 1830–1832. He became a professor of botany in Berlin.

me'yeri In honour of Carl Anton Meyer (1795–1855), German botanist in Russian employ, director of St. Petersburg botanic garden; or Ernst H. F. Meyer (1791–1858), German botanist chiefly celebrated for his very scholarly but never completed history of botany; or Frank Nicholas Meyer (1875–1918), Dutch-born agricultural explorer and collector in China, Manchuria, Korea, Mongolia and Turkistan for U.S. Department of Agriculture.

mezer'eum Latinized from a Persian name.

mi'cans Glittering.

Michau'xia f. Named for André Michaux (1746–1803), French collector, explorer, and plantsman who collected this, among many others, on an expedition to Persia, 1782–1785, and, in company with his son, François André Michaux (1770–1855), carried out remarkable journeys in North America (1785–1796) and sent back an extensive herbarium. When he himself returned to France, he took with him forty boxes of seeds in addition to the 60,000 living plants he had at various times already dispatched, most of which were lost by shipwreck. He died on an expedition to Madagascar. Much of his *Flora Boreali-Americana* (1803) was written by L. C. M. Richard (1754–1821). CAMPANULACEAE.

michauxioi'des Resembling *Michauxia*.

Miche'lia f. Named for Pietro Antonio Micheli (1679–1737), Florentine botanist, particularly celebrated for his remarkable pioneer researches on cryptogams, including fungi. MAGNOLIACEAE.

Mico'nia f. Tropical American trees and shrubs with showy foliage, named for Francisco Mico (b. 1528), Spanish physician and botanist. MELASTOMATACEAE.

Micran'themum n. Gr. *mikros*, small; *anthemon*, flower. SCROPHULARIACEAE.

micro- In compound words signifying small. Thus:
micracan'thus (small thorns); **micran'thus** (small-flowered); **microcar'pus** (small-fruited); **microce'phalus** (with a small head); **microchi'lus** (small-lipped); **micro'dasys** (small and shaggy); **mic'rodon** (small-toothed); **microglos'sus** (small-tongued); **micro'gynus** (with small ovary); **microle'pis** (small-scaled); **micro'meris** (with a small number of parts or small parts); **micrope'talus** (small-petalled); **microphyl'lus** (small-leaved); **micro'pterus** (small-winged); **microse'palus** (with small sepals); **microsper'mus** (with small seeds).

Microcit'rus m. Finger lime. Gr. *mikros*, small; *kitron*, lemon, etc. RUTACEAE.

Microcoe'lum n. Gr. *mikros*, small; *koilos*, hollow, hollowed; in allusion to a small hollow in the endosperm of the seed. PALMAE.

Microglos'sa f. Gr. *mikros*,small; *glossa*, a tongue; in reference to the short ray florets. COMPOSITAE.

Microle'pia f. Gr. *mikros*, small; *lepis*, a scale; a reference to the small indusia on these ferns. DENNSTAEDTIACEAE.

Microme'ria f. Gr. *mikros*, small; *meris*, a part. The flowers and leaves are very small. LABIATAE.

Micro'stylis f. Gr. *mikros*, small; *stylos*, a column. These orchids have small columns. ORCHIDACEAE.

middendor'ffii In honour of Alexander Theodor von Middendorff (1815–1894), zoologist who travelled and collected extensively in Siberia.

Mika'nia f. Climbing hempweed. Named in honour of Joseph Gottfried Mikan (1743–1814), professor of botany at Prague, father of Johann Christian Mikan (1769–1844), likewise a professor at Prague, who collected in Brazil. COMPOSITAE.

milia'ceus, -a, -um Pertaining to millet.

milita'ris, -is, -e Pertaining to soldiers; like a soldier.

Mil'la f. Named for Julian Milla, 18th-century gardener to the king of Spain. LILIACEAE (ALLIACEAE).

millefolia'tus, -a, -um; millefo'lius, -a, -um Many-leaved; literally, with a thousand leaves.

mil'leri In honour of Philip Miller (1691–1771), English gardener and botanist, curator of the Chelsea Physic Garden 1722–1770, author of the celebrated *Gardener's Dictionary* (1731; 8th ed. 1768).

Milto'nia f. Named for Viscount Milton, later Lord Fitzwilliam (1786–1851), patron of gardening. ORCHIDACEAE.

mim'etes Gr. *mimētēs*, an imitator.

Mimo'sa f. Gr. *mimos*, a mimic; in reference to the sensitive collapse, when touched, of the leaves of some species. LEGUMINOSAE.

mimosifo'lius, -a, -um With leaves like *Mimosa*.

mimosoi'des Like *Mimosa*.

Mi'mulus m. Monkey-flower. Latin diminutive of *mimus*, a mimic. The corolla looks somewhat like the face of a monkey. SCROPHULARIACEAE.

Mimu'sops f. L. *mimus*, a mimic; Gr. *ōps*, eye, face, appearance. SAPOTACEAE.

Mi'na f. In honour of Joseph X.F.X. Mina, a Mexican (fl. 1824). CONVOLVULACEAE.

mi'nax Threatening; forbidding.

minia'tus, -a, -um Cinnabar-red.

mi'nimus, -a, -um Smallest.

mi'nor; mi'nus, -a, -um Smaller.

Minuar'tia f. In honour of Juan Minuart (1693–1768), Spanish apothecary and botanist at Barcelona, then Madrid. CARYOPHYLLACEAE.

minu'tus, -a, -um Very small. Also:
minutiflo'rus (minute-flowered); **minutifo'lius** (minute-leaved); **minutis'simus** (most minute).

miquelia'nus, -a, -um; mique'lii In honour of Friedrich Anton Wilhelm Miquel (1811–1871), Dutch botanist of Hanoverian origin, professor of botany at Utrecht and Leiden, author of works on flora of Japan, Indonesia, etc.

Mira'bilis f. Four-o'clock. Latin word meaning wonderful. NYCTAGINACEAE.

mira'bilis, -is,-e Wonderful; remarkable.

Miscan'thus m. Gr. *miskos*, a stem; *anthos*, a flower; an allusion to the stalked spikelets. GRAMINEAE.

mischtchenkoa'nus, -a, -um Epithet of a *Scilla* named for P. I. Misczenko (1869–1938), Russian botanist.

mishimien'sis, -is, -e Of the Mishmi Hills, Assam, India.

missourien'sis, -is, -e; missu'ricus -a, -um Referring to Missouri, U.S.A.

Mitchel'la f. Partridge berry. An evergreen trailing herb named for John Mitchell (1711–1768), physician of Virginia, born in Lancaster county, who was a correspondent of Linnaeus. RUBIACEAE.

Mitel'la f. Bishop's cap, mitrewort. Diminutive of Gr. *mitra*, a cap; in allusion to the form of the young fruit. SAXIFRAGACEAE.

mi'tis, -is, -e Mild; gentle; without spines.

Mitra'ria f. Gr. *mitra*, a mitre or cap; from the shape of the seed pod. GESNERIACEAE.

mitra'tus, -a, -um Turbaned; mitred.

mitrifor'mis, -is, -e Caplike.

mix'tus, -a, -um Mixed.

mlokosewitschii In honour of Ludwik Franciszek Mlokosiewicz (1831–1909), Polish soldier, naturalist and forester stationed in the Caucasus at Lagodekhi; cf. *juliae*.

modes'tus, -a, -um Modest.

moesia'cus, -a, -um From *Moesia*, classical name of northern Balkan region.

Mogha'nia f. Latinized form of an Indian name. This genus is also known as *Flemingia*. LEGUMINOSAE.

Moh'ria f. In honour of Daniel Matthias Heinrich Mohr (d. 1808) of Kiel. SCHIZAEACEAE,

molda'vicus, -a, -um Of Moldavia, south-eastern Europe.

Moli'na f. Named for Juan Ignacio Molina (1740–1829), Jesuit historian, writer on the civil and natural history of Chile. GRAMINEAE.

mol'le Peruvian vernacular name.

molli'comus, -a, -um Soft-haired.

mol'lis, -is, -e Soft; with soft hairs.

mollis'simus, -a, -um Very soft.

mollu'go Specific name, derived from the genus *Mollugo* which has no garden interest except as a pernicious carpetweed.

Moloposper'mum n. Gr. *mōlops*, a stripe; *sperma*, a seed. The fruit appears to be striped. UMBELLIFERAE.

Mol'tkia f. Named for Count Joachim Godske Moltke (1746–1818), Danish statesman and patron of sciences. BORAGINACEAE.

molucca'nus, -a, -um Of Moluccas or Spice Islands of Indonesia.

Moluccel'la f. Shell-flower. Derivation of the name is uncertain, but is apparently a diminutive of *Molucca*. LABIATAE.

mo'ly Gr. *mōly*, name of a magic herb of uncertain identity, nothing whatever to do with *Allium moly*.

Momor'dica f. L. *mordeo*, to bite, possibly because the seeds of these tropical climbers appear bitten. CUCURBITACEAE.

monacan'thus, -a, -um One-spined.

monadel'phus, -a, -um From Gr. *mŏnŏs*, one; *adĕlphŏs*, brother; the filaments of the stamens being joined into one brotherhood.

monan'drus, -a, -um With one stamen.

Monan'thes f. Gr. *mŏnŏs*, one; *anthos*, a flower. The type-species was named from a solitary-flowered specimen but it can bear up to eight flowers on a stem. CRASSULACEAE.

Monar'da f. Bee-balm, bergamot, horse-mint. Named for Nicholas Monardes (1493–1588), physician and botanist of Seville who published in 1569, 1571 and 1574 a book on American products, translated into English in 1577 with the title *Joyfull Newes out of the newe founde Worlde*. The genus includes bee-balm, lemon mint, wild bergamot. LABIATAE.

Monardel'la. f. Diminutive of *Monarda* which this genus of aromatic herbs resembles. LABIATAE.

mon'do Japanese name used as a specific epithet.

Mo'neses f. One-flowered pyrola. Gr. *mŏnŏs*, single; *ĕsis*, a sending forth. This has a solitary flower. PYROLACEAE.

mongo'licus, -a, -um Mongolian.

moni'lifer, monili'fera, -um Bearing a necklace.

monilifor'mis, -is, -e In the form of a necklace with alternate swellings and constrictions.

mono- In compound words signifying single. Thus:
monoce'phalus (with one head); **mono'gynus** (with one pistil); **monope'talus** (single-petalled); **monophyl'lus** (one-leafed); **mono'pterus** (with a single wing); **monosema'tus** (with one mark or blotch); **monose'palus** (one-sepalled); **monosper'mus** (one-seeded); **monosta'chyus** (one-spiked).

monregalen'sis, -is, -e Of Mondovi, Piedmont, Italy; or of Monreale. Both are known in Latin as *Mons Regalis*.

Monso'nia f. In honour of Lady Ann Monson (*c*. 1714–1776) who collected at the Cape of Good Hope and in Bengal. GERANIACEAE.

monspelien'sis, -is, -e; monspessulan'us, -a, -um Of Montpellier in southern France, Latinized as *Mons Pessulanus*.

Monste'ra f. Derivation obscure, possibly contracted from *monstrifer*, monster-bearing, in allusion to the perforated leaves of these tropical evergreen climbers. ARACEAE.

monstro'sus, -a, -um Abnormal.

Monta'noa f. Named for Luis Montana, supposed to have been a mid-19th-century Mexican politician. COMPOSITAE.

monta'nus, -a, -um Pertaining to mountains.

Montbre'tia f. Named for Antoine François Ernest Conquebert de Montbret (1781–1801), one of the botanists accompanying Napoleon's invasion of Egypt (1798), where he died. Some plants formerly placed in *Montbretia* belong to *Tritonia* or *Crocosmia*. IRIDACEAE.

Mon'tia f. In honour of Giuseppe Monti (1682–1760), professor of botany at Bologna. PORTULACACEAE.

monti'cola A dweller on mountains.

monti'genus, -a, -um Mountain-born.

Monvi'llea f. In honour of Hippolyte Boissel (1794–1863), Baron de Monville, whose superb collection was sold by auction in 1846. CACTACEAE.

Morae'a (Morea) f. Butterfly iris. Named by Miller in honour of Robert More (1703–1780) of Shrewsbury, England, apparently a keen amateur botanist. IRIDACEAE.

Morican'dia f. In honour of Moise Étienne Moricand (1780–1854), Swiss botanist, commercial traveller in Swiss watches. CRUCIFERAE.

morifo'lius, -a, -um Mulberry-leaved.

Mori'na f. Whorl flower. Named for Louis Pierre Morin (1635–1715), French botanist and physician. DIPSACACEAE.

Morin'da f. Indian mulberry. L. *morus*, a mulberry; *indicus*, Indian. RUBIACEAE.

Morin'ga f. Latinized version of the Sinhalese name *murunga*. The seeds of these trees yield ben-oil used in perfumery and for lubricating watches and other fine mechanisms. MORINGACEAE.

Mori'sia f. In honour of Giuseppe Giacinto Moris (1796–1869), professor of botany at Turin, author of a flora of Sardinia. CRUCIFERAE.

Moriso'nia f. In honour of Robert Morison (1620–1683), a Scottish royalist who after exile in France became professor of botany at Oxford. CAPPARACEAE.

Mormo'des f. From Gr. *mŏrmō̄*, phantom; in allusion to the strange flowers of these orchids. ORCHIDACEAE.

morrisonen'sis, -is, -e Of Mount Morrison (Yushan, Niitakayama), Taiwan.

Mo'rus f. Mulberry. The Latin name. The tree is widely cultivated as food for silk-worms, most species being useful for this purpose. MORACEAE.

mosa'icus, -a, -um Coloured in a mosaic fashion; see *musaicus*.

Moscha'ria f. Gr. *mŏschŏs*, musk; in allusion to the fragrance. COMPOSITAE.

moscha'tus –a, –um Musky. (See under *Muscari*.)

moulmainen'sis, -is, -e Of Moulmein, Burma.

moupinensis, -is, -e Of Mupin, Sichuan, western China, a principality in which Armand David made large collections of plants.

moure'tii In honour of Lieutenant Marcellin Mouret (1881–1915),French soldier and amateur botanist who collected in North Africa and was killed in the First World War.

moy'esii In honour of the Rev. J. Moyes of the China Inland Mission, *c* 1906.
muco'sus, -a, -um Slimy.
mucrona'tus, -a, -um Pointed; mucronate.
mucronula'tus, -a, -um Having a short hard point.
Mucu'na f. From the Brazilian name for these vines, the pods of which often bear irritant hairs. LEGUMINOSAE.
Muehlenbe'ckia f. Named for Henri Gustave Muehlenbeck (1798–1845), physician at Mulhouse, France, investigator of the flora of Alsace. POLYGONACEAE.
Muhlenber'gia f. Named in honour of Gotthilf Henry Ernest Muhlenberg (1753–1815), Lutheran pastor in Pennsylvania, U.S.A., distinguished amateur botanist. GRAMINEAE.
Mulge'dium n. From *mulgeo*, to milk; by analogy with *Lactuca* to which this genus is closely related. COMPOSITAE.
mulien'sis, -is, -e From the Muli region of south-western Sichuan, China.
multi- In compound words signifies many. Thus:
multibractea'tus (with many bracts); **multicau'lis** (many-stemmed); **multica'vus** ((with many hollows); **mul'ticeps** (with many heads); **multi'color** (many-coloured); **multicosta'tus** (many-ribbed); **multi'fidus** (many times divided); **multiflo'rus** (many-flowered); **multi'jugus** (many yoked together, generally referring to leaves with many pairs of leaflets); **multilinea'tus** (many-lined); **multiner'vis** (many-nerved); **mul'tiplex** (much-folded, hence doubled); **multiradia'tus** (with many rays); **multiscapoi'deus** (with many scapes, that is leafless flower-stems); **multisec'tus** (much-cut).
mu'me Variant of the Japanese name *ume* for a species of *Prunus*.
mun'dulus, -a, -um Trim; neat.
mura'lis, -is, -e Growing on walls.
murica'tus, -a, -um Roughened, with hard points, like the shell of the mollusc (*Mūrex*) from which the celebrated purple dye of antiquity was obtained.
murie'lae In honour of Muriel Wilson (Mrs George L. Slate, d. 1976), daughter of E. H. Wilson; see *Sinowilsonia*.
Murray'a (also incorrectly **Mur'raea**) f. Curry-leaf, orange jessamine, satinwood tree. Named for Johann Andreas Murray (1740–1791), Swedish pupil of Linnaeus and professor of medicine and botany, Göttingen. RUTACEAE.
Mu'sa f. Banana. From Arabic *mouz* or *moz*, but accepted by Linnaeus because it could have been named for Antonius Musa (63–14 B.C.), physician to the first Roman emperor, Octavius Augustus. The Arabs regarded it as the 'Tree of Paradise', hence the Linnaean epithet *paradisiaca*. The genus is of Old World origin but quite early was established in the New. *M. paradisiaca*, in one of its many variants, provides most of the edible fruits. MUSACEAE.

musa'ceus, -a, -um *Musa*-like.

musa'icus, -a, -um Coloured or marked in a mosaic fashion; also spelled *mosaicus*.

Musca'ri n. Grape-hyacinth. Turkish name recorded by Clusius in 1583, the bulbs of *Muscari muscarimi* (*M. moschatum*) being received from Constantinople under the names *Muscari*, *Muschoromi* or *Muscurimi*, meaning musk of the Romans (i.e. Greeks), or *Muschio greco* (Greek musk), referring to the sweet aromatic scent of the flowers, hence from Persian *mushk*, Sanskrit *mushka*, testicle. The source of musk is a scent gland or 'pod' of the male musk-deer (*Moschus moschifer*). LILIACEAE (HYACINTHACEAE).

muscárius, -a, -um Relating to flies; from L. *musca*, fly.

musci'pula f. Mouse-trap; from L. *mus*, mouse.

muscito'xicus, -a, -um Poisonous to flies.

musci'vorus, -a, -um Fly-eating.

muscoi'des Fly-like.

musco'sus, -a, -um Resembling moss, mossy; from L. *muscus*, moss.

musifo'lius, -a, -um *Musa*-leaved.

Mussaen'da f. Latinization of Sinhalese vernacular name *mussenda*. RUBIACEAE.

mussi'nii Named for the Russian Count Apollos Mussin-Puschkin (d. 1815), who made an expedition to the Caucasus in 1802.

muta'bilis, -is, -e Changeable, especially as to colour.

muta'tus, -a, -um Changed.

mu'ticus, -a, -um Blunt; without a point.

mutila'tus, -a, -um Divided as though torn, as with some leaves.

Muti'sia f. Named for José Celestino Mutis (1732–1809), physician of Cadiz, who went to South America in 1760 and settled at Bogotá, whence he sent specimens to Linnaeus and his son. A team of artists worked under his direction making superb illustrations of plants, which were taken to Spain after his death and that of his assistant Caldas, and of which some were first published in 1955. COMPOSITAE.

Myce'lis f. Old name of obscure origin. COMPOSITAE.

myco'ni, Myconel'la f., **myco'nis** In honour of Francisco Mico (b. 1528), Spanish botanist and physician, not 'the mythical bear Mycon' ('nach dem mythischen Bären *Mycon*') as stated by Wittstein. See *Miconia*.

myiag'rus, -a, -um Fly-catching; in allusion to stickiness.

myoporoi'des Like *Myoporum*.

Myo'porum n. Gr. *myō*, to shut, close; *pŏrŏs*, opening, pore; referring to the transparent spots on the leaves filled (and thus closed) with a pellucid substance. MYOPORACEAE.

myosotidiflo'rus, -a, -um With flowers like *Myosotis*.

Myosoti'dium n. Chatham Island lily. This handsome plant, confined to the

Chatham Islands, is not a lily but a giant forget-me-not (*Myosotis*), hence the name. BORAGINACEAE.

Myoso'tis f. Forget-me-not. The classical Greek name, *myŏsōtis*, from *mus*, mouse; *ŏus*, *ōtŏs*, ear; applied to plants with short pointed leaves, later transferred to this genus. BORAGINACEAE.

Myosu'rus m. Mouse-tail. From Gr. *mus*, mouse; *oura*, tail; with reference to the slender elongated inflorescence. RANUNCULACEAE.

myri- In compound words signifying very many. Thus: **myriacan'thus** (with very many thorns); **myriocar'pus** (many-fruited); **myriophyl'lus** (many-leaved).

Myri'ca f. Gale, bog myrtle, bayberry, wax myrtle. Derived from the Greek name *myrikē* for tamarisk. The greasy covering of the fruit of these shrubs provides the aromatic tallow from which bayberry candles are made. MYRICACEAE.

Myrica'ria f. False tamarisk. From Gr. *myrikē*, tamarisk, to which this genus is related. TAMARICACEAE.

Myrioce'phalus m. Gr. *myrios*, many; *kĕphale*, a head; in allusion to the many one-flowered minor heads assembled into a single compound head. COMPOSITAE.

myriophylloi'des Resembling *Myriophyllum*.

Myriophyl'lum n. Water milfoil. Gr. *myrios*, many; *phyllon*, a leaf; with reference to the many divisions of the submerged leaves of these aquatic plants. HALORAGACEAE.

Myris'tica f. Nutmeg. Gr. *myristikos*, fit for anointing. The fruit of *M. fragans* is the source of nutmeg. MYRISTICACEAE.

myrmeco'philus, -a, -um Literally, ant-loving—an epithet applied to plants much frequented by ants as providing them with either food or shelter, as in the case of certain plants with hollow pseudo-bulbs, and many others.

Myroxy'lon n. Gr. *myrŏn*, a sweet-smelling oil; *xylon*, wood. The heartwood of these trees is resinous and fragrant. LEGUMINOSAE.

Myr'rhis f. Myrrh, sweet cicely. The Greek name for a plant often identified as *M. odorata* as well as for the true myrrh. *M. odorata* is an herbaceous plant of the carrot family useful as a salad plant for its aniseed flavour and aroma. Biblical myrrh is a fragrant gum resin once very highly esteemed for ceremonial, religious, and embalming purposes. It comes from *Commiphora myrrha*, a small East African and Arabian tree. UMBELLIFERAE.

Myr'sine f. From a Greek name for myrtle; transferred to these trees and shrubs. MYRSINACEAE.

myrsinifo'lius, -a, -um *Myrsine*-leaved.

myrsinoi'des Resembling *Myrsine*.

myrtifo'lius, -a, -um Myrtle-leaved.

Myrtillocac'tus m. L. *myrtillus*, a small myrtle. The allusion is to the small myrtle-like fruits, not unlike blueberries. CACTACEAE.

Myr'tus f. Myrtle. The Greek name. The Greeks held the plant sacred to Aphrodite (the Roman Venus) and in Christian religious painting it became one of the plants associated with the Virgin. MYRTACEAE.

mysoren'sis, -is, -e Of Mysore, southern India.

Mystaci'dium n. From Gr. *mystax*, moustache. ORCHIDACEAE.

N

Naege'lia f. Named for Karl Wilhelm Naegeli (1817–1891), Swiss botanist, successively a professor at Freiburg im Breisgau, Zürich and Munich, noted for his cytological research and speculations on heredity. GESNERIACEAE.

Na'jas f. Water nymph. NAJADACEAE.

nakotil'tus, -a, -um Having the wool plucked off.

Nandi'na f. Latinized form of the Japanese name *Nanten*. BERBERIDACEAE.

nanel'lus, -a, -um Very dwarf.

nankinen'sis, -is, -e From Nanking (Nanjing), Fukien, China.

nannophyl'lus, -a, -um, Small-leaved.

Nan'norhops f. A small tufted palm. From Gr. *nannŏs*, a dwarf; *rhōps*, a bush or low shrub. PALMAE.

na'nus, -a, -um Dwarf.

napel'lus m. Diminutive of *napus*, meaning a little turnip, in allusion to the tubers.

napifor'mis, -is, -e Turnip-shaped.

napobras'sica Specific epithet for swede or ruta-baga—literally, 'turnip-cabbage'.

napuliger, napuligera, -um With a little turnip-like root.

narbonen'sis, -is, -e Of Narbonne, southern France.

narcissiflo'rus, -a, -um Narcissus-flowered, i.e. with flowers in an umbel resembling some species of *Narcissus*.

narcissifo'lius, -a, -um Narcissus-leaved.

Narcis'sus m. Daffodil. Classical Greek name in honour of a beautiful youth who became so entranced with his own reflection that he pined away and the gods turned him into this flower. The word is possibly derived from an ancient Iranian language. AMARYLLIDACEAE.

Nardos'mia f. From Gr. *nardŏs*, spikenard; *ŏsmē*, smell. COMPOSITAE.

narino'sus, -a, -um Broad-nosed.

Narthe'cium n. A diminutive of *narthex*, rod. LILIACEAE (MELANTHIACEAE).

nastur'tium From L. *nasus tortus*, a twisted nose, due to the plant's pungent taste. Specific epithet of water-cress (*Rorippa nasturtium-aquaticum*). The plant familiarly known as nasturtium is *Tropaeolum* (*q.v*).

nasu'tus, -a, -um Large-nosed.

natalen'sis, -is, -e Of Natal, South Africa; discovered by Vasco da Gama on Christmas Day and named *Terra Natalis*.

nataliae In honour of Natalie Kesjko (1859–1900), Russian wife of Milan Karajorevich, prince, later king, of Serbia.

na'tans Floating.

Nauc'lea f. Said to be a variant of L. *naucula*, a little ship, with reference to valves of the fruit. RUBIACEAE.

nauseo'sus, -a, -um Nauseating.

Nautiloca'lyx m. L. *nautilus*, the nautilus mollusc; *calyx*, calyx. GESNERIACEAE.

navicula'ris, -is, -e Boat-shaped.

neapolita'nus, -a,-um Of Naples, Neapolitan.

nebroden'sis, -is, -e Of the Monti Nebrodi, north-eastern Sicily.

nebulo'sus, -a, -um Cloudlike.

Nectanoscordum n. From L. *nectar*, Gr. *nektar*, nectar; *skŏrŏdŏn*, garlic; with reference to the large nectaries on the ovary. LILIACEAE (ALLIACEAE).

neglec'tus, -a, -um Hitherto overlooked.

negun'do From Sanskrit and Bengali *nirgundi*, rendered as *negundo* by Acosta, name for *Vitex negundo*, and transferred on account of leaf resemblance to *Acer negundo*.

neilgherren'sis, -is, -e From the Nilgiri Hills, southern India; also rendered *nilagiricus* and *nilgiricus*.

Neil'lia f. Nine bark. Named for Patrick Neill (1776–1851), printer, of Edinburgh, Secretary of the Caledonian Horticultural Society. ROSACEAE.

nelumbifo'lius, -a, -um With leaves like *Nelumbo* or holy lotus.

Nelum'bo f. Holy lotus. Sinhalese name for this large and beautiful aquatic plant. NYMPHAEACEAE or NELUMBOACEAE.

Nema'stylis f. Gr. *nēma*, a thread; *stylos*, a column; from the slender style. IRIDACEAE.

Nematan'thus m. Gr. *nēma*, thread; *anthos*, flower; in allusion to the long thread-like flower-stalks. GESNERIACEAE.

Neme'sia f. From the Greek name *nĕmĕsiŏn* for a plant similar to these herbs. SCROPHULARIACEAE.

Nemopan'thus m. Mountain holly. Contracted from Gr. *nēma*, a thread; *pŏus*, foot; *anthos*, a flower; with reference to the long slender peduncles. AQUIFOLIACEAE.

Nemo'phila f. Gr. *nĕmŏs*, wooded pasture, glade; *phileo*, to love; in allusion to the habitat of some species. HYDROPHYLLACEAE.

nemora'lis, -is, -e; nemoro'sus, -a, -um Growing in groves or woods.

Neobentha'mia f. Gr. *neo*, new; *Benthamia*, in honour of George Bentham; see *benthamii*. The name *Benthamia* having been used earlier for another genus, this one had to be named anew. ORCHIDACEAE.

Neobessey'a f. Cactus. Named for Charles Bessey (1845–1915), professor of botany, University of Nebraska, author of an influential system of classification. CACTACEAE.

Neoglazio'via f. Named for Auguste F. M. Glaziou (1828–1906), French plant-collector in Brazil. BROMELIACEAE.

Neolloy'dia f. Named for Francis Ernest Lloyd (1868–1947), American plant physiologist, author of *The Carnivorous Plants*, for many years professor of botany at Montreal. The suffix *neo-*, new, was used owing to the existence of the name *Lloydia*. CACTACEAE.

Neomar'ica f. From Gr. *neo-*, new; *Marica*, a genus of *Iridaceae*; presumably named after the river nymph Marica. IRIDACEAE.

Neoporte'ria In honour of Carlos Porter, Chilean entomologist. CACTACEAE.

Neorege'lia f. In honour of E.A. von Regel; see *Aregelia*. BROMELIACEAE.

nepalen'sis, -is, -e From Nepal, central Himalayas; also rendered *napaulensis* and *nepaulensis*.

Nepen'thes f. Pitcherplants. From the name of a plant in Greek literature which could assuage grief and, classically, any plant capable of producing euphoria. Used here in connexion with its supposed medical qualities. NEPENTHACEAE.

Ne'peta f. Latin name for an aromatic plant; includes the genus catnip (*Nepeta cataria*). LABIATAE.

nepetoi'des Like *Nepeta*.

Nephro'dium n. From Gr. *nĕphrodēs*, kidney-like, referring to the indusium. ASPLENIACEAE.

Nephro'lepis f. Sword fern. From Gr. *nĕphrōs*, a kidney; *lĕpis*, a scale; in allusion to the form of the indusium. DAVALLIACEAE.

nephro'lepis, -is, -e With kidney-shaped scales.

Nephthy'tis f. From Nephthys, in Egyptian mythology, mother of Anubis and wife of Typhon. ARACEAE.

Neptu'nia f. From L. *Neptunus*, Neptune, god of the sea, rivers and fountains; in allusion to its watery habitat. LEGUMINOSAE.

neriifo'lius, -a, -um Oleander-leaved.

Neri'ne f. Named for the water nymph of that name, obviously in allusion to the legend that bulbs of the type-species, the Guernsey lily (*Nerine sarniensis*), were cast on to the shore of Guernsey after a shipwreck: it appears, however, that bulbs brought from South Africa were given to John de Saumarez, Dean of Guernsey, about 1655 by mariners whose ship was cast ashore there. AMARYLLIDACEAE.

Ne'rium n. Oleander, rose bay. The classical Greek name. Although oleanders are very poisonous, they are very popular, being very ornamental where the climate is suitable. Men have died from drinking tea stirred with oleander twigs. APOCYNACEAE.

neriiflo'rus, -a, -um Oleander-flowered.

Ner'tera f. Gr. *nerteros*, lowly; from the habit of growth of these creeping herbs. RUBIACEAE.

nerva'tus, -a, -um; ner'vis, -is, -e With nerves, i.e. fairly strong unbranched veins or ribs.

nervo'sus, -a, -um With well-developed or conspicuous nerves.

neurocar'pus, -a, -um With prominently nerved or veined fruits.

nevaden'sis, -is, -e Of the Sierra Nevada, Spain or California, or the state of Nevada, U.S.A.

Neviu'sia f. Snow wreath. Named for Ruben Denton Nevius (1827–1913), of Alabama, who discovered it there. ROSACEAE.

nicae'ensis, -is, -e Of Nice (formerly Nicaea Maritima), southern France, or Iznik (formerly Nicaea), Turkey.

Nican'dra f. Apple-of-Peru. Named for Nikander of Colophon, a poet who wrote of plants and their medical uses, *c.* 100 B.C. SOLANACEAE.

Nicode'mia f. In honour of Gaetano Nicodemo (d. 1803), Italian botanist from Naples, curator of the Lyons botanic garden from 1799 to his death by murder or suicide in 1803. LOGANIACEAE.

Nicola'ia f. Torch ginger. In honour of the Czar Nicholas I (1796–1855) of Russia. ZINGIBERACEAE.

Nicotia'na f. Tobacco. Named for Jean Nicot (1530–1600), French ambassador to Lisbon who introduced tobacco into France. SOLANACEAE.

nic'titans Blinking; moving.

Nidula'rium n. Diminutive of L. *nidus*, a nest; in fanciful reference to rosettes of leaves resembling nests of birds. BROMELIACEAE.

ni'dus A nest.

Nierember'gia f. Cup-flower. Named for Juan Eusebio Nieremberg (1595–1658), Spanish Jesuit and author of a book on the marvels of nature published in 1635. SOLANACEAE.

Nigel'la f. Love-in-a-mist. Fennel-flower. From the diminutive of L. *niger*, black; in allusion to the colour of the seeds. RANUNCULACEAE.

ni'ger, ni'gra, -um Black. Often applied to roots, as in the case of the Christmas rose, *Helleborus niger*, as well as to flowers.

ni'gratus, -a, -um; nigres'cens Blackish.

ni'gricans Blackish.

nigropuncta'tus, -a, -um With small black spots or dots.

nilo'ticus, -a, -um Of the valley of the Nile.

Ni'pa f. From the Moluccan name for this palm. PALMAE.

Niphae'a f. From Gr. *niphas*, snow; in allusion to the white unspotted flowers of the type-species; *Phinaea* is an anagram. ORCHIDACEAE.

nippo'nicus, -a, -um Japanese.

ni'tens; nitidus, -a, -um Shining.

nitidifo'lius, -a, -um With glossy leaves.

niva'lis, -is, -e; niv'eus, -a, -um; nivo'sus, -a, -um Snow-white; growing near snow.

Nive'nia f. In honour of James Niven (*c.* 1774–1826), Scottish gardener and collector in South Africa, 1798–1803. IRIDACEAE.

no'bilis, -is, -e Notable, famous, renowned, excellent.

noctiflor'us, -a, -um; noctur'nus, -a, -um Night-flowering.

nodiflor'us, -a, -um Flowering at the nodes.

nodo'sus, -a, -um Having conspicuous joints or nodes.

nodulo'sus, -a, -um With small nodes.

Nola'na f. Chilean bellflower. L. *nola*, a small bell; from the form of the corolla. NOLANACEAE.

Noli'na f. Bear-grass. Short-trunked desert plants named for P. C. Nolin, French agricultural writer *c.* 1755. AGAVACEAE.

noli-tan'gere Touch-me-not. Specific name of *Impatiens*.

Nol'tea f. Soap-bush. Named for Ernst Ferdinand Nolte (b. 1791), early 19th-century professor of natural history at Kiel. RHAMNACEAE.

Nomocha'ris f. From Gr. *nŏmŏs*, meadow; *charis*, grace, loveliness. LILIACEAE.

non-scrip'tus, -a, -um Without markings.

nootkaten'sis, -is, -e Of Nootka Sound, British Columbia.

Nopa'lea f. From the Mexican name for these cacti, upon a species of which the cochineal insect is cultured. CACTACEAE.

Nopalxo'chia f. Reputedly from the Aztec name. CACTACEAE.

nor'thiae In honour of Marianne North (1830–1890), prolific English botanical painter and wide-ranging traveller whose paintings are exhibited in the North Gallery at the Royal Botanic Gardens, Kew.

norve'gicus, -a, -um Norwegian.

nota'tus, -a, -um Spotted, marked.

Nothofa'gus f. Southern beech. From Gr. *nŏthŏs*, false; L. *fagus*, beech; being closely related to beech and from the southern hemisphere. FAGACEAE.

Notholae'na f. From Gr. *nŏthŏs*, false; *chlaina*, cloak; in allusion to the incomplete indusium. POLYPODIACEAE.

Notholi'rion n. From Gr. *nŏthŏs*, false; *lĕiriŏn*, lily. This group of lily-like plants has been included both in *Lilium* and *Fritillaria* but is now considered distinct from both. LILIACEAE.

Notho'panax m. From Gr. *nŏthŏs*, false; *panax*, ginseng. ARALIACEAE.

Nothoscor'dum n. Gr. *nŏthŏs*, false; *skŏrdon*, garlic; being related to *Allium* (garlic, onion, etc.) but lacking its flavour and odour. ALLIACEAE.

no'thus, -a, -um False.

Noto'nia f. In honour of Benjamin Noton, assay master in Bombay, 1812–1835, who collected in southern India. COMPOSITAE.

Notospar'tium n. Gr. *nŏtŏs*, southern; *spartiŏn*, broom. LEGUMINOSAE.

Noty'lia f. Gr. *nōtŏn*, back; *tylŏs*, hump. ORCHIDACEAE.

novae-ang'liae Of New England, U.S.A. (from *Anglia*, England).

novae-caesar'eae Of New Jersey, U.S.A. (from Roman *Caesara*, Jersey, Channel Islands).

novae-zelan'diae Of New Zealand.

noveboracen'sis, -is, -e Of New York (*Novum Eboracum*, from the Roman name for York).

novi-belgii Of New York, established by Dutch settlers, who named it from the ancient Belgic tribe inhabiting the Low Countries (Netherlands and Belgium).

nubi'cola Dweller among the clouds.

nubi'genus, -a, -um Born among the clouds.

nu'cifer, nuci'fera, -um Bearing nuts.

nuda'tus, -a, -um; nu'dus, -a, -um Naked; bare. Also: **nudicau'lis** (bare-stemmed); **nudiflo'rus** (with flowers coming before the leaves).

numi'dicus, -a, -um Of Algeria (ancient Numidia).

nummula'rius, -a, -um Resembling a coin (*nummus*) usually applied to plants with small almost circular leaves.

Nu'phar f. Spatter-dock. Yellow water-lily. Ultimately from Persian *nufar*. NYMPHAEACEAE.

nu'tans Nodding.

nutka'nus Of Nootka Sound, British Columbia.

nuttal'lii In honour of Thomas Nuttall (1786–1839), English botanist, who travelled extensively in U.S.A. in 1811–1834.

nyctagi'neus, -a, -um Night-blooming.

Nyctan'thes f. Gr. *nyx, nyktos*, night; *anthos*, a flower. The flowers open at nightfall and drop at dawn. OLEACEAE.

Nyctoce'reus m. Gr. *nyx, nyktos*, night; *Cereus*, cactus. They are night-blooming. CACTACEAE.

nymansen'sis Raised at Nymans Gardens, Handcross, Sussex, England.

Nymphae'a f. Water-lily. Gr. *nymphaia*, referring to a water nymph. NYMPHAEACEAE.

Nymphoi'des f. Resembling *Nymphaea*. MENYANTHACEAE.

Nys'sa f. Tupelo, sour gum. Named for Nysa or Nyssa, one of the water nymphs. The cotton gum, *Nyssa aquatica*, grows in swamps. NYSSACEAE.

O

obcon'icus, -a, -um Shaped like an inverted cone.

obe'sus, -a, -um Fat; succulent.

obfusca'tus, -a, -um Confused; clouded.

obla'tus, -a, -um Flattened at the ends, hence used of rounded shapes broader than long.

obli'quus, -a, -um Lopsided, oblique.

oblitera'tus, -a, -um Erased, suppressed.

oblonga'tus, -a, -um Oblong.

oblongifo'lius, -a, -um With oblong leaves.

oblon'gus, -a,-um Oblong, i.e. with almost parallel sides.

obova'tus, -a,-um Inverted ovate, i.e. egg-shaped with broadest end uppermost.

obscu'rus, -a, -um Dark, dusky, indistinct, uncertain.

obsole'tus, -a, -um Rudimentary, scarcely apparent.

obtusa'tus, -a, -um Blunted.

obtusifo'lius, -a, -um Blunt-leaved.

obtu'sior Blunter.

obtu'sus, -a, -um Blunt.

obvallar'is, -is, -e; obvalla-tus, -a, -um Apparently surrounded by a wall, applied to the corona of *Narcissus obvallaris*.

occidenta'lis, -is, -e Western.

occita'nus, -a, -um Relating to Occitania, i.e. Languedoc, southern France.

oceanen'sis, -is, -e Growing near the sea; of the sea.

ocella'tus, -a, -um With an eye; having a spot enclosed within another spot of a different colour, or with a circular patch of colour.

Ochaga'via f. In honour of Sylvestri Ochagavia, minister of education in Chile from 1853 to 1854. BROMELIACEAE.

Och'na f. Greek name for the wild pear to which the foliage of this genus has some resemblance. OCHNACEAE.

ochra'ceus, -a, -um Ochre-coloured.

ochroleu'cus, -a, -um Yellowish-white.

ocimoi'des Resembling *Ocimum*.

O'cimum n. Basil. From the Greek name *ōkimŏn* for an aromatic herb, possibly this one. LABIATAE.

Oco'tea f. Name coined by Aublet, presumably Latinized from a French Guiana vernacular name. LAURACEAE.

octan'drus, -a, -um Having eight stamens.

Octome'ria f. Gr. *okto*, eight; *meris*, a part; in allusion to the eight pollen-bearing masses. ORCHIDACEAE.

octope'talus, -a, -um Having eight petals.

ocula'tus, -a, -um With an eye, provided with a circular patch of colour; see *ocellatus*.

oculiro'seus, -a, -um With a rose-coloured eye.

ocymoi'des Resembling basil (*Ocimum*).

odessa'nus, -a, -um From Odessa, the Black Sea port.

Odontade'nia f. From Gr. *ŏdŏus*, *ŏdŏntŏs*, tooth; *adĕn*, gland. APOCYNACEAE.

Odon'tia f. Collective name for hybrids between the orchid genera *Miltonia* and *Odontoglossum*.

Odontio'da f. Collective name for hybrids between the orchid genera *Cochlioda* and *Odontoglossum*.

odontochi'lus, -a, -um Having a toothed lip.

Odontoglos'sum n. Gr. *ŏdŏus, ŏdŏntŏs*, a tooth; *glōssa*, a tongue; in allusion to the toothed lip. ORCHIDACEAE.

Odontone'ma n. Gr. *ŏdŏus*, *ŏdŏntŏs*, a tooth; *nēma*, a thread. The stamens have toothed filaments. ACANTHACEAE.

Odontoso'ria f. Gr. *ŏdŏus, ŏdŏntŏs*, a tooth; *sorus*, a spore case. DENNSTAEDTIACEAE.

odora'tus, -a, -um; odori'fer, -a, -um; odo'rus, -a, -um Fragrant. Also *odoratissimus*, very fragrant.

Oemleria f. In honour of Augustus Gottlieb Oemler, German-born pharmacist and naturalist at Savannah, Georgia, U.S.A. The name *Oemleria* (1841) has priority over *Osmaronia* (1891), both being replacement names for *Nottallia* Torrey and Gray (1838) not Rafinesque (1817). ROSACEAE.

Oenan'the f. Gr. name for a plant smelling of wine, from *oinos* (*oenos*), wine. UMBELLIFERAE.

oenensis, -is, -e Of the river Inn, Austria. *Oenipons* is Innsbruck.

Oenocar'pus m. Wine palm. From Gr. *oinos*, wine; *karpos*, fruit. PALMAE.

Oenothe'ra f. Evening primrose. The Greek name *ŏinōthēras*, which is sometimes supposed to derive from *oinos*, wine; *thera*, booty, is more likely a corruption of *ŏnŏthēras*, from *ŏnŏs*, ass; *thēra*, hunting, chase, pursuit, or *thēr*, wild beast, comparable to the alternative name *ŏnagra*. Originally, of course, it had nothing to do with the yellow-flowered American plants to which it has been transferred. ONAGRACEAE.

oertendah'lii Specific epithet for a Natal species of *Pleitranthus* named in honour of Ivan Anders Örtendahl (1870–1935), Swedish horticulturist for many years in charge of the Uppsala botanic garden.

officina'lis, -is, -e Sold in shops; applied to plants with real or supposed medicinal properties.

officina'rum Of shops, generally those of apothecaries.

olbius -a, -um; olbiensis, olbiense, -e Of the Iles d'Hyères, southern France.

oldha'mii In honour of Richard Oldham (1837–1864), Kew gardener who collected in China.

O'lea f. Olive. The Latin name (cognate with Gr. *ĕlaio*) for this most long-lived and economically important fruit tree which from antiquity has been a symbol of peace and good will. OLEACEAE

olean'der Specific epithet of a shrub belonging to the genus *Nerium q.v.*, derived from the Italian *oleandro* and referring to the olive-like leaves.

Olean'dra f. In allusion to the oleander-like fronds. DAVALLIACEAE.

Olea'ria f. Tree aster, daisy bush. Australian and New Zealand trees and shrubs named in honour of Johann Gottfried Ölschläger (1635–1711), German horticulturist, whose family name was latinized as Olearius, author of works on hyacinths and the flora of Halle. COMPOSITAE.

ole'ifer, olei'fera, -um Oil-bearing.

oleifo'lius, -a, -um With leaves like the olive.

oleoi'des Resembling olive.

olera'ceus, -a, -um Of the vegetable garden; applied to vegetables and potherbs.

oligan'thus, -a, -um With few flowers. Also:
oligocar'pus (with few fruits); **oligophyl'lus** (with few leaves);
oligosper'mus (with few seeds).

Oli'nia f. In honour of a Swedish medical man Johan Henrik Olin (1769–1824) of Växjö, who was a student of Thunberg's. OLINIACEAE.

olitor'ius, -a, -um Pertaining to culinary herbs.

oliva'ceus, -a, -um Greenish-brown; olive-coloured.

Oliveran'thus m. Named for George Watson Oliver (1857–1923), horticultural expert in United States Department of Agriculture. The genus is separated from *Echeveria*. CRASSULACEAE.

olive'ri Of Daniel Oliver (1830–1916), English botanist, from 1864 to 1890 keeper of the Kew Herbarium.

olivie'ri Of Guillaume Antoine Olivier (1756–1814), French entomologist and plant collector, traveller in the Orient from 1794 to 1798.

olusa'trum n. L. name *ŏlus atrum*, black herb, from *ŏlus*, garden herb; *ater*, black; in allusion to the black seeds of this umbellifer.

olym'picus, -a, -um Of Mount Olympus, a name used for a number of high mountains in the Balkan Peninsula and Asia Minor and elsewhere.

omeien'sis, -is, -e Of Mount Omei (Emei Shan), western Szechwan (Sichuan), China.

omor'ika Local name of the Serbian spruce, of which this is the specific epithet.

Omphalo'des f. Gr. *ŏmphalōdēs*, navel-like, from *ŏmphalŏs*, a navel. The nutlet hollowed out on one side suggests the human navel. BORAGINACEAE.

Omphalogram'ma n. From Gr. *ŏmphalŏs*, navel; *gramma*, letter, marking; in allusion to the seeds. PRIMULACEAE.

Onci'dium n. Diminutive of Gr. *ŏnkŏs*, a tumour. The lip bears swellings. ORCHIDACEAE.

On'coba f. Latinized form of the Arabic name, *onkob*. FLACOURTIACEAE.

Onobry'chis f. Sainfoin. Gr. *ŏnŏbrychis*, apparently from *ŏnŏs*, an ass, and *brycho*, eat greedily. The English name means holy clover (sometimes saintfoin) because it was reputed to have been the hay in Christ's manger. LEGUMINOSAE.

Ono'clea f. Greek name, *ŏnŏklĕia*, for another plant, possibly borage, and connected with *ŏnŏs*, an ass (meaning 'good for asses only'), but applied here because of an alternative meaning: *onos*, a vessel; *kleio*, to close; refers to the closely rolled fertile fronds of these ferns. ASPLENIACEAE.

Ono'nis f. Rest-harrow. The classical Greek name, *ŏnōnis*. LEGUMINOSAE.

Onopor'dum n. Scotch thistle. The Greek name *ŏnŏpŏrdŏn*, from *ŏnŏs*, an ass; *pŏrdē*, fart; supposedly referring to its effect on donkeys. COMPOSITAE.

Onos'ma f. Golden drop. Gr. plant name from *ŏnŏs*, an ass, *ŏsmē*, a smell, referring to an obscure plant not of the present genus. BORAGINACEAE.

Onosmo'dium n. False gromwell. Resembles *Onosma* (*q.v.*) to which it is closely related. BORAGINACEAE.

Ony'chium n. Gr. *ŏnyx*, a claw; in allusion to the shape of the lobes of the fronds. ADIANTACEAE.

ophioglossifo'lius, -a, -um With leaves like adder's tongue fern.

Ophioglos'sum n. Adder's tongue fern. Gr. *ŏphis*, a snake; *glōssa*, a tongue; referring to the slender fertile spike of *O. vulgatum*. OPHIOGLOSSACEAE.

Ophiopo'gon m. From Gr. *ŏphis*, snake; *pōgōn*, beard; allusion obscure. LILIACEAE.

Oph'rys f. The Greek name for a two-leaved orchid. ORCHIDACEAE.

Ophthalmophyl'lum n. From Gr. *ophthalmŏs*, eye; *phyllŏn*, leaf. AIZOACEAE.

Oplis'menus m. From Gr. *hŏplismos*, equipment for war, weapon. The spikelets of these grasses have awns. GRAMINEAE.

Oplo'panax m. From Gr. *hŏplŏn*, weapon; *panax*, ginseng, a related genus; in allusion to its ferociously spiny habit, which has earned it the name of 'Devil's Club'. ARALIACEAE.

Opo'panax m. Gr. *ŏpŏs*, a milky juice especially that of the fig-tree; *panax*, a remedy, hence used for ginseng. The sap was once used medicinally. Today, gum opopanax is obtained from the root for use in perfumery. UMBELLIFERAE.

opori'nus, -a, -um Autumnal; from Gr. *ŏpōra*, autumn.

oppositifo'lius, -a, -um With leaves growing opposite to each other on each side of a stem.

opuloi'des Resembling the guelder-rose, *Viburnum opulus*. Also: *opuliflorus* and *opulifolius*, with flowers and leaves, respectively, like the guelder-rose; *ŏpŭlus*, however, was a kind of maple.

Opun'tia f. Prickly pear. Greek name for a different plant which grew around the ancient town of Opus in Greece, the chief town of a tribe called from it the Locri Opuntii; the name *Opus* is supposed to mean the town of figs (see *Opopanax*). CACTACEAE.

Orbea f. From L. *orbis*, a circular shape, disc. ASCLEPIADACEAE.

orbicula'ris, -is, -e; orbicula'tus, -a, -um Round and flat; disc-shaped.

orchi'deus, -a, -um; orchioi'des; orchoi'des Orchid-like.

orchidiflo'rus, -a, -um With orchid-like flowers.

Or'chis f. Classical Greek name, from *orchis*, a testicle; in allusion to the paired rounded tubers, thus *herba orchis*. For that reason, *Orchis* has been regarded since antiquity as an aphrodisiac. ORCHIDACEAE.

orega'nus, -a, -um; orego'nus, -a, -um From the State of Oregon, U.S.A., or the old Oregon territory of the Hudson's Bay Company, which included the later States of Washington and Oregon.

Oreodo'xa f. From Gr. *ŏrŏs*, mountain; *doxa*, glory; in allusion to the stately habit of these palms. PALMAE.

Oreo'panax m. Gr. *ŏrŏs*, mountain; plus *Panax*. ARALIACEAE.

oreo'philus, -a, -um Mountain-loving.

Oreo'pteris From *ŏrŏs*, mountain; *ptĕris*, fern. THELYPTERIDACEAE.

ores'bius, -a, -um Gr. *ŏrĕsbiŏs*, living on mountains.

organen'sis, -a, -um Of the Organ Mountains of Brazil.

orgya'lis, -is, -e The distance between the tips of the fingers when the arms are extended, i.e. about six feet; from *orgya*, a classical and medieval measure corresponding to the fathom, hence tall.

orienta'lis, -is, -e From the Orient; Eastern.

origanifo'lius, -a, -um With leaves like marjoram or *Origanum*.

origanoi'des Resembling marjoram.

Ori'ganum n. Marjoram. The classical Greek name *ŏrīganŏn* for these aromatic herbs. LABIATAE.

Ori'xa f. From the Japanese name. RUBIACEAE.

Or'menis f. Presumably named for Ormenis, granddaughter of Ormenus, a king of Thessaly. COMPOSITAE.

Ormo'sia f. Gr. *hŏrmos*, a necklace. The scarlet seeds of *O. coccinea* are strung as beads in Guyana. LEGUMINOSAE.

ornatiss'imus, -a, -um Very showy.

orna'tus, -a, -um Ornamental, showy.

Ornithi'dium n. Gr. *ornis*, a bird; *eidos*, a shape. The upper lip of the stigma is beaklike. ORCHIDACEAE.

ornithoce'phalus, -a, -um Bird-headed.

Ornithochi'lus m. Gr. *ornis*, a bird; *cheilos*, a lip; in allusion to the shape of the lip. ORCHIDACEAE.

Ornitho'galum n. Gr. *Ornis*, a bird; *gala*, milk. The flowers of these bulbous plants are usually white. The bulbs of star-of-Bethlehem (*O. umbellatum*) are supposed by some to have been the 'dove's dung' of the Bible of which a 'cab' measure was sold for a shekel during the Babylonian siege of Jerusalem. LILIACEAE (HYACINTHACEAE).

ornitho'podus, -a, -um; orni'thopus Resembling a bird's foot.

Oroban'che f. Broomrape. From Gr. *ŏrŏbŏs*, a kind of vetch; *anche*, strangle. OROBANCHACEAE.

Oron'tium n. Golden club. The Greek name, *ŏrŏntiŏn*, now applied to a North American aquatic, is said to have belonged to some plant growing in the Syrian river Orontes. ARACEAE.

Oro'ya f. Name of a village in Peru, in the Andes above Lima, where the type-species grows. CACTACEAE.

Oroxy'lon n. Gr. *ŏrŏs*, a mountain; *xylon*, wood; the tree grows also on plains. BIGNONIACEAE.

Orphanide'sia f. In honour of Theodoros Georgios Orphanides (1817–1886), professor of botany at Athens, mid-19th century. Now united with *Epigaea*. ERICACEAE.

ortho- In compound words signifying upright or straight. Thus:
ortho'botrys (with upright clusters); **orthocar'pus** (with upright fruit); **orthocla'dus** (with straight branches); **orthoglos'sus** (with a straight tongue); **ortho'pterus** (straight-winged); **orthose'palus** (straight-sepalled).

Orthocar'pus m. Gr. *orthos*, upright; *karpos*, fruit; from the upright pods. SCROPHULARIACEAE.

Orthrosan'thus m. From Gr. *ŏrthrŏs*, morning; *anthos*, flower. 'From the flowers expanding in the morning, and fading before noon'. IRIDACEAE.

orvala Latinized from French vernacular name *orvale* for *Salvia sclarea*.

Orychophrag'mus m. From Gr. *ŏrychō*, dig up; *phragma*, fence, partition; in allusion to the minutely pitted septum of the pod. CRUCIFERAE.

Ory'za f. Rice. The English name *rice* is ultimately derived from the Latin and Greek *ŏr̄yza*, which in turn came from an Asiatic name. Rice (*O. sativa*) is one of the staple cereals. With the exception of the so-called 'hill rice' which is grown on dry land, all varieties need to grow in water for part of their development. GRAMINEAE.

Osbe'ckia f. In honour of Pehr Osbeck (1723–1805), Swedish clergyman, a student of Linnaeus, who travelled to India and China. MELASTOMATACEAE.

Osman'thus m. Sweet olive. Gr. *ŏsmē*, fragrance; *anthŏs*, a flower. All the species are very fragrant. OLEACEAE.

osman'thus, -a, -um With fragrant flowers.

Osmaro'nia f. Osoberry. Gr. *ŏsmē*, fragrance; *Aronia*, choke-cherry. See *Oemleria*.

Osmun'da f. Derivation unknown, often said to be named for Osmundus or

Asmund, *c.* 1025, a Scandinavian writer of runes who helped prepare the way for the Swedish acceptance of Christianity. OSMUNDACEAE.

Osteome'les f. Gr. *ŏstĕŏn*, a bone; *mēlŏn*, an apple; an allusion to the hard nutlets. ROSACEAE.

Ostrow'skia f. Giant bellflower. Named for Michael Nicholazewitsch von Ostrowsky, Russian Minister of Imperial Domains *c.* 1884 and a patron of botany. CAMPANULACEAE.

Os'trya f. Hop-hornbeam. From the Greek name, *ŏstrys*, for this tree. BETULACEAE.

Otan'thus m. From *ŏus*, *ōtŏs*, ear; *anthŏs*, flower. COMPOSITAE.

Ota'tea f. From the vernacular name otate. GRAMINEAE.

Othon'na f. Greek name for a plant other than this genus, some members of which are succulent. COMPOSITAE.

Ottelia f. Latinized from the Malabar name *ottel-ambel.* HYDROCHARITACEAE.

otto'nis In honour of Eduard Otto (1812–1885), botanical collector in Cuba and Venezuela, later curator of Hamburg botanic garden, or his father Friedrich Otto (1782–1856), curator of the Berlin botanic garden.

ouletri'chus, -a, -um With curly hair.

Ouri'sia f. Named for Ouris, governor of the Falkland Islands, where Commerson, Bougainville's naturalist, collected it on their voyage round the world, 1766–1768. SCROPHULARIACEAE.

ova'lis, -is, -e Oval, broadly elliptic.

ova'tus, -a, -um Ovate; egg-shaped, with the broad end at the base.

o'vifer, ovi'fera, -um; o'viger, ovi'gera, -um Bearing eggs, i.e. egg-shaped structures.

ovi'nus, -a, -um Relating to sheep; providing fodder for sheep.

Oxa'lis f. From Gr. *oxys*, acid, sour, sharp. *O. acetosella*, wood-sorrel, is one of the several plants with 3 leaflets called shamrock (see p. 316). OXALIDACEAE.

Oxe'ra f. Gr. *ŏxys*, acid; in allusion the sour sap. VERBENACEAE.

oxonia'nus, -a, -um; oxonien'sis, -is, -e Of Oxford, England.

oxyacan'thus, -a, -um Sharp-spined.

Oxycoc'cos m. From Gr. *ŏxys*, acid; *kokkos*, a round berry. Now included in *Vaccinium*. ERICACEAE.

Oxyden'drum n. Sour-wood. Gr. *ŏxys*, acid; *dendron*, a tree. The foliage is bitter. ERICACEAE.

oxygo'nus, -a, -um With sharp angles.

Oxylo'bium n. From Gr. *oxys*, sharp; *lobos*, pod. LEGUMINOSAE.

oxy'lobus, -a, -um With sharp-pointed lobes.

Oxype'talum n. From Gr. *ŏxys*, sharp; *petalum*, in botanical L. a petal. ASCLEPIADACEAE.

oxy'philus, -a, -um Loving acid soil.
oxyphyl'lus, -a, -um With sharp-pointed leaves.
oxy'pterus, -a, -um With sharp-pointed wings.
Oxyria f. Gr. *ŏxys*, referring to the sharp or bitter taste of this northern herb with antiscorbutic properties. POLYGONACEAE.
Oxy'tropis f. Gr. *ŏxys*, sharp; *trŏpis*, a keel; an allusion to the form of the flower. LEGUMINOSAE.

P

pabula'ris, -is, -e Providing pasture or fodder.
Pachi'ra f. From the native Guyanese name. BOMBACACEAE.
Pachis'tima f. From Gr. *pachys*, thick; *stigma*, a stigma; an irregularly formed name like many of Rafinesque's coinages. CELASTRACEAE.
pachy- In compound words signifies thick. Thus:
pachyan'thus (thick-flowered); **pachycar'pus** (with a thick pericarp); **pachyphloe'us** (with thick bark); **pachyphyl'lus** (thick-leaved); **pachy'podus** (with a thick foot, i.e. stalk); **pachy'pterus** (with thick wings); **pachy'trichus** (with thick hair).
Pachyce'reus m. Gr. *pachys*, thick; *Cereus*, cactus. These cacti have very stout stems. CACTACEAE.
Pachyphrag'ma n. From Gr. *pachys*, thick; *phragma*, fence, partition; in allusion to the stout ribbed septum of the pod. CRUCIFERAE.
Pachy'phytum f. Gr. *pachys*, thick; *phyton*, a plant. Both stems and leaves are thickened on these succulent plants. CRASSULACEAE.
Pachyrhi'zus m. Gr. *pachys*, thick; *rhiza*, a root. The roots of these twining herbs are thick and tuberous. LEGUMINOSAE.
Pachysan'dra f. Mountain spurge. Gr. *pachys*, thick; *aner*, *andros*, in the sense of stamen; with reference to the thickened white filaments. BUXACEAE.
Pachy'stachys Gr. *pachys*, thick; *stachys*, ear of corn, spike; in allusion to the dense flower clusters. ACANTHACEAE.
paci'ficus, -a, -um Of the Pacific Ocean.
padus Greek name of a wild cherry.
Paede'ria f. From L. *paedor*, filth, stench; with reference to the unpleasant smell of the leaves, etc. RUBIACEAE.
Paeo'nia f. Peony. The classical Greek name said to commemorate Paeon (Paiōn), physician of the gods and reputed discoverer of its medicinal properties, which, according, however, to one version, were revealed to him by Leto. The name is possibly derived from the word *paio* (I strike) and the incantation *iēpaiēon* uttered by the medical practitioner. We still

speak of a paeon or song of praise, originally a hymn to Apollo. PAEONIACEAE.

palaesti'nus, -a, -um Of Palestine.

Palafo'xia f. In honour of an unspecified Spaniard, either the prelate Juan de Palafox y Mendoza (1600–1659), or the general José de Palafox y Melzi (1780–1847). COMPOSITAE.

Pala'quium n. Latin version of the Philippine name. SAPOTACEAE.

Pa'laua f. In honour of Antonio Palau (d. 1793), Spanish botanist. MALVACEAE.

palhin'hae In honour of Ruy Telles Palhinha (1871–1957), Portuguese professor of botany, author of a catalogue of plants of the Azores (1966).

Paliso'ta f. Named for Baron A. M. F. J. Palisot de Beauvois (1752–1820), French botanist and traveller, author of *Flore d'Oware et Benin* (1804–1821) and works on grasses and mosses. COMMELINACEAE.

Paliu'rus m. The ancient Greek name for the terribly spiny Christ-thorn (*Paliurus spina-christi*) from which the crown of thorns is traditionally supposed to have been made. RHAMNACEAE.

pal'lens; pal'lidus, -a, -um Pale. Also *pallescens*, rather pale.

pallia'tus, -a, -um Shrouded, cloaked.

palma'ris, -is, -e A hand's breadth (about 3 in.) high or wide.

palma'tus, -a, -um Palmate, i.e. lobed like a hand with fully outspread fingers.

palmifo'lius, -a, -um Palm-leaved.

paludo'sus, -a, -um; palus'tris, -is, -e Marsh-loving.

Pamian'the f. In honour of Major Albert Pam (1875–1955) of Wormleybury, Hertfordshire, author of *Adventures and Recollections* (1945), a financier and specialist in the cultivation of bulbous plants; who received this fine member of the *Amaryllidaceae* in 1926 among bulbs collected for him in Peru. AMARYLLIDACEAE.

Pa'nax m. Ginseng. From Gr. *panakēs*, all-healing, a panacea; in allusion to the high value placed on it by the Chinese for aphrodisiacal and medicinal purposes. American ginseng (*P. quinquefolius*) is very closely related to the Asiatic ginseng (*P. ginseng* from which the English name derives). The word *-panax* has been used to form several generic names in the *Araliaceae*, e.g. *Acanthopanax*, *Echinopanax*, *Neopanax* and *Tetrapanax*. ARALIACEAE.

Pancra'tium n. Mediterranean lily, sea daffodil. Greek name for a bulbous plant. AMARYLLIDACEAE.

Pan'danus m. Screw-pine. Latinized version of the Malayan name, *pandan*. PANDANACEAE.

Pando'rea f. Named for Pandora. In Greek mythology she was the woman sent to earth by Zeus in subtle revenge for the theft of fire by Prometheus. The gods endowed her with their choicest gifts, while Zeus himself gave her a box which was to be given to the man she married, whom Zeus intended to be Prometheus. He, however, was too crafty to be caught in this way, but his brother married Pandora and opened the box, releasing all the evils Zeus intended for Prometheus. The type-species was associated with an insect plague on Norfolk Island, whence the name. *Podranea* is an anagram of *Pandorea*. BIGNONIACEAE.

pandura'tus, -a, -um Fiddle-shaped.

panicula'tus, -a, -um With flowers arranged in panicles.

Pani'cum n. Millet. The Latin name. GRAMINEAE.

panno'nicus, -a, -um Of the Roman province Pannonia, which comprised part of modern Austria, Hungary and what was northern Jugoslavia.

panno'sus, -a, -um Tattered.

panormita'nus, -a, -um Of Palermo (ancient Panŭrmŭs), Sicily.

panteumor'phus, -a, -um Altogether beautiful.

Papa'ver n. Poppy. The Latin name. PAPAVERACEAE.

papavera'ceus, -a, -um Poppy-like.

Paphi'nia f. From Gr. *paphia*, an epithet of Aphrodite (Venus) whose chief seat of worship was at Paphos on the island of Cyprus. ORCHIDACEAE.

Paphiope'dilum From Gr. *paphia* (see above); *pĕdilŏn*, sandal. see *Cypripedium*. ORCHIDACEAE.

papi'lio Latin word for butterfly. Specific epithet for the tropical butterfly orchid (*Oncidium papilio*).

papil'liger, papilli'gera, -um; papillo'sus, -a, -um Having papillae, soft protuberances on a surface.

papyra'ceus, -a, -um Papery.

papy'rifer, papyri'fera, -um Paper-bearing.

Papy'rus m. Greek name of the paper made in rolls from the pith of *Cyperus papyrus* in Ancient Egypt. CYPERACEAE.

Paradi'sea f. St. Bruno's lily. Named for Count Giovanni Paradisi (1760–1826) of Modena. LILIACEAE (ASPHODELACEAE).

paradi'si; paradisi'acus, -a, -um Belonging to Paradise.

parado'xus -a, -um Paradoxical, contrary to expectation.

Parahe'be f. Gr. *para*, near, alongside; *Hebe*, from which genus this has been cut off. SCROPHULARIACEAE.

Paraqui'lega f. Gr. *para*, near, alongside; *Aquilegia*, columbine, to which this genus is akin. RANUNCULACEAE.

parasi'ticus, -a, -um Parasitic; an epithet sometimes used by early authors for epiphytes as well as parasites.

pardalian'thes Greek name of a plant used for poisoning leopards and other beasts of prey, hence the English name Leopardsbane.

pardali'nus, -a, -um; pardi'nus, -a, -um Relating to leopards; spotted like a leopard.

pardanthi'nus, -a, -um Spotted like the flower of *Pardanthus* (Gr. *pardos*, male panther; *anthos*, flower), a synonym of *Belamcanda*.

Parieta'ria f. Pellitory. The name derives from the L. *pariētārius*, of walls, from *paries*, a wall, where the plant likes to grow, as Pliny knew when he described it. URTICACEAE.

Pa'ris f. Herb-Paris. L. *par*, equal; alluding to the regularity of the parts. LILIACEAE (TRILLIACEAE).

pari'shii In honour of the Rev. Charles Samuel Pollock Parish (1822–1897), chaplain at Moulmein, Burma, from 1852 to 1878, if a Burmese plant; of Samuel Bonsall Parish (1838–1928), American botanist, if a Californian plant.

par'keri In honour of Richard Neville Parker, (1884–1958) conservator of forests and forest botanist in India, author of *Common Indian Trees* (1933).

Parkinso'nia f. Named for John Parkinson (1567–1650), apothecary of London and author of two important books, *Paradisi in Sole Paradisus Terrestris* (1629) and *Theatrum Botanicum* (1640). Having perpetrated a gross pun in the title of the first (it can be translated as 'Park in sun's earthly paradise'), he proceeds to extol the joys of a garden and to describe in vivid detail the plants available. LEGUMINOSAE.

Parmentie'ra f. Panama candle-tree. Named for Antoine Augustin Parmentier (1737–1813), French agricultural economist who promoted cultivation of the potato in France. Potage Parmentier (potato soup) is named after him. BIGNONIACEAE.

parmula'tus, -a, -um Armed with a little round shield (*parmula*).

par'nassi Of Mount Parnassus, Greece.

Parnas'sia f. Grass-of-Parnassus. Shortened version of the 16th-century name *Gramen Parnassi*, referring to Mount Parnassus in Greece. SAXIFRAGACEAE.

parnas'sicus, -a, -um Of Mount Parnassus.

parnassifo'lius, -a, -um With leaves like *Parnassia*.

Paro'chetus m. Blue oxalis. Gr. *para*, near; *ŏchĕtŏs*, a brook; from the normal habitat in moist places of these trailing perennials, which occur on high mountains in both Asia and Africa. LEGUMINOSAE.

Paro'dia f. In honour of Lorenzo Raimundo Parodi (1895–1966), Argentinian botanist especially noted for his studies of South American grasses. CACTACEAE.

Parony'chia f. Whitlow-wort. Gr. *parōnychia*, a whitlow (from *para*, near; *onyx*, nail), for which these herbs were thought to be a cure. CARYOPHYLLACEAE.

Parro'tia f. Named for F. W. Parrot (1792–1841), German naturalist and traveller who climbed Mount Ararat in 1834. HAMAMELIDACEAE.

Parrotiopsis f. From *Parrotia* and *opsis*, resemblance. HAMAMELIDACEAE.

par'ryi In honour of Charles Christopher Parry (1823–1890), American botanical explorer born in England.

Parthe'nium n. American feverfew. Greek name, *parthĕniŏn*. COMPOSITAE.

parthe'nium Specific epithet for *Chrysanthemum* (see *Parthenium*).

Parthenocis'sus n. Boston ivy, Virginia creeper, woodbine, etc. From Gr. *parthenos*, a virgin; *kissos*, ivy; in allusion to its English vernacular name. The state of Virginia was named after England's virgin queen Elizabeth I. *P. quinquefolia*, Virginia creeper, is often listed under *Ampelopsis*. VITACEAE.

parti'tus, -a, -um Parted.

parvibractea'tus, -a, -um With small bracts.

par'vus, -a, -um Small. In compound words *parvi-*, as in *parviflorus*, with small flowers; *parvifolius*, with small leaves; also *parvissimus*, *parvulus*, very small.

Pas'palum n. Millet. The Greek name. GRAMINEAE.

Passiflo'ra f. Passion-flower. L. *passio*, passion; *flos*, a flower. The name was given by the early missionaries in South America who thought they saw in the parts of the flower various signs of Christ's crucifixion. The corona became the crown of thorns, the five anthers represented five wounds, the three styles three nails, and so on. PASSIFLORACEAE.

Pastina'ca f. Parsnip. From L. *pastus*, food; in allusion to the edible root. UMBELLIFERAE.

pastora'lis, -is, -e Relating to shepherds, hence growing in pastures.

patago'nicus, -a, -um Of Patagonia, region in Chile and Argentina, South America.

patavi'nus, -a, -um Of the neighbourhood of Padua (earlier *Patavium*), Italy.

patell'aris, -is, -e; patelliformis, -is, -e Dish-shaped, saucer-shaped.

pa'tens; pat'ulus, -a, -um Spreading.

Patri'nia f. In honour of Eugène L. M. Patrin (1724–1815), French mineralogist who travelled from 1780 to 1787 in Siberia and elsewhere in Asia and collected plants as well as minerals. VALERIANACEAE.

pau'cus, -a, -um Few. In compound words *pauci-*. Thus:
pauciflo'rus (with few flowers); **paucifo'lius** (with sparse leaves);
pauciner'vus (with few nerves).

Paulli'nia f. Named by Linnaeus for Simon Paulli (1603–1680), professor of anatomy, surgery and botany at Copenhagen, as stated by Linnaeus himself in his *Critica botanica* no. 238 (1737), and not Christian Franz Paullini (1643–1742), a German botanist at Eisenach, as stated by many authors. SAPINDACEAE.

Paulow'nia f. Named for Princess Anna Paulowna (1795–1865), daughter of Czar Paul I of Russia. SCROPHULARIACEAE.

pauper'culus, -a, -um Poor.

Pavet'ta f. Sinhalese vernacular name. RUBIACEAE.

Pavo'nia f. Tropical herbs and shrubs named for José Antonio Pavón (1754–1840), Spanish botanist and traveller in South America, part-author of the incomplete *Flora Peruviana*, of whose unfortunate history an account will be found in A. R. Steele's *Flowers for the King* (1964). MALVACEAE.

pavoni'nus, -a, -um Peacock-blue or with a distinct eye recalling that on a peacock's tail feather.

paxto'nii In honour of Sir Joseph Paxton (1801–1865), gardener and architect. Starting as a poor apprentice in the Chiswick garden of the Horticultural Society of London, he became head gardener to the Duke of Devonshire at Chatsworth when aged only 25; he was manager of the Duke's princely estates and his general adviser and close friend. Having designed and built the huge Chatsworth conservatory of nearly 300 feet in length, he used this experience in glasshouse construction to help him to design the great Crystal Palace, the basic plan of which was prepared in nine days, for the Great Exhibition of 1851. Later, he designed many other important buildings, became a company director and a Member of Parliament. In gardening, he dispatched collectors to many parts of the world and introduced a number of new plants. The plant of the great water lily, *Victoria amazonica* (*V. regia*), which he brought into bloom at Chatsworth in 1849, was obtained from the Royal Botanic Gardens, Kew.

pectina'tus, -a, -um Comblike.

pecti'nifer, pectini'fera, -um Having a comb.

pectora'lis, -is, -e Relating to the chest, e.g. used in cough mixtures.

peda'lis, -is, -e About one foot (30 cm) long.

pedati'fidus, -a, -um Cut like a bird's foot, i.e. with a few divisions radiating from the same centre.

peda'tus, -a, -um Like a bird's foot, see above.

pedemonta'nus, -a, -um Of Piedmont (Piemonte), Italy.

Pedicula'ris f. Louse-wort, wood-betony. From L. *pedicularis*, relating to lice, from *pediculus*, a louse. The presence of the plant in fields was supposed to produce lice in sheep. SCROPHULARIACEAE.

Pedilan'thus m. Slipper-flower. From Gr. *pĕdilon*, sandal; *anthos*, a flower; from the shape of the flowers. EUPHORBIACEAE.

Pediocac'tus m. Gr. *pĕdion*, a plain; plus *Cactus*; from its habitat on the Great Plains, U.S.A. CACTACEAE.

peduncula'ris, -is, -e; peduncula'tus, -a, -um With a flower stalk or peduncle, usually a well-developed one.

pedunculo'sus, -a, -um With many or especially well-developed flower stalks or peduncles.

Pelargo'nium n. The geranium of the florists. From Gr. *pelargos*, a stork. The fruit has a beak not unlike that of a stork. Sometimes called stork's-bill; see *Erodium*, *Geranium*. GERANIACEAE.

Pelecy'phora f. From Gr. *pĕlĕkys*, a hatchet; *phoreo*, to bear; with reference to the strongly flattened tubercles. CACTACEAE.

pelisseria'nus, -a, -um In honour of Guillaume Pellicier (1490–1568), bishop of Montpellier, the moral force behind the university's academic tolerance when Clusius, Platter, Lobel, J. Bauhin and Rabelais were students there. The tropical genus *Pelliciera* commemorates him.

Pella'ea f. Cliff-brake. From Gr. *pellaios*, dark; in allusion to the stalks of this fern which are generally dark. ADIANTACEAE.

Pellio'nia f. Named for Alphonse Odet Pellion (1796–1868), later a French admiral, who when young accompanied the French navigator Freycinet on his second voyage around the world in 1817 to 1820. URTICACEAE.

pellu'cidus, -a, -um Transparent; can be seen through.

Peltan'dra f. Arrow-arum. Gr. *pĕltē*, a shield; *aner*, *andros*, man, hence a stamen. ARACEAE.

Pelta'ria f. Shield-wort. Gr. *pĕltē*, a shield; from the form of the leaf. CRUCIFERAE.

pelta'tus, -a, -um Shield-shaped, i.e. attached on the lower surface away from and not at the margin.

peltifo'lius, -a, -um With peltate leaves.

Peltiphyl'lum n. Umbrella plant. Gr. *pĕltē*, a shield; *phyllon*, a leaf; from the form of the leaves. The name *Peltophyllum* having been used earlier for a genus of *Triuridaceae*, this has been renamed *Darmera* (which see).

Pelto'phorum n. Gr. *pĕltē*, a shield; *phoreo*, to bear; from the shape of the stigma. LEGUMINOSAE.

pelvifor'mis, -is, -e Forming a shallow cup.

pemakoen'sis, -is, -e Of the province of Pemako, Tibet.

penduliflo'rus, -a, -um With pendulous flowers.

penduli'nus, -a, -um; pen'dulus, -a, -um Hanging, pendulous.

penicilla'tus, -a, -um Having a tuft of hair like a paint-brush.

peninsula'ris, -is, -e Growing on a peninsula.

Penioce'reus m. Gr. *pēnē*, a thread; *Cērĕus*, cactus. The stems are very slender. CACTACEAE.

penna'tus, -a, -um Feathered, pinnate.

pen'niger, penni'gera, -um Bearing feathers, as featherlike leaves.

Pennise'tum n. From L. *penna*, a feather; *seta*, a bristle. The flower has long, feathery bristles. GRAMINEAE.

pennive'nius, -a, -um Pinnately veined.

pen'silis, -is, -e Hanging down.

Penste'mon (also, incorrectly, **Pentstemon**) m. Beard-tongue. Gr. *pĕnte*, five; *stēmōn*, a stamen. The flowers have five stamens, one of them sterile. SCROPHULARIACEAE.

pensylva'nicus, -a, -um Of Pennsylvania, U.S.A.

penta- In compound words signifying five. Thus:
pentade'nius (with five glands); **pentago'nus** (five-angled); **penta'gynus** (with five pistils); **pentan'drus** (with five stamens); **pentaphyl'lus** (with five leaves); **penta'pterus** (five-winged).

Pentaglot'tis From Gr. *pĕntĕ*, five; *glōtta*, tongue; in allusion to the five tongue-like scales of the corolla. BORAGINACEAE.

Pentaptery'gium n. Gr. *pĕntĕ*, five; *pterygion*, a small wing; the calyx has five small wings. ERICACEAE.

Pen'tas f. Gr. *pentas*, a series of five; with reference to the flower parts being in fives instead of fours as in related genera. RUBIACEAE.

Pepero'mia f. Gr. *pĕpĕri*, pepper; *hŏmŏiŏs*, resembling. The plants resemble and are closely related to true pepper; see *Piper*. PIPERACEAE.

peramoe'nus, -a, -um Very pleasing.

perbel'lus, -a, -um Very beautiful.

percus'sus, -a, -um Perforated, or apparently so.

peregri'nus, -a, -um Exotic; immigrant.

peren'nans; peren'nis, -is, -e Perennial.

Peres'kia f. Barbados gooseberry. Named for Nicholas Claude Fabre de Peiresc (1580–1637), French naturalist and archaeologist. CACTACEAE.

Pereskiop'sis f. *Pereskia* (*q.v.*) plus Gr. *opsis*, like. The two genera resemble each other. CACTACEAE.

Pere'zia f. Named for Lorenzo Perez, 16th-century Spanish apothecary, author of a history of drugs, *Libro de Theriaca* (1573). COMPOSITAE.

perfolia'tus, -a, -um; perfos'sus, -a, -um With the leaf surrounding or embracing the stem; perfoliate.

perfora'tus, -a, -um Having or appearing to have small holes.

pergra'cilis, -is, -e Very slender.

pericly'menum n. Greek name, *pĕriklymĕnon*, for a honeysuckle.

Perilep'ta f. Gr. *pĕrileptos*, embraced, with reference to the stem-clasping leaf bases. ACANTHACEAE.

Peril'la f. Beefsteak plant. Derivation obscure; probably a diminutive of L. *pēra*, bag, wallet, in allusion to the fruiting calyx. LABIATAE.

Peri'ploca f. Silk vine. Gr. *pĕriplŏkē*, a twining around, from *pĕri*, around; *plŏkē*, something twisted or woven. ASCLEPIADACEAE.

Periste'ria f. Dove plant. Gr. *pĕristĕra*, a dove. The column has a dovelike appearance. ORCHIDACEAE.

Peri'strophe f. Gr. *pĕristrŏphē*, a turning around, from *pĕri*, around; *strŏphē*, turning, twist; in allusion to the twisted corolla tube. ACANTHACEAE.

permix'tus, -a, -um Much-mixed.

Pernet'tya (originally **Pernettia**) f. Ornamental shrubs of the heath family named for Antoine Joseph Pernetty (1716–1801), who visited the Falkland

Islands and South America with Bougainville, 1763–1764, and wrote about it. ERICACEAE.

per'nyi In honour of Paul Hubert Perny (1818–1907), French Roman Catholic missionary in China.

Perov'skia f. Named for V. A. Perovski (1794–*c.* 1857), Russian general. LABIATAE.

perralderia'nus, -a, -um In honour of Henri René le Tourneux de la Perraudière (1831–1861), French naturalist who collected in the Canary Islands and North Africa.

Per'sea f. Avocado or alligator pear. From the Greek name, *pĕrsĕa*, for an Egyptian tree (*Cordia myxa*). LAURACEAE.

Persica'ria f. Medieval name of a knotweed, from *Persica*, peach; in allusion to the shape of the leaves. POLYGONACEAE.

persicifo'lius, -a, -um With leaves like a peach.

per'sicus, -a, -um Persian. The peach, the *pĕrsike* or *persica malus* (Persian apple) of the Ancients, reached Europe from China by way of Iran (Persia).

persim'ilis, -is, -e Very similar to another species.

persis'tens Persistent.

Persoo'nia f. Named for Christian Hendrick Persoon (1761–1836), South African botanist, especially celebrated as a mycologist, who worked and died in Paris. PROTEACEAE.

perspi'cuus, -a, -um Transparent.

pertu'sus, -a, -um Perforated; thrust through.

perula'tus, -a, -um Provided with perules (bud scales); mostly used when these are conspicuous or persistent.

peru'tilis, -is, -e Very useful.

peruvia'nus, -a, -um Peruvian.

pes m. L. foot, hence *pes-caprae*, foot of goat, for leaflet of an *Oxalis*, leaf of an *Ipomoea*.

Pescato'ria f. In honour of Monsieur Pescatore, a French orchid-specialist, *c.* 1852. ORCHIDACEAE.

petaloi'deus, -a, -um Petal-like.

Petaloste'mum n. Prairie-clover. Gr. *pĕtalŏn*, a petal; *stēmōn*, a stamen. Petals and stamens are joined. LEGUMINOSAE.

Petasi'tes m. Butter-bur. The Greek name, from *pĕtasos*, a hat with a broad brim; with reference to the large leaves. COMPOSITAE.

petectica'lis, -is, -e Spotted, from Italian *petecchia*, spot produced by a fever.

petiolar'is, -is, -e; petiola'tus, -a, -um Furnished with a leaf-stalk or a particularly long one.

Petive'ria f. Named for James Petiver (1665–1718), apothecary of London, an ardent collector of botanical specimens now in the British Museum (Natural History). PHYTOLACCACEAE.

petr- In Gr. compounds can refer either to rock forming cliffs, ledges (*pĕtra*) or pieces of rock (*pĕtrŏs*).

petrae'us, -a, -um Rock-loving.

Pet'rea f. Named for Lord Robert James Petre (1713–1743), English patron of botany and horticulture, called 'the Phenix of this age' by Collinson; both he and Collinson helped to introduce many plants into European gardens. VERBENACEAE.

pet'riei In honour of Donald Petrie (1846–1925), Scottish botanist who became New Zealand inspector of schools, a tireless amateur investigator of the flora of New Zealand.

Petrocal'lis f. From Gr. *pĕtrŏs*, rock; *kallis*, beauty; from the habitat and the beauty of the flowers. CRUCIFERAE.

Petrocop'tis f. From Gr. *pĕtrŏs*, a rock; *kopto*, to break. These plants grow in cracks and broken crevices of rock. CARYOPHYLLACEAE.

Petroma'rula f. Cretan vernacular name meaning 'rock lettuce'. CAMPANULACEAE.

Petro'phila f. Gr. *pĕtrŏs*, rock; *phileo*, to love; from the habitat. PROTEACEAE.

Petrophy'ton n. Gr. *pĕtrŏs*, rock; *phyton*, a plant; from the rocky habitat. ROSACEAE.

Petrorha'gia f. From Gr. *pĕtrŏs*, a rock; *rhagas*, rent, chink, from *rhēgnymi*, break asunder. Equivalent to Latin *saxifraga*. CARYOPHYLLACEAE.

Petroseli'num n. Parsley. Gr. *pĕtrŏs*, a rock; *sĕlinŏn*, parsley or celery; i.e. wild parsley. UMBELLIFERAE.

Pette'ria f. Named for Franz Petter (1798–1858), Austrian high school teacher of German at Spalato, Dalmatia, who wrote on the plants of Dalmatia. LEGUMINOSAE.

Petu'nia. f. Latinized form of the Brazilian name *petun* for tobacco, to which these annuals are allied. SOLANACEAE.

Peuce'danum n. Gr. *pĕukĕdanŏn*, parsnip. See *Pastinaca*. UMBELLIFERAE.

Peu'mus f. Chilean boldo tree. The Chilean name Latinized. MONIMIACEAE.

pexa'tus, -a, -um, Clothed in a garment with a nap on it, referring to hair-covering.

Pfei'ffera f. In honour of Louis Carl Georg Pfeiffer (1805–1877), many-sided, extremely industrious German medical man, translator, botanist and conchologist, botanically renowned for his work on cacti and his monumental *Nomenclator botanicus* (1873–1874). CACTACEAE.

Phace'lia f. California-bluebell. Gr. *phakĕlos*, a bundle; from the bunched flowers of the original species. HYDROPHYLLACEAE.

Phaedranas'sa f. Gr. *phaidrŏs*, gay; *anassa*, queen. AMARYLLIDACEAE.

Phaedran'thus m. Gr. *phaidrŏs*, gay; *anthos*, a flower; with reference to the showy flowers of this Mexican climber. BIGNONIACEAE.

phaeocar'pus, -a, -um With dark fruit.

pha'eus, -a, -um Dusky.

Phaio'phleps f. From Gr. *phaios*, dark; *phleps*, vein; in allusion to the veining of the flowers. IRIDACEAE.

Phai'us m. Gr. *phaios*, dusky. The flowers of the type-species are dark-coloured. ORCHIDACEAE.

Phalaeno'psis f. Moth orchid. Gr. *phalaina*, a moth; *ŏpsis*, like. ORCHIDACEAE.

Phala'ris f. Greek name for a grass of this genus: *P. arundinacea picta* is a ribbon-grass, *P. canarienis* canary grass, sometimes grown for bird feed. GRAMINEAE.

Phase'olus m. Latin diminutive *phăsĕŏlus* or *făsĕŏlus* or *phăsĕlus*, Gr. *phasēlŏs*, the name for a kind of bean (*Vigna*). Cultivated all over the world, this genus includes the French or string bean, limas, and scarlet runners, all of which being of American origin were unknown in antiquity. LEGUMINOSAE.

Pheba'lium n. Said to be from a Greek name for myrtle used by ancient Greek comic poets. RUBIACEAE.

Phego'pteris f. From *phēgŏs*, oak; *ptĕris*, fern. THELYPTERIDACEAE.

Phelloden'dron n. Cork-tree. Gr. *phĕllos*, cork; *dĕndrŏn*, a tree; from the corky bark. Commercial cork comes from the cork-oak (*Quercus suber*). RUBIACEAE.

Phellosper'ma n. Gr. *phĕllos*, cork; *sperma*, a seed; because of the corky base of the seed. CACTACEAE.

philadel'phicus, -a, -um Of Philadelphia, U.S.A.

Philadel'phus m. Mock-orange. Gr. *philadĕlphŏs*, loving one's brother or sister; a Grecian and Roman family name. In some parts of England the flowers were often used to take the place of orange blossoms in the country bride's wedding bouquet. It is sometimes (wrongly) called syringa as a vernacular name. HYDRANGEACEAE.

Phile'sia f. Gr. *phileo*, to love. SMILACACEAE (PHILESIACEAE).

Philly'rea f. The classical Greek name, apparently derived from a non-Greek name. OLEACEAE.

Philoden'dron n. Gr. *phileo*, to love; *dendron*, a tree; from the tree-climbing habits. ARACEAE.

phleban'thus, -a, -um With veined flowers; from *phleps*, vein; *anthos*, flower.

Phlebo'dium n. From Gr. *phlĕps*, *phlĕbŏs*, vein; with reference to the numerous veins and veinlets of the fronds. POLYPODIACEAE.

phleioi'des *Phleum*-like or resembling a reed.

Phle'um n. Timothy. Gr. *phlēŏs*, a kind of grass. GRAMINEAE.

phlogiflo'rus, -a, -um With flame-coloured flowers; with flowers like *Phlox*.

phlogifo'lius, -a, -um With leaves like *Phlox*.

Phlo'mis f. Jerusalem sage. Greek name, *phlŏmis*, for some plant possibly not of this genus. LABIATAE.

Phlox f. Gr. *phlox*, a flame. Also the Greek name for some plant with flame-coloured flowers. POLEMONIACEAE.

phoeni'ceus, -a, -um Purple-red; from Phoenicia, where the cities of Tyre and Sidon produced the celebrated Tyrian purple dye in antiquity, from a secretion of a sea-snail (*Murex*); see *Punica*.

phoenicola'sius, -a, -um With purple hairs.

Phoe'nix f. Date palm. The Greek name. PALMAE.

Pholido'ta f. Rattlesnake orchid. Gr. *phŏlidōtos*, scaly, clad in scales; in allusion probably to the scaly bracts. ORCHIDACEAE.

Phor'mium n. New Zealand flax. Gr. *phŏrmiŏn*, mat. The plant provides a very strong fibre which can be woven and was extensively used by the Maoris for their traditional garments and basketry. Sir Joseph Banks brought seeds of it back from Captain Cook's first voyage. It is an important cordage plant in New Zealand. AGAVACEAE (PHORMIACEAE).

Photi'nia f. Apparently from Gr. *phōs*, *phōtŏs*, light, in allusion to the shining leaves. ROSACEAE.

Phragmipe'dium n. From Gr. *phragma*, a fence or partition; name formed by analogy with *Cypripedium*. ORCHIDACEAE.

Phragmi'tes m. Reed, Gr. *phragmītēs*, of fences, from *phragma*, a fence or screen; abbreviated from Gr. *kalamos phragmitēs*, reed of hedges. GRAMINEAE.

phry'gius, -a, -um From Phrygia, western Asia Minor.

phu Gr. *phŏu*, a kind of valerian.

Phuo'psis f. From a resemblance (*ŏpsis*) to *Valeriana phu*. RUBIACEAE.

Phyge'lius m. Probably from Gr. *phygē*, flight, avoidance; 'in consequence of its having so long escaped the researches of botanists', according to W.J. Hooker (1855). SCROPHULARIACEAE.

Phyla f. Gr. *phylē*, tribe, union; probably from the flowers being tightly clustered in heads. VERBENACEAE.

Phy'lica f. Gr. *phyllikos*, leafy; in allusion to the abundant foliage. RHAMNACEAE.

Phylla'gathis f. From Gr. *phyllon*, a leaf; *agathis*, ball of thread; in allusion to the large leafy bracts below the compact flower-heads. MELASTOMATACEAE.

Phyllan'thus m. Gr. *phyllon*, a leaf; *anthos*, a flower. In some species the flowers are produced on the edges of the leaf-like branches. EUPHORBIACEAE.

Phylli'tis f. Hart's-tongue fern. Gr. *phyllitis*, from *phyllon*, leaf. ASPLENIACEAE.

Phyllocac'tus m. Gr. *phyllon*, a leaf; *Cactus*; in allusion to the flattened leaf-like stems. CACTACEAE.

Phyllo'cladus m. Gr. *phyllon*, a leaf; *klados*, a branch. The branches are flattened like leaves. PHYLLOCLADACEAE.

Phyllo'doce f. Heathlike shrubs named for a sea-nymph, Phyllŏdŏcē (mentioned by Virgil), in accordance with the custom of naming genera of *Ericaceae* after nymphs and goddesses initiated by Linnaeus when he named *Andromeda*. ERICACEAE.

phyllomani'acus, -a, -um Producing leafy growth in excessive abundance.

Phyllo'stachys f. Gr. *phyllon*, a leaf; *stachys*, ear of corn, a spike; referring to the leafy inflorescence of these bamboos. GRAMINEAE.

-phyllus In Gr. compounds means -leaved, e.g. *macrophyllus*, large-leaved; *microphyllus*, small-leaved; *polyphyllus*, many-leaved.

Phy'salis f. Cape gooseberry, ground-cherry. From Gr. *physa*, a bladder; from the inflated calyx. SOLANACEAE.

Physocar'pus m. Nine bark. From Gr. *physa*, a bladder; *karpos*, fruit; from the inflated follicles. ROSACEAE.

Physochlai'na f. Gr. *physa*, bladder; *chlaina*, cloak, outer garment; with reference to the inflated calyx covering the capsule. SOLANACEAE.

Physoplexis f. From Gr. *physa*, bladder; *plĕxis*, plaiting, weaving, presumably with reference to the divisions of the basally swollen corolla being joined together at the top. CAMPANULACEAE.

Physoste'gia f. Obedient plant, false dragon-head. Gr. *physa*, a bladder; *stĕgē*, roof covering. The fruits of these herbaceous perennials are covered by an inflated calyx. LABIATAE.

Phyte'lephas m. Ivory-nut palm. From Gr. *phyton*, a plant; *ĕlĕphas*, elephant; because the nut of this palm is vegetable ivory which in some ways resembles true ivory. It can be carved into various articles as large as billiard balls. PALMAE.

Phyteu'ma n. Rampion. The Greek name for *Reseda phyteuma*. CAMPANULACEAE.

Phytolac'ca f. Pokeweed, pokeberry, inkberry. From Gr. *phyton*, a plant; modern Latin *lacca*, from Hindi *lakh*, referring to the dye extracted from the lac insect. The allusion is to the staining qualities of the fruit which has been used to redden wine. PHYTOLACCACEAE.

Pi'cea f. Spruce. L. *pĭcĕa*, pitch-pine, from *pix*, pitch. PINACEAE.

pictura'tus, -a, -um Variegated.

pic'tus, -a, -um Painted; brightly coloured.

Pi'eris f. Named for one of the Pierides, the generic name of the Muses. ERICUCEAE.

Pi'lea f. L. *pīlĕus*, a cap; from the shape of the female flowers. URTICACEAE.

pilea'tus, -a, -um Furnished with a cap.

pi'lifer, pili'fera, -um With soft short hairs.

Pilosoce'reus m. L. *pilosus*, shaggy; *Cereus*, *q.v.* CACTACEAE.

pilo'sus, -a, -um Covered with long soft hairs.

pilula'ris, -is, -e; pilu'lifer, piluli'fera, -um With little balls, referring to the globular fruits, notably of nettle.

Pimele'a f. Rice-flower. Gr. *pimělē*, fat; from its richness in oil. THYMELEACEAE.

pimeleoi'des Like *Pimelea.*

Pimen'ta f. Allspice. From Spanish *pimento.* Not to be confused with pimiento (*Capsicum*). MYRTACEAE.

Pimpinel'la f. Anise. Derivation of the name of these herbs is uncertain. The fruit of one species yields oil of anise with a strong, aromatic flavour. UMBELLIFERAE.

pimpinellifo'lius, -a, -um With leaves like *Pimpinella.*

Pinan'ga f. From the Malayan name *pinina.* PALMAE.

pinas'ter Latin for a wild pine, indicating by use of the ending *-aster* one inferior to the cultivated pine (*Pinus pinea*).

Pinel'lia f. In honour of Giovanni Vincenzo Pinelli (1535–1601), who had a botanic garden in Naples. ARACEAE.

pineto'rum Of pine forests.

pi'neus, -a, -um Relating to the pine.

Pingui'cula f. Butterwort. L. *pinguis*, fat; in allusion to the greasy appearance of the leaves. LENTIBULARIACEAE.

pinguifo'lius, -a, -um With fat leaves.

pinifo'lius, -a, -um With leaves like pine.

pinna'tus, -a, -um Featherlike; having leaflets arranged on each side of a common stalk. Also:
pinnati'fidus (pinnately cut); **pinnatifo'lius** (with pinnate leaves); **pinnati'frons** (with pinnate fronds); **pinnatiner'vis** (with pinnately arranged nerves).

Pi'nus m. Pine. The Latin name, particularly for *Pinus pinea. P. strobus*, the magnificent white pine of North America, now often called Weymouth pine in England, became known early in the 18th century as 'Lord Weymouth's pine' because of the large and successful plantings on his Longleat estate. In New England the arbitrary marking of all pines over 24 inches diameter as Crown property for mast timber was one of the causes of discontent which led to the American Revolution. As a consequence it is seldom possible to find pine boards more than 23 inches wide in old houses in New England. PINACEAE.

Pi'per n. Pepper. The Latin name from Gr. *pěpěri*, itself derived from an Indian name. Black pepper is made from the whole fruit of *Piper nigrum.* PIPERACEAE.

piperi'tus, -a, -um Pepperlike, tasting like pepper, hot and sharp.

Piptade'nia f. Gr. *pipto*, to fall; *adēn*, a gland. LEGUMINOSAE.

Piptan'thus m. From Gr. *pipto*, to fall; *anthos*, a flower. The calyx, corolla and stamens fall off together, leaving the young pod without a calyx at base. LEGUMINOSAE.

Pique'ria f. Named for Andrés Piquer, 18th-century Spanish physician. COMPOSITAE.

Pisci'dia f. From L. *piscis*, fish; *caedo*, cut down, slaughter, kill; in allusion to use as a fish-poison; formed by analogy with *hŏmĭcīda*, man-killer, murderer. LEGUMINOSAE.

pi'sifer, pisi'fera, -um Pea-bearing.

pisocar'pus, -a, -um With pea-like fruit.

Pista'cia f. Pistachio. From *pistakē*, the Greek word for the nut, derived from a Persian name. ANACARDIACEAE.

Pis'tia f. Water-lettuce. From Gr. *pistos*, water; in allusion to the floating habit. ARACEAE.

Pi'sum n. Pea. The Latin name. Peas, known since prehistoric times, have been found in Swiss lakeside dwellings of the Bronze Age. LEGUMINOSAE.

Pitcair'nia f. Named for William Pitcairn (1711–1791), London physician, who had a botanic garden at Islington. BROMELIACEAE.

Pithecello'bium n. From Gr. *pithēkŏs*, ape, monkey; *ĕllŏbiŏn*, ear-ring. The name of this genus was originally published as *Pithecollobium* then amended to *Pithecellobium*. The Brazilian name, Brincos de Sahoy, means monkey's ear-ring. LEGUMINOSAE.

Pithecocte'nium n. From Gr. *pithēkŏs*, ape, monkey; *ctĕniŏn*, comb; in allusion to the spiny fruits. BIGNONIACEAE.

Pittos'porum n. Parchment bark, Australian laurel. From Gr. *pitta*, pitch; *spŏra*, seed. The seed has a sticky resinous coating. PITTOSPORACEAE.

Pityrogram'ma f. Gold fern, silver fern. From Gr. *pityrŏn*, chaff, bran, scurf; *gramma*, writing; from the powder on the fronds. ADIANTACEAE.

placa'tus, -a, -um Quiet, calm, gentle.

Plagian'thus m. Ribbon wood. Gr. *plagiŏs*, oblique; *anthos*, a flower, the flowers have asymmetrical petals. MALVACEAE.

Plagioboth'rys f. From Gr. *plagiŏs*, oblique, placed sideways; *bŏthrŏs*, pit. In allusion to the scar on the nutlet. BORAGINACEAE.

Pla'nera f. Water-elm. Named for Johann Jakob Planer (1743–1789), German botanist, professor in Erfurt. ULMACEAE.

plane'tus, -a, -um Wandering.

plantagin'eus, -a, -um Resembling a plantain, usually like leaves of *Plantago major*.

Planta'go f. Plantain. The Latin name. PLANTAGINACEAE.

pla'nus, -a, -um Flat. Also:

planiflo'rus (with flat flowers); **planifo'lius** (with flat leaves); **pla'nipes** (with a flat stalk).

platanifo'lius, -a, -um With leaves like the plane tree (*Platanus*).

platanoi'des Resembling a plane tree (*Platanus*).

Pla'tanus f. Plane tree. The Greek name, *platanŏs*, for the long-lived oriental plane (*P. orientalis*). The London plane (*Platanus hybrida*, syn. *P. acerifolia*) is the most widely planted city tree in London, Paris, New York, and other northern cities. It originated as a cross (*P. occidentalis* × *P. orientalis*). Handel's well-known *Largo* is an ode to a plane tree. In the wild *P. orientalis* grows only along water courses. PLATANACEAE.

platy- In Greek compound words usually signifies broad, rarely flat. Thus: **platyacan'thus** (with broad spines); **platyan'thus** (with broad flowers); **platycar'pus** (broad-fruited); **platycau'lis** (with a broad (flat) stem); **platycen'trus** (with a broad (flat) spur); **platy'cladus** (flat-branched); **platyglos'sus** (broad-tongued); **platype'talus** (with broad petals); **platyphyl'lus** (broad-leaved); **platy'podus** (broad-stalked); **platyspa'thus** (with a broad spathe); **platysper'mus** (broad-seeded).

Platyca'rya f. From Gr. *platys*, broad; *karyŏn*, a nut; from the winged nutlets. JUGLANDACEAE.

Platyce'rium n. Staghorn fern. Gr. *platys*, broad; *kĕras*, a horn; in allusion to the branched fertile fronds. POLYPODIACEAE.

Platyco'don m. Balloon flower, Japanese bellflower. Gr. *platys*, broad; *kōdōn*, a bell; from the form of the corolla. CAMPANULACEAE.

Platycra'ter f. From Gr. *platys*, broad; *kratēr*, bowl; in allusion to the expanded saucer-like calyces of the sterile flowers. HYDRANGEACEAE.

Platyste'mon m. Gr. *platys*, broad; *stemon*, a stamen. This has broad stamens. PAPAVERACEAE.

plebe'ius, -a, -um Common.

Plectran'thus m. From Gr. *plēctrŏn*, spur; *anthos*, flower. LABIATAE.

Pleiobla'stus m. From Gr. *pleios*, more; *blastos*, bud; with reference to the several buds, later shoots at each node of these bamboos. GRAMINEAE.

Plei'one f. Indian crocus. Asiatic. Named for *Plēiŏnē*, the mother of the seven Pleiades. ORCHIDACEAE.

pleioner'vis, -is, -e With many nerves.

Pleiospi'los m. Gr. *pleios*, many; *spilos*, a spot, fleck. The leaves are conspicuously spotted. AIZOACEAE.

pleniflo'rus, -a, -um With double flowers.

plenis'simus, -a, -um Entirely double; superlative of *plenus*.

ple'nus, -a, -um Double or full, e.g. with more than the usual number of petals.

pleuro'stachys With spikes at the side.

Pleurothal'lis f. Gr. *pleuron*, side, rib; *thallŏs*, branch. ORCHIDACEAE.

plica'tus, -a, -um Pleated.

plumar'ius, -a, -um; pluma'tus, -a, -um Plumed; feathered.

plumbaginioi'des Like *Plumbago*.

Plumba'go f. Leadwort. The Latin name derived from *plumbum*, lead; *-ago*, a termination of many Latin plant-names used to indicate a resemblance or a property. PLUMBAGINACEAE.

plum'beus, -a, -um Relating to lead.

Plume'ria (also, incorrectly, **Plumi'era**) f. Frangipani. Named for Charles Plumier (1646–1704), French monk of the Franciscan order, botanist and traveller, who in 1689 and 1690 visited Martinique, Guadeloupe and Haiti in the West Indies and made remarkably detailed and accurate drawings and descriptions of their flowering-plants and ferns. He revived the ancient custom of naming genera after persons. APOCYNACEAE.

plumo'sus, -a, -um Feathery.

pluriflo'rus, -a, -um Many-flowered.

Po'a f. Gr. *pŏa*, grass. The genus comprises a variety of useful lawn and fodder grasses. GRAMINEAE.

poco'phorus, -a, -um Fleece-bearing; from Gr. *pŏcŏs*, wool, a fleece.

poculifor'mis, -is, -e Cup-shaped.

Podachae'nium n. Gr. *pŏus, pŏdŏs*, a foot; modern Latin *achaenium*, an achene or dry one-seeded fruit. The allusion is to the stalked achenes (*cypsalae*). COMPOSITAE.

podag'ricus, -a, -um Gouty; coarsely swollen at base; relating to gout as a remedy; from Gr. *pŏdagra*, gout in the feet.

Podaly'ria f. Named for Podalyrius, son of Aesculapius, celebrated in Greek mythology as a skilful physician. LEGUMINOSAE.

Podocar'pus m. From Gr. *pŏus, pŏdŏs*, a foot; *karpŏs*, a fruit. The fruits are borne on a fleshy stalk. PODOCARPACEAE.

Podo'lepis From Gr. *pŏus, pŏdŏs*, a foot; *lĕpis*, a scale. The pedicels are scaly. COMPOSITAE.

podo'licus, -a, -um Of Podolia, old Polish province divided between Austria and Russia in 1793.

Podophyl'lum n. May-apple. A contraction of *Anapodophyllum*, from L. *anas*, duck; Gr. *pŏus, pŏdŏs*, a foot; *phyllon*, a leaf; with reference to the shape of the leaf in the American species *P. peltatum*. The rhizomes have medicinal properties. BERBERIDACEAE.

podophyl'lus, -a, -um With stout-stalked leaves.

Podra'nea f. Anagram of *Pandorea*, name of the genus of *Bignoniaceae* to which this is related. BIGNONIACEAE.

poeciloder'mus, -a, -um With spotted skin.

Poellni'tzia f. In honour of Karl von Poellnitz (1896–1945), German specialist on succulent plants. ALOACEAE.

poeta'rum Of the poets, an epithet applied usually to plants associated with ancient Greek and Roman poets.

poe'ticus, -a, -um Pertaining to poets; see above.

pogonan'thus, -a, -um With bearded flowers.

Pogonathe'rum n. From Gr. *pōgōn*, beard; *athēr*, awn, barb. GRAMINEAE.

Pogo'nia f. Adder's mouth. Gr. *pōgōnias*, bearded. In most species of these terrestrial orchids the lip is fringed. ORCHIDACEAE.

pogonopetalus, -a, -um With long-haired petals.

pogonosty'lus, -a, -um With a bearded style.

Pogoste'mon m. From Gr. *pōgōn*, beard; *stēmōn*, thread, stamen. LABIATAE.

pohuashanen'sis, -is, -e From Po hua shan (Hundred Flower Hill), west of Peiping (Beijing), China.

Poincia'na f. Flamboyant. Named for M. de Poinci, 17th-century French Governor of Martinique; *Poinciana regia*, the flamboyant, is now placed in the genus *Delonix* and *P. pulcherrima*, Barbados pride, in *Caesalpinia*. LEGUMINOSAE.

Poinset'tia f. Although usually included in *Euphorbia* it will probably always retain the popular name of *Poinsettia*. Joel R. Poinsette (1775–1851), gardener, botanist, and diplomat of South Carolina, was the first American ambassador to Mexico in 1824; he introduced the plant to Charleston friends in 1833. EUPHORBIACEAE.

Polemo'nium n. The Greek name, *pŏlĕmōniŏn*, originally applied to a medicinal plant associated with Polemon of Cappadocia. Jacob's ladder is *P. caeruleum*. POLEMONIACEAE.

Polian'thes f. Tuberose. Gr. *pŏliŏs*, grey, whitish; *anthos*, a flower. AGAVACEAE.

polifo'lius, -a, -um Grey-leaved like *Teucrium polium*.

Poliothyr'sis f. Gr. *pŏliŏs*, grey, whitish; *thyrsos*, a panicle; in allusion to the whitish panicles of this tree. FLACOURTIACEAE.

poli'tus, -a, -um Elegant, polished, neat.

poly- In Greek compound words signifies many. Thus:
polyacan'thus (with many thorns); **polyan'drus** (with many stamens); **polyan'themos** (with many flowers); **polyan'thus** (with many flowers); **polybot'ryus** (with many clusters); **polycar'pus** (many-fruited); **polyce'phalus** (many-headed); **polychro'mus** (of many colours); **poly'lepis** (with many scales); **polymor'phus** (of many forms; variable); **polype'talus** (with many petals); **polyphyl'lus** (with many leaves); **polyrrhi'zus** (many-rooted); **polyse'palus** (with many sepals); **polysper'mus** (many-seeded); **polysta'ch'yus** (many-spiked); **polystic'tus** (many-spotted).

Polybot'rya f. Gr. *polys*, many; *botrys*, bunch; with reference to the many clusters of sporangia on the fertile fronds. ASPLENIACEAE.

Poly'gala f. Milkwort. Latin name from the Greek name, *pŏlygalŏn*, from *pŏlys*, much; *gala*, milk; they were reputed to aid the secretion of milk. POLYGALACEAE.

polygaloi'des Resembling *Polygala*.

Polygo'natum m. Solomon's seal. The Greek name *pŏlygŏnatŏn*, from *polys*, many; *gony*, the knee-joint; in allusion to the many joints of the rhizomes. LILIACEAE (CONVALLARICEAE).

Poly'gonum n. Smart or knotweed. From the Greek name, *pŏlygŏnon*, from *polys*, many, much; and either *gŏnŏs*, offspring, seed, in allusion to the numerous seeds, or *gŏny*, knee-joint, in allusion to the swollen joints of the stems. POLYGONACEAE.

Polypo'dium n. Polypody fern. Gr. *pŏlypŏdiŏn*, from *polys*, many; *pous*, a foot. The rhizomes are much branched. POLYPODIACEAE.

Polys'cias f. From Gr. *pŏlys*, many; *skias*, canopy like a sunshade, umbel; because of the main umbel being divided into numerous lesser umbels. ARALIACEAE.

Polysta'chya f. Gr. *pŏlys*, many; *stachys*, ear of corn, hence a spike; with reference to the many branches or spikes on a single inflorescence. ORCHIDACEAE.

polysta'chyus -a, -um With many spikes.

Polys'tichum n. Christmas fern. Gr. *pŏlys*, many; *stichos*, a row. The sori of these ferns are in many rows. ASPLENIACEAE.

poma'ceus, -a, -um Applelike.

Pomader'ris f. Gr. *pōma*, a lid; *dĕrris*, a skin, leathern covering; from the membranous covering of the capsule. RHAMNACEAE.

pomeridia'nus, -a, -um Of the afternoon.

po'mifer, pomi'fera, -um Apple-bearing.

pompo'nius, -a, -um Meaning obscure, possibly connected with the Roman name Pomponius.

Ponci'rus m. Hardy orange. Fr. *poncire*, a kind of citron. This shrub is extremely thorny and hence has been used for hedging. RUTACEAE.

pondero'sus, -a, -um Heavy.

Ponga'mia f. From the Malabar name, *pongam*. LEGUMINOSAE.

Pontede'ria Named for Guilio Pontedera (1688–1757), professor of botany at Padua. *P. cordata* is pickerel-weed. PONTEDERIACEAE.

pon'ticus, -a, -um Of the south shore of the Black Sea, the north coast of Asia Minor.

populifo'lius, -a, -um Poplar-leaved.

popul'neus, -a, -um Relating to poplar.

Po'pulus f. Poplar. The Latin name. Includes aspen, cottonwood, popple, and so on. SALICACEAE.

Pora'na f. Derivation obscure. CONVOLVULACEAE.

porci'nus, -a, -um Relating to pigs; pig-food.

porophyl'lus, -a, -um Having holes in the leaves or seeming to.

porphy'reus, -a, -um Warm reddish colour.

porphyrophyl'lus, -a, -um Purple-leaved.

porrifo'lius, -a, -um With leaves like leek. L. *porrum*, leek.

Por'tea f. In honour of Marius Porte (d. 1866), French explorer and plant-collector in Brazil, Malaya and Philippines.

Porteran'thus m. In honour of Thomas Conrad Porter (1822–1901), American botanist. (See *Gillenia.*) ROSACEAE.

portenschlagia'nus, -a, -um In honour of Franz von Portenschlag-Ledermayer (1777–1827), Austrian botanist.

Portlan'dia f. Named for Margaret Cavendish Bentinck (1715–1785), Duchess of Portland, bluestocking, friend and correspondent of Jean Jacques Rousseau, a keen botanist who had a private botanic garden and museum at Bulstrode, Buckinghamshire. RUBIACEAE.

Portula'ca f. Purslane. The Latin name for *P. oleracea*. PORTULACACEAE.

portulaca'ceus, -a, -um *Portulaca*-like.

Portulaca'ria f. Closely related to *Portulaca*. PORTULACACEAE.

poscharskya'nus, -a, -um In honour of Gustav Adolf Poscharsky (1832–1914), head gardener in Dresden.

Posoque'ria f. From the native Guianan name. RUBIACEAE.

Potamoge'ton m. Pondweed. Variant of L. *pŏtamogiton*; Gr. *pŏtamŏgeitōn*, from *pŏtamŏs*, river; *geitōn*, neighbour. POTAMOGETONACEAE.

potamo'philus, -a, -um Loving rivers or wet places.

potani'nii In honour of Grigori Nikolaevich Potanin (1835–1920), Russian explorer of central and eastern Asia, who made enormous botanical collections in western China.

potato'rum Of drinkers or topers; applied to plants used to make beers, etc.

Potentil'la f. Cinquefoil. L. *potens*, powerful; from the reputed medicinal properties. ROSACEAE.

poteriifo'lius, -a, -um With leaves like *Poterium*.

Pote'rium n. Greek name for another plant (burnet), not these herbs and shrubs. ROSACEAE.

Po'thos m. From the Sinhalese name *potha* for these climbers, not connected with L. and Gr. *pŏthos*, which was applied to a blue-flowered larkspur and a white-flowered asphodel. ARACEAE.

Potina'ra f. Multigeneric hybrid group derived from *Brassavola* × *Cattleya* × *Laelia* × *Sophronitis*, named in honour of M. Potin, president of the French orchid society in 1922.

pot'tsii In honour of John Potts (d. 1822), gardener, who collected in China and Bengal for the Horticultural Society of London, or C. H. Potts (*fl.* 1877) of Lasswade near Edinburgh.

Poute'ria f. Latinized from the Guiana (Galibi) vernacular name *pourama-pouteri*. SAPOTACEAE.

praeal'tus, -a, -um Very tall.
prae'cox Very early.
praemor'sus, -a, -um Appearing to be bitten off at the end; see p. 320.
prae'stans Distinguished; excelling.
praete'ritus, -a, -um Past and gone, passed over, omitted.
praeter'missus, -a, -um Neglected, omitted, overlooked.
praetex'tus, -a, -um Bordered.
praever'nus, -a, -um Coming very early, before the spring.
pra'sinus, -a, -um Leek-green; from Gr. *prasŏn*, leek.
praten'sis, -is, -e Of the meadows.
Pra'tia f. Named for a French naval officer, Ch. L. Prat-Bernon, who sailed with Freycinet on his voyage to the South Pacific in 1817 but who died within a few days of the expedition's setting out. CAMPANULACEAE.
prattii In honour of Antwerp E. Pratt, English zoologist and traveller in China, 1887–1890.
praviss'imus, -a, -um Very crooked.
precator'ius, -a, -um Relating to prayer, used of seeds made into rosaries.
Prem'na f. Gr. *prĕmnŏn*, the stump of a tree. VERBENACEAE.
Prenan'thes f. Rattlesnake root. From Gr. *prēnēs*, with face downwards; *anthos*, a flower; with reference to the drooping flower-heads. COMPOSITAE.
prep'tus, -a, -um Distinguished, eminent.
Pri'mula f. Primrose. Contraction of medieval name *primula veris* for the daisy, meaning 'firstling of spring', being a diminutive of L. *primus*, first, used for early-flowering herbs. PRIMULACEAE.
primuliflo'rus, -a, -um Primrose-flowered.
primulifo'lius, -a, -um Primrose-leaved.
primuli'nus, -a, -um Primrose-coloured, i.e. a pale greenish yellow.
primuloi'des Resembling a primrose.
prin'ceps The most distinguished.
Prinse'pia f. Named for James Prinsep (1799–1840), Secretary of the Asiatic Society of Bengal. ROSACEAE.
prisma'ticus, -a, -um Shaped like a prism, i.e. with flat surfaces separated by angles.
Pritchar'dia f. Named for W. T. Pritchard, 19th-century British official in Polynesia; see *Eupritchardia*. PALMAE.
Probosci'dea f. From Gr. *prŏbŏskis*, elephant's trunk, in allusion to the elongated curved ends of the fruit. PEDALIACEAE.
probosci'deus, -a, -um Snoutlike; with an elongated proboscis.
proce'rus, -a, -um Tall.
procum'bens Prostrate.
procur'rens Spreading out, extending, projecting.

produc'tus, -a, -um Lengthened, stretched out.

profu'sus, -a, -um Abundant.

pro'lifer, proli'fera, -um Proliferous, i.e. free-flowering or producing side shoots or buds in order to increase.

proli'ficus, -a, -um Very fruitful.

Promenae'a f. Named after a Greek priestess of Dodona, Epirus, known to Herodotus. ORCHIDACEAE.

pro'nus, -a, -um Bending or hanging forwards.

propen'sus, -a, -um Hanging down.

propin'quus, -a, -um Related.

Proserpina'ca f. L. plant-name of obscure application, probably from *proserpo*, creep along. HALORHAGACEAE.

Proso'pis f. Mesquite. The Greek name for the burdock, but why it was given to this genus is obscure. LEGUMINOSAE.

Prostanthe'ra f. Mint-bush. Gr. *prŏstithemi*, to append or add to; *prŏsthĕma*, appendage; *anthera*, anther; in allusion to a spur-like appendage on the back of the anther. LABIATAE.

prostra'tus, -a, -um Flat on the ground, prostrate.

Pro'tea f. South African plants. Named for the Greek sea-god, Proteus, with the power of assuming a diversity of shapes; an allusion to the diversity within the genus. PROTEACEAE.

protru'sus, -a, -um Protruding.

provincia'lis, -is, -e From Provence in southern France.

pruina'tus, -a, -um; pruino'sus, -a, -um Glistening as though frosted over.

Prunel'la f. Self-heal. In the 15th and 16th centuries German herbalists used the names *Prunella* and *Brunella* interchangeably; they may be derived from the south German word *braun* (from L. *prunum*) meaning purple and referring to the flower-colour, or from the German *Bräune*, meaning quinsy, which these herbs were supposed to cure: the north German *braun* means brown. LABIATAE.

prunelloi'des Resembling *Prunella*.

prunifo'lius, -a, -um With leaves like plum.

Pru'nus f. Plum, cherry. The Latin name. ROSACEAE.

pruhonicia'nus, -a, -um Of Pruhonice, near Prague, Czechoslovakia.

pru'riens Causing an itch.

przewal'skii In honour of Nicolai Mikhailovich Przewalski (1839–1888), celebrated Russian explorer of central Asia.

Psammo'phora f. From Gr. *psammŏs*, sand; *phŏrŏs*, bearing; in allusion to their sanded appearance. AIZOACEAE.

pseuda'corus Gr. *pseudo*, false; *Acorus*, sweet flag; the rhizomes of *Iris pseudacorus* are rather similar in appearance though not in smell.

Pseuderan'themum n. Gr. *pseudo*, false; plus *Eranthemum*. Some of these plants were formerly placed in *Eranthemum*. ACANTHACEAE.

Pseudofumaria Gr. *pseudo*, false; *Fumaria*. FUMARIACEAE

Pseudo'larix f. Golden larch. Gr. *pseudo*, false; *Larix*, larch, which it resembles. PINACEAE.

pseudonarcissus Gr. *pseudo*, false; *Narcissus*, poet's narcissus, *N. poeticus*.

Pseudo'panax m. Gr. *pseudo*, false; *Panax*, ginseng; to which this genus is related. ARALIACEAE.

Pseudophoe'nix f. Gr. *pseudo*, false; *Phoenix*, the date palm, to which these palms bear a general resemblance. PALMAE.

Pseudosa'sa f. Gr. *pseudo*, false; *Sasa*, a Japanese genus of bamboos to which this is related. GRAMINEAE.

Pseudotsu'ga Gr. *pseudo*, false; *Tsuga*, hemlock. The Douglas fir is *Ps. menziesii*. PINACEAE.

Pseudowin'tera Gr. *pseudo*, false; *Wintera*, in honour of Captain John Wynter or Winter, Elizabethan sea-captain who saved his crew from scurvy at the Straits of Magellan by discovering in 1578 the antiscorbutic property of the bark of *Drimys winteri*, known from 1582 as *Cortex Winteranus* or Winter's Bark. WINTERACEAE.

Psi'dium n. Guava. Gr. *psidion*, a pomegranate. MYRTACEAE.

psilan'drus, -a, -um With glabrous stamens.

psilos'tachys With glabrous spike.

psilos'temon With glabrous stamens.

Psilotum n. Gr. *psilos*, naked; with reference to the leafless branches. PSILOTACEAE.

psittaci'nus, -a, -um Parrotlike, i.e. with green or contrasting colours.

psittacor'um Of parrots.

Psophocar'pus m. Gr. *psŏphŏs*, a noise; *karpos*, fruit. The seed capsules explode noisily when ripe. LEGUMINOSAE.

Psora'lea f. Scurfy pea. Gr. *psōralĕos*, scabby. These herbs are covered with glandular dots. LEGUMINOSAE.

Psycho'tria f. Wild coffee. Variant coined by Linnaeus from Gr. *psychŏtrŏphŏn*, applied by Patrick Browne to Jamaican plants but by the Greeks to betony, because found in cold places. RUBIACEAE.

psyco'des Butterfly-like.

ptar'mica Gr. name *ptarmikē* for a plant which caused sneezing.

Pte'lea f. Gr. name *ptĕlĕa* for an elm tree, presumably applied to this genus on account of its flattened, almost circular fruit having a wing all around like that of an elm. RUTACEAE.

pteran'thus, -a, -um With winged flowers.

Pteri'dium n. Bracken, brake. Diminutive of Gr. *ptĕris*, fern, from *ptĕron*, wing, in allusion to the form of the fronds. Commonly known as *Pteris*. DENNSTAEDTIACEAE.

pteridoi'des Resembling *Pteris*.

Pteridophyl'lum n. From Gr. *ptĕris*, fern; *phyllŏn*, leaf. FUMARIACEAE.

Pte'ris f. Gr. *ptĕris*, fern; from the feathery fronds. ADIANTACEAE.

Pteroca'rya f. Wing-nut. Gr. *ptĕron*, a wing; *karyon*, a nut. JUGLANDACEAE.

Pteroce'phalus m. Gr. *ptĕron*, a wing; *kephale*, a head. The fruiting head appears to be covered with feathers. DIPSACACEAE.

pteroneu'rus, -a, -um With winged nerves.

Pterosper'mum n. Gr. *ptĕron*, a wing; *sperma*, a seed; with reference to the winged seeds. STERCULIACEAE.

Pterosty'rax f. Gr. *ptĕron*, a wing; plus *Styrax*. The seeds of one species of these *Halesia*-like trees and shrubs have winged seeds like *Styrax*. STYRACACEAE.

Ptychosper'ma n. Gr. *ptyche*, a fold; *sperma*, a seed; in allusion to a technical characteristic of the seed of these palms. PALMAE.

pubens; pubes'cens; pu'biger, pubi'gera, -um Downy.

pud'dum Variant of the Hindi vernacular name *padam* for a beautiful Himalayan cherry.

pudero'sus Very bashful.

pu'dicus, -a, -um Bashful.

pue'lii After Timothée Puel (1812–1890), French botanist and physician, who published on French and Syrian plants.

Puera'ria f. Named for Marc Nicolas Puerari (1766–1845), Swiss botanist, born at Geneva, who spent most of his life as a professor at Copenhagen, but gave his important herbarium to De Candolle of Geneva. LEGUMINOSAE.

pugionifor'mis, -is, -e Dagger-shaped.

pulchel'lus, -a, -um; pul'cher, pul'chra, pulchrum Pretty.

pule'gium n. Latin name, from L. *pulex*, flea; reputedly a flea repellent.

Pulica'ria f. Felabane. Latin name, from L. *pulex*, flea; reputedly a flea repellent. COMPOSITAE.

pul'lus, -a, -um Dark-coloured.

Pulmona'ria f. Lungwort. L. *pulmo*, the lung. In accordance with the old Doctrine of Signatures the plant was considered to be an effective remedy for diseases of the lung because the spotted leaves were supposed to resemble diseased lungs. BORAGINACEAE.

Pulsatil'la f. Pasqueflower. From L. *pulso*, strike, set in violent motion, its relevance uncertain. RANUNCULACEAE.

Pultena'ea f. Named for Richard Pulteney (1730–1801), English botanist and physician, author of *A General View of the Writings of Linnaeus* (1781) and *Historical and Biographical Sketches of the Progress of Botany in England* (1790). LEGUMINOSAE.

pulverulen'tus, -a, -um Powdered as with dust.

pulvina'tus, -a, -um Cushionlike.

pu'milus, -a, -um Dwarf.

puncta'tus, -a, -um Spotted.

pun'gens Sharp-pointed.

Pu'nica f. Pomegranate. The Latin name contracted from *punicum malum*, Carthaginian apple, in turn derived from *Poenus*, Carthaginians, *Phoinikes*, Phoenicians; see *phoeniceus*. PUNICACEAE.

puni'ceus, -a, -um Reddish-purple.

pur'gans Purgative, purging.

purpura'tus, -a, -um Made purple.

purpu'reus, -a, -um Purple. Also *purpurascens*, tending to purple.

purpu'sii In honour of Carl Albert Purpus (1853–1941), or his brother Joseph Anton Purpus (1860–1932), German plant-collectors in central America.

Pur'shia f. Named for Frederick Traugott Pursh (1774–1820), German explorer, collector, horticulturist, author, who lived in the United States from 1799 to 1811 and wrote an important *Flora Americae Septentrionalis* (1814). ROSACEAE.

Puschki'nia f. Striped squill. Named for the Russian Count Mussin-Puschkin (d. 1805), who collected in the Caucasus; *Nepeta mussinii* also commemorates him. LILIACEAE (HYACINTHACEAE).

pusil'lus, -a, -um Very small.

pustula'tus, -a, -um Appearing as though blistered.

Putoria f. From L. *putor*, foul smell. RUBIACEAE.

Puy'a f. The Chilean vernacular name. BROMELIACEAE.

Pycnan'themum n. Mountain mint. From Gr. *pyknŏs*, dense; *anthŏs*, a flower. The flowers are densely crowded. LABIATAE.

Pycno'stachys f. From Gr. *pyknŏs*, dense; *stachys*, ear of corn, a spike; referring to the dense flower spikes. LABIATAE.

Pyg'mea f. A genus of dwarf plants so-named from their habit of growth. SCROPHULARIACEAE.

pygma'eus, -a, -um Pygmy; the Pygmies to the Ancient Greeks and Romans were an African tribe of dwarfs reputedly at war with the cranes.

pylzowia'nus, -a, -um Named for Mikhail Alexandrovich Pyltsov, Russian army officer, who accompanied Przewalski in China 1820–1873.

Pyracan'tha f. Fire-thorn. Greek name *pyrakantha*, from *pyr*, fire, *akantha*, a thorn; in allusion to the thorny branches and the showy crimson fruit. ROSACEAE.

pyramida'lis, -is, -e; pyramida'tus, -a, -um Pyramid-shaped.

pyrenae'us, -a, -um; pyrena'icus, -a, -um Of the Pyrenees; Pyrenean.

Py'rethrum n. Feverfew. Gr. *pyrĕthrŏn*, from *pyr*, fire; in allusion to the bitter roots of *Anacyclus pyrethrum* which were regarded as a remedy for fevers and by their hotness promoted a flow of saliva, whence the alternative Latin name *salivaris*. Now classified under *Tanacetum*, but the other name remains in familiar use. COMPOSITAE.

pyrifo'lius, -a, -um Pear-leaved.

pyrifor'mis, -is, -e Pear-shaped.

Py'rola f. Shinleaf. The Latin diminutive of *Pyrus*, pear, from the pear-like leaves. PYROLACEAE.

Pyroste'gia f. From Gr. *pyr*, fire; *stĕgē*, a roof; from the crimson-orange colour and form of the upper lip of the flowers. BIGNONIACEAE.

pyrrhoan'thus, -a, -um With flaming red flowers.

Pyrro'sia f. Gr. *pyrrŏs*, flame-coloured, reddish yellow, tawny; in allusion to the fronds. POLYPODIACEAE.

Py'rus f. Pear. The Latin name. ROSACEAE.

Pyxidanthe'ra f. Flowering moss. Gr. *pyxis*, a small box; *anthera*, anther. The anthers open by a tranverse slit, thus as if like the lid of a box. DIAPENSIACEAE.

pyxida'tus, -a, -um Provided with a pyxis or a lid.

Q

quadra'tus, -a, -um In fours or in four. Thus:
quadrangula'ris (with four angles); **quadrangula'tus** (with four angles);
quadriauri'tus (with four ears); **quadri'color** (four-coloured);
quadridenta'tus (four-toothed); **quadri'fidus** (cut into four);
quadrifolius (four-leaved); **quadriparti'tus** (parted four ways);
quadriradia'tus (with four ray florets); **quadrival'vis** (with four valves).

Qua'moclit f. Derivation of the name is obscure. Possibly Gr. *kuamos*, bean. CONVOLVULACEAE.

Quas'sia f. Name given by Linnaeus to honour Graman Quasi, a Negro slave of C. G. Dahlberg, who used the bark of this tree of Surinam as a remedy for fever. It was used in medicine as a bitter tonic and a vermifuge. The name like the West Indian Quashie is from the (Tui) *Kwasi*, a personal name given to a boy born on the first day of the week, *Kwasida* or Sunday. SIMAROUBIACEAE.

querce'torum Of oak woods.

quercifo'lius, -a, -um With leaves like *Quercus* or oak.

querci'nus, -a, -um Relating to the oak.

Quer'cus f. Oak. The Latin name. The oak has long been considered a symbol of strength and is celebrated in legend and mythology in many lands. It was sacred to Jupiter and Thor, while the Druids thought that it inspired prophecy and built their altars under its branches. The timber of most species is valuable, the bark of some is used in tanning, while the acorn is used both for animal and human food. FAGACEAE.

Quilla'ja f. Soap-bark tree. The Chilean name. ROSACEAE.

quina'tus, -a, -um In fives. Thus:
quinqueflo'rus (with five flowers); **quinquefo'lius** (five-leaved); **quinquener'vis** (five-nerved); **quinquepuncta'tus** (five-spotted); **quinquevul'nerus** (with five marks).

Quisqua'lis f. Rangoon Creeper. L. *quis*, who? *qualis*, what? This curious name is the result of a play on words by Rumphius (1627–1702). The Malay name for this plant is *Udani*, rendered as *hoedanig*, meaning in Dutch 'how, what', which Rumphius translated into Latin as *quis qualis*, expressive of his surprise at the variability of the plant's growth. The flowers open white, become pink and end blood-red. COMBRETACEAE.

quiten'sis, -is, -e Referring to Quito, Ecuador.

R

racemiflo'rus, -a, -um; racemo'sus, -a, -um With flowers in racemes, as in lily-of-the-valley.

raddei; raddeanus, -a, -um In honour of Gustav Ferdinand Richard Radde (1831–1903), German naturalist who made made large natural history collections in the Caucasus and Amur region.

raddii; raddianus, -a, -um In honour of Giuseppe Raddi (1770–1829), Italian botanist who collected in Brazil and Egypt and gave special attention to cryptogams.

rad'ians Radiating outward.

radia'tus, -a, -um With rays or of radiating form.

ra'dicans Having rooting stems. Also:
radica'tus (having conspicuous roots); **radico'sus** (with many roots); **rad'icum** (of roots).

radio'sus, -a, -um Having many rays.

rad'ula Rasp, file.

Raffenal'dia f. In honour of Alire Raffeneau Delile (1778–1850), professor of botany at Montpellier, one of the scientists who accompanied Napoleon to Egypt, also commemorated by the genus *Lilaea*. CRUCIFERAE.

raffil'lii. In honour of Charles Raffill (1876–1950), Kew gardener.

Raffle'sia f. Appropriately named in honour of Sir Thomas Stamford Raffles (1781–1826), founder of Singapore, a great patron of science as well as statesman, *Rafflesia* has the world's largest flower but, owing to its parasitic habit, unfortunately no horticultural potentialities. RAFFLESIACEAE.

rafinesquianus, -a, -um. In honour of Constantine Samuel Rafinesque (1783–1840), naturalist of French-German parentage who spent most of his life in Sicily and U.S.A. and published over 900 publications abounding in new names.

ragusi'nus, -a, -um Relating to Dubrovnik (formerly Ragusa).

ramiflo'rus, -a, -um Flowering on older branches.

Ramon'da f. Stemless herbs named for Louis Francis Ramond, Baron de Carbonnière (1753–1827), French botanist and traveller in the Pyrenees. GESNERIACEAE.

ramondioi'des Resembling *Ramonda*.

ramo'sus, -a, -um Branched. Also, *ramulosus*, twiggy; *ramosissimus*, much branched.

Ran'dia f. After Isaac Rand (d. 1743), London apothecary, director of the Chelsea Physic Garden. RUBIACEAE.

Rane'vea f. Named for Louis Ranevé. PALMAE.

ranunculi'nus, -a, -um; ranunculoi'des Resembling *Ranunculus*.

Ranun'culus, m. Buttercup, crowfoot. The Latin name from the diminutive of *rana*, a frog, because many species grow in damp places. RANUNCULACEAE.

Ranza'nia f. In honour of Ono Ranzan (1729–1810), celebrated Japanese naturalist. BERBERIDACEAE.

Raou'lia f. Tufted or creeping perennial herbs named for Edouard Raoul (1815–1852), French naval surgeon who collected in New Zealand from 1840 to 1846 and published on New Zealand plants in 1846. COMPOSITAE.

rapa'ceus, -a, -um Relating to turnips.

Ra'phanus m. Radish. The Latin name, from Gr. *rhaphanis*, for this vegetable which has been known from antiquity. CRUCIFERAE.

Ra'phia f. Raffia—the source of the very useful garden tying fibre. From Gr. *haphis*, a needle; in allusion to the beaked fruit of these palms. PALMAE.

Raphio'lepis f. See *Rhaphiolepis*.

rapunculoi'des Specific epithet, resembling *Rapunculus*, the now obsolete name for a group of bellflowers, meaning a little turnip and referring to the roots.

rariflo'rus, -a, -um With scattered flowers.

ra'rus, -a, -um Rare, uncommon.

Rathbu'nia f. Cacti named for Dr. Richard Rathbun (1852–1918), Assistant Secretary of the Smithsonian Institute, Washington, D.C. CACTACEAE.

Rati'bida f. Meaning obscure. See *Lepachys*. COMPOSITAE.

Rauvol'fia f. Named for Leonhart Rauwolf (d. 1596), German physician and adventurous observant traveller in the Orient. Long familiar to Oriental medicine, certain species have now achieved respectability by their acceptance in

Western medicine, especially in the treatment of nervous disorders. APOCYNACEAE.

Ravena'la f. Traveller's tree. From the native name in Madagascar. STRELITZIACEAE.

ra'vus, -a, -um Greyish or tawny.

Rebu'tia f. Named for P. Rebut, French dealer in cacti, *c.* 1900. CACTACEAE.

Rechsteine'ria In honour of a Swiss-German clergyman and botanist, Rechsteiner (1797–1858). GESNERIACEAE.

reclina'tus, -a, -um Bent backward.

rectifo'lius, -a, -um With erect or ascending leaves.

rec'tus, -a, -um Upright.

recurva'tus, -a, -um; recur'vus, -a, -um Curved backward.

recuti'tus, -a, -um Circumcised.

redivi'vus, -a, -um Brought back to life, used of plants reviving after a period of desiccation.

reduc'tus, -a, -um Drawn back, reduced, hence made small.

reevesia'nus, -a, -um In honour of John Reeves (1774–1856), from 1812 to 1831 tea inspector at Macao and Canton, who was instrumental in introducing Chinese plants into British gardens.

refle'xus, -a, -um; refrac'tus, -a, -um Bent sharply backward.

reful'gens Shining brightly.

rega'lis, -is, -e Royal; of outstanding merit.

regi'nae Of the queen. As an epithet in *Tradescantia* honouring Queen Marie of the Belgians, wife of Leopold II (1815–1909).

re'gius, -a, -um Royal, princely.

reh'deri In honour of Alfred Rehder (1863–1949), American dendrologist of German origin, from 1898 until his retirement in 1940 at the age of 76 on the staff of the Arnold Arboretum, Massachusetts; author of the celebrated *Manual of Cultivated Trees and Shrubs* (1940) and, after retirement, of the *Bibliography of Cultivated Trees and Shrubs* (1949).

Rehman'nia f. Named for Joseph Rehmann (1753–1831), German physician, who settled in St. Petersburg. SCROPHULARIACEAE.

Reine'ckea f. Named for Johann Heinrich Julius Reinecke (1799–1871) of Berlin. LILIACEAE.

Reinwar'dtia f. Named for Caspar Georg Carl Reinwardt (1773–1854), professor at Harderwijk, then Leyden, founder of Bogor (Buitenzorg) botanic garden, Java. LINACEAE (CONVALLARIACEAE).

religio'sus, -a, -um Used for religious ceremonies; sacred.

remo'tus, -a, -um Scattered.

Remusa'tia f. In honour of Abel Remusat (1788–1832), celebrated French scholar, professor of Chinese at Paris. ARACEAE.

Renanthe'ra f. From L. *rēnes*, kidneys; *anthera*, anther; in reference to the kidney-shaped pollinia. ORCHIDACEAE.

renar'dii In honour of Charles Claude (Karl Ivanovich) Renard (1809–1886) of Moscow, second secretary of the Imperial Society of Naturalists of Moscow.

Reneal'mia f. In honour of Paul de Reneaulme (1560–1624), French physician and author of *Specimen Historiae Plantarum* (1611). ZINGIBERACEAE.

renifor'mis, -is, -e Kidney-shaped.

repan'dus With slightly wavy margins.

re'pens; rep'tans Creeping.

replica'tus, -a, -um Doubled back; folded.

Rese'da f. Mignonette. The Latin name derived from *resedo*, to heal. RESEDACEAE.

resedifo'lius, -a, -um With leaves like Mignonette.

resi'nifer, resini'fera, -um; resino'sus, -a, -um Resinous.

resupina'tus, -a, -um Bent back, put on its back, applied to organs turned upside down by a twist in their support.

Reta'ma f. Latinized form of the Arabic name *retem*. LEGUMINOSAE.

reticula'tus, -a, -um Netted; with a net-like pattern.

retrofle'xus, -a, -um; retor'tus, -a, - um; retrofrac'tus, -a, -um Twisted back.

retu'sus, -a, -um With a rounded, slightly notched tip.

rever'sus, -a, -um Reversed; turned around.

revolu'tus, -a, -um Rolled backward, as some leaves are at the edges.

rex King; epithet for an outstanding species.

Reynoutria f. In honour of Reynoutre, 16th-century French naturalist, acquaintance of Lobel. POLYGONACEAE.

Rhabdotham'nus m. Gr. *rhabdŏs*, a rod; *thamnŏs*, a bush. A New Zealand shrub with a twiggy habit. GESNERIACEAE.

rhabdo'tus, -a, -um Striped, with rod-like marking.

rhae'ticus, -a, -um Of the Rhaetian Alps of Switzerland and Austria.

Rhago'dia f. Salt-bush. Gr. *rhagos*, a berry; in reference to the fruit. Sheep delight in this drought-resistant Australian shrub. Few plants, however, could appear less inviting. CHENOPODIACEAE.

rhamnifo'lius, -a, -um With leaves like *Rhamnus*.

rhamnoi'des Resembling *Rhamnus*.

Rham'nus m. Buckthorn. The Greek name of various spiny shrubs. Many species have medicinal value, the dried bark of *R. purshianus* being the source of *Cascara sagrada*. RHAMNACEAE.

Rhaphido'phora f. Gr. *rhaphis*, needle; *phŏrŏs*, carrying; in allusion to the needle-like points on the fruit. ARACEAE.

Rhaphio'lepis f. Gr. *rhaphis*, a needle; *lĕpis*, a scale; from the very narrow persistent bracteoles on the inflorescence. ROSACEAE.

Rhaphitham'nus m. Gr. *rhaphis*, a needle; *thamnos*, a shrub; with reference to the spiny habit. VERBENACEAE.

Rhapidophyl'lum n. Blue palmetto, needle palmetto. Dwarf fan palms like *Rhapis* in producing suckers freely, with needle-like appendages on leaf sheaths. PALMAE.

Rha'pis f. Ground rattan. Gr. *rhapis*, a needle; with reference to the needlelike leaf segments of these sucker-producing palms. PALMAE.

Rhazy'a f. In honour of Abu Bekr-el-Rasi (d. *c.* 932), Arabian physician, author of a medical encyclopaedia, who was known in the West as Rhazes. APOCYNACEAE.

Rhektophyl'lum n. From Gr. *rhēktŏs*, rent, penetrable; *phyllon*, a leaf. The leaves are cut and perforated. ARACEAE.

Rhe'um n. Rhubarb. The Greek name for the roots and rhizome imported from Iran was *rhēŏn* or *rha*, whence the names *Rhaponticum*, meaning the *rha* of Pontus, and *Rhabarbarum*, the *rha* of the foreigners or barbarians. These names probably derive from Iranian *rēwās*. Many species have considerable medicinal value. In the palmy days of the China trade (1800–1850), the best medicinal rhubarb was exported from Canton. POLYGONACEAE.

Rhe'xia f. Meadow-beauty, deer-grass. Name used by Pliny, of uncertain origin, for some other plant. MELASTOMATACEAE.

Rhipo'gonum See *Ripogonum.*

Rhipsalidop'sis f. From *Rhipsalis; opsis*, likeness. CACTACEAE.

Rhip'salis Gr. *rhips*, wicker-work; from the pliant interlacing twigs. CACTACEAE.

rhizophyl'lus,-a,-um With rooting leaves.

rhodan'thus,-a,-um With rose-coloured flowers.

rhoden'sis, -is, -e; rhodius, -a, -um Of the island of Rhodes (Rodos).

Rhod'iola f. L. diminutive of Gr. *rhŭdon*, rose, referring to the rose-scented roots of the type species known as *Rhudia Radix* or *Rosea Radix*, by the old herbalists. CRASSULACEAE.

Rhodochi'ton m. From Gr. *rhŏdŏ-*, red; *chitōn*, a tunic cloak; the calyx is large and red. SCROPHULARIACEAE.

Rhododen'dron Gr. *rhŏdŏdĕndrŏn*, a name for the rose-flowered oleander (*Nerium oleander*), from *rhŏdŏn*, a rose; dĕndrŏn, a tree; transferred to this genus. ERICACEAE.

Rhodohypo'xis f. Gr. *rhŏdŏ-*, red; *Hypoxis*, the yellow-flowered genus to which this genus with red, pink or white flowers is related. HYPOXIDACEAE.

Rhodomyr'tus f. Gr. *rhŏdŏ-*, red; *myrtos*, myrtle; from the rose-coloured, myrtle-like flowers of these shrubs. MYRTACEAE.

rhodo'pensis, -is, -e Of the Rhodope Mountains (Gr. *Rhŏdŏp)*, southern Bulgaria.

Rhodospa'tha f. Gr. *rhŏdŏ-*, red; *spatha*, spathe. ARACEAE.

Rhodotham'nus m. Gr. *rhŏdŏ-*, red; *thamnos*, a shrub; from the rosy flowers of this alpine shrub. ERICACEAE.

Rhodo'typos m. Gr. *rhŏdŏn*, a rose; *typŏs*, a type. ROSACEAE.

Rho'eo f. Derivation of the name unknown. COMMELINACEAE.
Rhoicis'sus f. Presumably from L. *rhŏicus*, belonging to Sumach (Rhus); *cissus*, ivy. VITACEAE.
rhom'bicus, -a, -um; rhomboi'deus, -a, -um Diamond-shaped; rhomboidal.
Rhombophyl'lum n. A genus once included in *Mesembryanthemum*, named for its rhomboid leaves. AIZOACEAE.
Rhopalosty'lis f. Gr. *rhŏpalŏn*, a club; *stylis*, a small pillar; from the club-shaped spadix of these palms. PALMAE.
Rhus f. Sumach. The Greek name for one species, *Rhus coriaria*. *R. typhina*, stag-horn sumach, was also known to the early North American colonists as dyer's sumach from the yellow dye which could be made from it. ANACARDIACEAE.
Rhynchely'trum n. From Gr. *rhynchos*, beak; *ĕlytron*, cover, husk; in allusion to the beaked flower-parts (upper glume and lower lemma) of these grasses. GRAMINEAE.
Rhyncho'sia f. Gr. *rhynchos*, a beak; from the beaked keel of the flowers. LEGUMINOSAE.
Rhynchosty'lis f. Gr. *rhynchos*, a beak; *stylis*, a small pillar; in allusion to the beaked column of these epiphytic orchids. ORCHIDACEAE.
Rhytico'cus f. Gr. *rhytis*, a wrinkle; *Cocos*, coconut. Closely related to *Cocos*; the name derives from the seed. PALMAE.
rhytidophyl'lus, -a, -um With wrinkled leaves.
Ri'bes n. Currant, gooseberry, etc. A large and useful genus, the name of which apparently derives from the Arabic or Persian name *ribas*, acid-tasting. GROSSULARIACEAE.
ricasolia'nus, -a, -um In honour of Vincenzo Ricasoli, 19th-century Italian general and patron of horticulture.
Richar'dia f. In honour of Richard Richardson (1663–1741), English physician and botanist with garden at North Brierley, Yorkshire. RUBIACEAE.
Richea f. In honour of A. Riche (d.1791), French naturalist on the D'Entrecasteaux voyage. EPACRIDACEAE.
ricinifo'lius, -a, -um With leaves like *Ricinus*.
ricinoi'des Resembling *Ricinus*.
Ric'inus m. Castor bean. L. *ricinus*, a tick; from the appearance of the seeds. The seeds themselves are very poisonous, and must be harvested with care; as soon as the capsules are ripe they dehisce audibly, shooting the seeds many feet. In the Middle Ages this was known as *Palma Christi* (the Hand of Christ) from the usually 5-lobed leaves but the same name was also used for *Dactylorhiza* orchids because of their finger-like roots. EUPHORBIACEAE.
Rico'tia f. Named for Ricot, a little-known French botanist. CRUCIFERAE.
ri'gens; rig'idus, -a, -um Rigid; stiff.
rigidifo'lius, -a, -um Rigid-leaved.

rin'gens Gaping, e.g. the mouth of an open two-lipped corolla.

ripa'rius, -a, -um Of the banks of rivers.

Ripo'gonum n. Gr. *rhips*, wicker-work; *gonu*, a knee; from the many-jointed stalks on these climbers that make formidable tangles in the New Zealand bush and are used in basketry. LILIACEAE (RHIPOGONUMACEAE).

riva'lis, -is, -e Growing by streams.

Rivi'na f. Rouge plant, bloodberry. Herbs with attractive berries named for August Quirinus Rivinus (A.Q. Bachmann, 1652–1722), professor of botany, Leipzig. PHYTOLACCACEAE.

rivula'ris, -is, -e Brook-loving.

rob'biae In honour of Mrs Mary Anne Robb (1829–1912) of Liphook, Hampshire, England, who introduced *Euphorbia robbiae* from Turkey by bringing back rooted plants in her hat-box in 1891; whence the name 'Mrs Robb's Bonnet'.

Robi'nia f. Yellow locust, false acacia. Named for Jean Robin (1550–1629), of Paris, gardener to Henri IV and Louis XIII of France, who received new plants from Canada. LEGUMINOSAE.

ro'bur Latin word for oak-wood, also strength. Specific epithet of English oak (*Quercus robur*).

robustispi'nus, -a, -um With strong spines.

robus'tus, -a, -um Stout; strong in growth.

Ro'chea f. Named for Daniel de la Roche (1743–1814), Swiss physician, part-author of *Pharmacopaea Genevensis* (1780), who settled in Paris in 1782 and became medical officer to the Swiss Guard. His son François de la Roche (1782–1814), likewise a physician in Paris, was also a distinguished botanist and wrote a monograph on *Eryngium*. CRASSULACEAE.

ro'ckii In honour of Joseph F. C. Rock (1884–1962), originally Josef Franz Karl Rock, American botanist of Austrian origin, plant-collector, explorer and orientalist, noted for his work on plants of Hawaii, western China and eastern Tibet.

Rodger'sia f. Named for Rear-Admiral John Rodgers (1812–1882), distinguished American naval officer who commanded in 1852–1856 a Paciifc expedition during which the first species of this genus was discovered. SAXIFRAGACEAE.

Rodrigue'zia f. Named for Manuel Rodriguez, 18th-century Spanish physician and botanist. ORCHIDACEAE.

Roel'la f. In honour of G. Roelle, 18th-century professor of anatomy in Amsterdam. CAMPANULACEAE.

Roeme'ria f. Named for Johann Jakob Roemer (1763–1819), professor of botany, Zürich. PAPAVERACEAE.

rog'ersii Named for George L. Coltman-Rogers (1854–1929), landowner and dendrologist in Powys, Wales, author of *Conifers* (1920); if South African, William Moyle Rogers (1835–1920) or his son Frederick Arundel Rogers (1876–1944).

Roh'dea f. Named for Michael Rohde (1782–1812), physician of Bremen. LILIACEAE (CONVALLARIACEAE).

roma'nus, -a, -um Roman.

Romanzof'fia f. Named for Prince Nicholas Romanzoff, who financed a round-the-world expedition, 1816–1817. HYDROPHYLLACEAE.

Romne'ya f. Matilija poppy. In honour of Thomas Romney Robinson (1792–1882), celebrated Irish astronomer. PAPAVERACEAE.

Romu'lea f. In honour of Romulus, legendary king of the Romans and founder of Rome. IRIDACEAE.

Rondele'tia f. Named for Guillaume Rondelet (1507–1566), chancellor of the University of Montpellier, France, and a very influential professor of natural history and medicine. RUBIACEAE.

Rorip'pa f. Water-cress. Latinized from a Saxon (East German) vernacular name, Rorippen, mentioned by Euricius Cordus. CRUCIFERAE.

Ro'sa f. Rose, The Latin name. ROSACEAE.

rosa'ceus, -a, -um Roselike.

Rosche'ria Derivation of the name is obscure. PALMAE.

Rosco'ea f. In honour of William Roscoe (1753–1831) of Liverpool, lawyer, opponent of the slave-trade, historian, promoter of the arts, a founder of the Liverpool botanic garden and author of a magnificent book on the ginger family (*Zingiberaceae*). ZINGIBERACEAE.

ro'sea Rose-like. As a feminine noun used as a specific epithet it does not have to agree with the gender of the genus, e.g. *Sedum rosea*, the roseroot, which has yellow flowers and rose-scented roots; see *Rhodiola*.

ro'seus, -a, -um Rose-coloured.

rosmarinifo'lius, -a, -um With leaves like rosemary.

Rosmari'nus m. Rosemary. The Latin name derived from *ros*, dew; *marinus*, maritime. It is found wild on sea cliffs in southern Europe. LABIATAE.

ros'sii. In honour of Sir John Ross (1777–1856), British naval officer and Arctic explorer.

rostra'tus, -a, -um Beaked.

rosula'ris, -is, -e Having rosettes; hence **Rosularia** f. CRASSULACEAE.

Rota'la. f. From *rŏtalis*, wheeled, wheel-like; with reference to the whorled leaves. LYTHRACEAE.

rota'tus, -a, -um Wheel-shaped; used of circular, almost flat, or saucer-like flowers.

rothschi'ldianus, -a, -um. In honour of British members of the Rothschild banking family, notably Lionel Walter Rothschild (1868–1937) and

Ferdinand James de Rothschild (1839–1898) who had a large collection of orchids at Waddesdon, Buckinghamshire.

rotunda'tus, -a, -um; rotun'dus, -a, -um Rounded.

rotundifo'lius, -a, -um With round leaves.

Rou'pala f. From the native Guyanese name. PROTEACEAE.

rowleyi; rowleyanus, -a, -um In honour of Gordon Rowley (b. 1921), lecturer in horticultural botany, University of Reading 1961–1981, cactus enthusiast and prolific author on succulent plants.

roxburghii In honour of William Roxburgh (1751–1815), Scottish doctor, and superintendent of Calcutta botanic garden 1793–1814.

Roye'na f. Named for Adrian van Royen (1705–1779), professor of botany, Leyden, and friend of Linnaeus. This genus is now included in the large genus *Diospyros*. EBENACEAE.

roylei In honour John Forbes Royle (1800–1858), British botanist and physician, in India 1819–1831, then professor of materia medica in London 1837–1856.

Roysto'nea f. Named for General Roy Stone (1836–1905), American army engineer in Puerto Rico. PALMAE.

rubel'lus; -a, -um; rubes'cens Pale red, becoming red.

ru'bens; ru'ber, rub'ra, rub'rum Red. In compound words, *rubri-*.

Ru'bia f. Madder. L. *ruber*, red; in allusion to the reddish dye obtained from the roots. RUBIACEAE.

rubigino'sus, -a, -um Rusty.

rubioi'des Resembling *Rubia*.

rubri'caulis, -is, -e Red-stemmed.

rubropilo'sus, -a, -um Red-haired.

Ru'bus m. Blackberry, bramble, raspberry. The Latin name. ROSACEAE.

rudbaren'sis, -is, -e; rudba'ricus, -a, -um Of Rudbar, north-west Iran, near Resht.

Rudbe'ckia f. Cone-flower, black-eyed Susan. Named for Olof Rudbeck (1630–1702), 'one of the most remarkable Swedes of all time', anatomist, botanist and antiquarian, founder of the Uppsala botanic garden. The woodblocks illustrating some 11,000 plants and intended for a great work on which he had spent much of his life were nearly all destroyed in the terrible Uppsala fire of 1702, and he did not long survive this disaster. His son Olof Rudbeck the younger (1660–1740), also a professor in Uppsala, befriended Linnaeus when he was a poverty-stricken student. *Rudbeckia* commemorates both. COMPOSITAE.

ru'dis, -is, -e Rough, untilled, coarse.

Ruel'lia f. Herbs and shrubs named for Jean Ruel (1474–1537). herbalist to François I of France. ACANTHACEAE.

rufes'cens Becoming reddish.

ru'fus, -a, -um Red.

rugo'sus, -a, -um Wrinkled, rugose.

Ruhmoh'ra f. In honour of Karl von Ruhmohr (1785–1843), German art expert. DAVALLIACEAE.

Ru'mex f. Sorrel, dock. The Latin name. POLYGONACEAE.

runcina'tus, -a, -um Saw-toothed, with the teeth pointing towards the base.

rupes'tris, -is, -e Rock-loving.

rupi'cola A dweller among rocks.

rupi'fragus, -a, -um Rock-breaking, i.e. growing in clefts of the rock.

Rus'chia f. Named after a South African farmer, Ernst Julius Rusch (1867–1957). *Ruschianthus* commemorates his son Ernst Franz Theodor Rusch (1897–1964). AIZOACEAE.

ruscifo'lius, -a, -um Having leaves in appearance resembling the leaf-like shoots (phylloclades) of butcher's broom (*Ruscus*).

Rus'cus m. Butcher's broom. The Latin name. LILIACEAE (RUSCACEAE).

Ruspo'lia f. In honour of Prince Eugenio Ruspoli (1866–1893), Italian explorer of Somaliland killed by an elephant. ACANTHACEAE.

russa'tus, -a, -um Russet.

Russe'lia f. In honour of Dr. Alexander Russell (*c.* 1715–1768), physician to the English Factory at Aleppo (*c.* 1740 to 1753) and the author of a *Natural History of Aleppo* (1756). SCROPHULARIACEAE.

russellia'nus, -a, -um In honour of John Russell (1766–1839), 6th Duke of Bedford, a zealous promoter of botany, gardening, forestry and agriculture at Woburn Abbey, responsible for the publication of works on grasses, heaths, willows and conifers.

russotin'ctus, -a, -um Red-tinged.

rustica'nus, -a, -um; rus'ticus, -a, -um Pertaining to the country.

Ru'ta f. Rue. The Latin word. As in English, the same word means bitterness or unpleasantness. Juice on the skin exposed to sunlight may cause dermatitis. RUTACEAE.

ruthen'icus, -a, -um From Ruthenia, a region of European Russia inhabited by the Ruthenians; the name was, however, used in a general sense for much of European Russia.

ru'tilans Reddish.

Rut'tya f. In honour of John Rutty (1697–1775), physician in Dublin and author of *The Natural History Of County of Dublin* (1772). The name *Ruttyruspolia* covers the hybrids of *Ruspolia* and *Ruttya*. ACANTHACEAE.

S

Sa'bal f. Palmetto. Possibly from the South American name for these spineless palms. PALMAE.

Saba'tia f. Rose pink, American centaury. Named for an Italian botanist and gardener, Liberato Sabbati (b. *c.* 1714), curator of the Rome botanic garden. GENTIANACEAE.

saba'tius, -a, -um Of Savona, Riviera, north-west Italy.

sabau'dus, -a, -um Of Savoy, south-eastern France.

sacca'tus, -a, -um Resembling a bag; provided with pronounced sacs or nectar-producing pits; see *saccifer*.

sacchara'tus, -a, -um Sugared or sugary, i.e. sweet-tasting or looking as if sprinkled with sugar.

saccha'rifer, sacchari'fera, -um Sugar-producing.

sacchari'nus, -a, -um Sugary.

saccharoi'des Resembling sugarcane.

Sac'charum n. Sugarcane. Gr. *sakcharon*, the sweet juice of sugarcane, from an Asiatic word, seemingly the Malay *singkara*. Sugarcane has been known since antiquity in Asia whence it passed to Africa. It was introduced into Sicily around A.D. 800 and by the Moors into Spain before A.D. 1000. The cane was taken by Columbus to the New World on his second voyage where it quickly became widely cultivated and led to the establishment of negro slavery. GRAMINEAE.

sac'cifer, sacci'fera, -um Bag-bearing, i.e. provided with hollowed-out organ or sac.

Saccola'bium n. From L. *saccus*, bag; *labium*, lip; in allusion to the lip of the flower. ORCHIDACEAE.

sachalinen'sis, -is, -e Of the island of Sakhalin, north of Japan.

sacro'rum Of sacred places.

sadda'riffa. Epithet of a *Hibiscus*, possibly a Turkish name.

Sadle'ria f. Named for Joseph Sadler (1791–1841), professor of botany, Budapest. BLECHNACEAE.

Sagi'na f. Pearlwort. L. *sagina*, fodder; from the fattening qualities of spurrey (*Spergula sativa* syn. *Sagina spergula*) on which sheep quickly thrive. CARYOPHYLLACEAE.

sagitta'lis, -is, -e; sagitta'tus, -a, -um Arrow-shaped.

Sagitta'ria f. Arrowhead. L. *sagitta*, an arrow; from the form of the leaves. ALISMATACEAE.

sagittifo'lius, -a, -um Arrow-leaved.

Saintpau'lia f. African violet. Named for a German, Baron Walter von Saint Paul-Illaire (1860–1910), who collected the first African violet (*S. ionantha*) in East Africa. GESNERIACEAE.

salicifo'lius, -a, -um Willow-leaved.

salici'nus, -a, -um Willow-like.

Salicor'nia f. Glasswort, marsh samphire. L. *sal*, salt; *cornu*, a horn; from the hornlike branches of these salt-marsh herbs. CHENOPODIACEAE.

salicornioi'des Resembling *Salicornia*.

salig'nus, -a, -um Resembling willow.

sali'nus, -a, -um Growing in salty places.

Sa'lix f. Willow, osier, sallow. The Latin name. Willow bark was, until the early part of the 20th century, the sole source of salicylic acid which has great medicinal value, especially in the treatment of rheumatism. German chemists discovered before 1914 how to synthesize the acid which became the principal ingredient of aspirin. In many country places in England, pussy willows were eagerly sought by children for the silky catkin-bearing twigs called 'palms' to be used for church decoration on Palm Sunday. SALICACEAE.

salmia'nus, -a, -um; sal'mii In honour of Prince Joseph M. F. A. H. I. Salm-Reifferscheid-Dyck (1773–1861), Rhineland German prince who became a leading authority on succulent plants; see also *Dyckia*.

Salpi'chroa f. From Gr. *salpinx*, a trumpet; *chroma*, colour; in allusion to the pretty trumpet-shaped flower. SOLANACEAE.

Salpiglos'sis f. From Gr. *salpinx*, a trumpet; *glōssa*, a tongue; with reference to the elongated trumpet-shaped style. SOLANACEAE.

Sal'sola f. Russian thistle. From L. *sal*, salt. They are generally salt-soil plants, often prickly but not true thistles. CHENOPODIACEAE.

salsugino'sus, -a, -um Growing in salt marshes.

saluenen'sis, -is, -e Of the Salween (Salwin) River (Nu Jiang) of Burma and China.

Sal'via f. Sage. The Latin name, presumably derived from *salvus*, safe, well, sound; from the supposed medicinal values of the plants. S. *sclarea*, clary or clear-eye, was particularly valued for affections of the eye, hence its English name. LABIATAE.

salviifo'lius, -a, -um *Salvia*-leaved.

Salvin'ia f. A floating plant named for Antonio Maria Salvini (1633–1729), professor of Greek at Florence, who helped Micheli in his botanical studies. SALVINIACEAE.

Sama'nea f. Rain-tree or monkey-pod. From the native South American name, *saman*. It is called rain-tree because the leaflets fold up in cloudy weather and at night. LEGUMINOSAE.

sambucifo'lius, -a, -um With leaves like elder or *Sambucus*.

sambu'cinus, -a, -um Resembling *Sambucus* or elder, particularly in smell.

Sambu'cus f. Elder. The Latin name, perhaps connected with *sambuca*, a kind of harp. CAPRIFOLIACEAE.

Sa'molus m. The Latin name, probably of Celtic origin. PRIMULACEAE.

Samue'la f. Date-yucca. Named for Samuel Farlow Trelease (1892–1958), American botanist nicknamed 'Selenium Sam', son of William Trelease (1857–1945). Both flowers and fruit are eaten locally in Mexico. AGAVACEAE.

Sanche'zia f. Named for José Sanchez, a 19th-century professor of botany at Cadiz. ACANTHACEAE.

sanc'tus, -a, -um Holy.

sanderi In honour of Henry Frederick Conrad Sander (1847–1920), German-born British nurseryman with orchid nurseries at St. Albans, England and Bruges, Belgium.

Sanderso'nia f. In honour of John Sanderson (1820 or 1821–1881), who discovered this remarkable plant in Natal. LILIACEAE (COLCHICACEAE).

Sanguina'ria f. Bloodroot. From L. *sanguis*, blood. All parts have copious yellowish-red sap. PAPAVERACEAE.

sanguin'eus, -a, -um Blood-red.

Sanguisor'ba f. Burnet. From L. *sanguis*, blood; *sorbeo*, to soak up. The tannin-rich rootstock has styptic qualities and was used as an infusion to stop bleeding from dysentery, etc. ROSACEAE.

Sansevie'ria f. Bow-string hemp and, in the florists' shops, snake plant or leopard lily. Named for an 18th-century Italian patron of horticulture. AGAVACEAE (DRACAENACEAE).

San'talum n. Sandalwood. Gr. *santalon*, sandalwood tree, derived via the form *sandanon* from Sanskrit *chandana*, fragrant. Fragrant carved boxes and fans from India are made from sandalwood which is also employed in Buddhist funeral and religious rites. Ground to a paste, it is one of the pigments used by Brahmans in making their caste marks. SANTALACEAE.

Santoli'na f. Lavender cotton. L. *sanctum linum*, holy flax, an old name for the species *S. virens*. COMPOSITAE.

Sanvita'lia f. Apparently named for Federico Sanvitali (1704–1761), professor at Brescia, Italy. COMPOSITAE.

sa'pidus, -a, -um Pleasant to taste.

sapien'tum Pertaining to wise men; used of the better forms of banana in contrast to those called *troglodytarum* considered fit only for apes or cave-dwellers.

Sapin'dus m. Soap-berry. From L. *sapo*, soap; *indicus*, Indian. The pulp of the fruits lathers like soap and was used as such by Indians. SAPINDACEAE.

Sa'pium m. Latin name for a resinous pine. The stems of these trees, which are not pines, exude a sticky sap. *S. sebiferum*, the Chinese tallow tree, has seeds with a waxy covering which can be used in making soap and candles. EUPHORBIACEAE.

sapona'ceus, -a, -um Soapy.

Sapona'ria f. Soapwort, bouncing Bet. L. *sapo*, soap. The roots of *S. officinalis* lather and were once used as a substitute for soap. CARYOPHYLLACEAE.

sapo'ta f. From native South American name. The sapodilla or naseberry (*Manilkara zapota*, syn. *Sapota achras*) has a delicious fruit but commercially it is most important as the source of chicle, the basis of chewing gum. It should not be confused with the sapote or mammee sapote (*Pouteria sapota*, syn. *Calocarpum sapota*) which has a much larger seed.

Sa'raca f. Apparently a corruption of an East Indian name *Asoka*, commonly applied to this tree which is remarkable among *Leguminosae* in having brightly coloured sepals but no petals. LEGUMINOSAE.

Sarcan'thus m. From Gr. *sarx*, *sarkos*, flesh; *anthos*, a flower. The flowers are fleshy. ORCHIDACEAE.

Sarcocau'lon n. From Gr. *sarx*, *sarkos*, flesh; *caulŏs*, stem; in allusion to the fleshy stems. GERANIANACEAE.

Sarcochi'lus m. Gr. *sarx*, *sarkos*, flesh; *cheilos*, a lip; in allusion to the fleshy lip of the flower. ORCHIDACEAE.

Sarcococ'ca f. Sweet box. Gr. *sarkos*, flesh; *kokkos*, a berry; in allusion to the fleshy fruits of these ornamental shrubs. BUXACEAE.

sarco'des Fleshlike.

Sarco'stemma n. From Gr. *sarx*, *sarkos*, flesh; *stĕmma*, wreath, garland; in allusion to the fleshy corona. ASCLEPIADACEAE.

sargentia'nus, -a, -um; sargen'tii In honour of Charles Sprague Sargent (1841–1927), leading American dendrologist, first director of the Arnold Arboretum of Harvard University.

Sargento'doxa f. In honour of C.S. Sargent (see above); Gr. *doxa*, repute, glory. SARGENTODOXACEAE.

sarma'ticus, -a, -um From Sarmatia, i.e. eastern Poland and south-western European Russia.

sarmento'sus, -a, -um Having or bearing runners.

sarnien'sis, -is, -e Of Guernsey (ancient Sarnia), Channel Islands.

Sarotham'nus m. From Gr. *sarŏn*, broom; *thamnŏs*, shrub. LEGUMINOSAE.

Sarrace'nia f. Pitcherplant. Named for Michel Sarrasin (1659–1734), French physician and botanist at Quebec, who sent the first of these carnivorous plants to Europe. SARRACENIACEAE.

Sa'sa f. The Japanese name for various bamboos. GRAMINEAE.

sasan'qua Japanese name *sasankwa* for a popular native species of *Camellia*.

Sas'safras n. Derivation of the name for these handsome trees is dubious, but it probably comes from an American Indian name used in Florida. The roots and bark are aromatic and are the source of oil of sassafras which is used in perfumery. LAURACEAE.

sati'vus, -a, -um Cultivated.

satura'tus, -a, -um Saturated.

Sature'ja (also at one time, **Satureia**) f. Savory. The Latin name for this herb which was well known to the ancients. It is an excellent bee plant and was recommended for planting around hives by Virgil in the *Georgics*. LABIATAE.

sauroce'phalus, -a, -um Lizard-headed.

Sauro'matum n. An old name used by Dodoens and derived from Gr. *sauros*, a lizard. The dried tubers of *S. venosum* sold as Monarch of the East or Voodoo lily, produce their ill-smelling inflorescences without being watered; the leaves come later. ARACEAE.

Sauru'rus m. Gr. *sauros*, a lizard; *oura*, a tail; from the dense spicate inflorescence. SAURURACEAE.

Saussu'rea f. Perennial mountain herbs named for Horace Bénédict de Saussure (1740–1799), Swiss philosopher and scientist, for many years professor of philosophy at Geneva, a pioneer investigator of the geology, meteorology and botany of the high Alps, celebrated for his ascent in 1787 of Mont Blanc, Europe's highest mountain, first conquered in 1786 by Paccard and Balmat through his encouragement. COMPOSITAE.

savann'arum Of savannas, grassy plains with few trees.

saxa'tilis, -is, -e Found among rocks.

Saxegothae'a f. Prince Albert's yew. Named for Prince Albert (1819–1861) of Saxe-Coburg-Gotha, Prince Consort of Queen Victoria, and a zealous promoter of the arts and sciences. PODOCARPACEAE.

saxi'cola A grower among rocks.

Saxif'raga f. Saxifrage. L. *saxum*, a rock; *frango*, to break. Growing in rock crevices this herb was supposed to be capable of breaking rocks. Hence, by deduction, it was accorded a medicinal quality of breaking up stone in the bladder. SAXIFRAGACEAE.

saxo'sus, -a, -um Full of rocks, hence growing among rocks.

sca'ber, scab'ra, scab'rum Rough.

Scabio'sa f. Mourning bride, pincushion flower, scabious. L. *scabies*, the itch, which, by the analogy of the roughness of the leaves, the plant was supposed to cure. DIPSACEAE.

scabiosifo'lius, -a, -um With leaves like *Scabiosa*.

scabrifo'lius, -a, -um Rough-leaved.

Scadoxus m. Compounded by Rafinesque from Gr. *skiadion*, parasol, umbel; *doxa*, glory. AMARYLLIDACEAE.

scala'ris, -is, -e. Relating to a ladder (*scalae*, staircase, ladder), hence with reference to a pinnate leaf with many leaflets.

scan'dens Climbing.

sca'piger, scapi'gera, -um Scape-bearing.

scapo'sus, -a, -um With scapes, i.e. with leafless flower-stems.

scario'sus, -a, -um Shrivelled, used of thin, dry organs.

scarlati'nus, -a, -um Scarlet.

scelera'tus, -a, -um Wicked, hurtful.

Scele'tium n. From Gr. *skĕlĕtŏs*, dried up; with reference to the persistent skeletonized dead leaves. AIZOACEAE.

scep'trum A sceptre.

Schaue'ria f. Named for Johann Conrad Schauer (1813–1848), professor at Greifswald, Germany, a specialist in the study of *Myrtaceae* and *Verbenaceae*. ACANTHACEAE.

Schee'lea f. Named for Karl Wilhelm Scheele (1742–1786), celebrated Swedish chemist of German origin, the discoverer of many substances, including Scheele's Green and prussic acid. PALMAE.

Scheffle'ra f. Trees or shrubs named for J. C. Scheffler, 19th-century botanist of Danzig (Gdansk, Poland). ARALIACEAE.

schiedean'us, -a, -um Named for Christian Wilhelm Schiede (1780–1836), German botanist and plant collector in Mexico from 1828 to 1836.

schillin'gii Named for Antony Schilling (b. 1935), English gardener and plant collector in Nepal, Bhutan and China, Deputy Curator of Royal Botanic Gardens, Kew, in charge of Wakehurst Place 1963–1991.

Schi'ma f. Derivation obscure; according to Backer, from Gr. *skiasma*, shadow, shelter; with reference to the thick leafy crown of the tree. THEACEAE.

schin'seng Variant of *ginseng*.

Schi'nus m. Pepper-tree. Gr. *schinos*, the mastic tree, which this genus resembles in that the trees exude a mastic-like juice. ANACARDIACEAE.

schipkaen'sis, -is, -e Referring to the Shipka Pass, Bulgaria.

Schisan'dra f. Gr. *schizo*, to divide; *aner*, *andros*, male; in allusion to the well separated anther cells. SCHISANDRACEAE.

Schismatoglot'tis f. Gr. *schismas*, cleft, division; *glōtta*, a tongue. The tongue-like blade of the spathe becomes cut off from the more persistent tubular basal part. ARACEAE.

schisto'calyx With split calyx.

Schivere'ckia f. Named for S. B. Schivereck (1782–1815) of Krakow, Poland. CRUCIFERAE.

Schizae'a f. Comb or rush fern. Gr. *schizo*, to divide. The fronds are divided into a fan shape. SCHIZAEACEAE.

Schizan'dra See *Schisandra*.

Schizan'thus m. Butterfly or fringe flower. Gr. *schizo*, to divide; *anthos*, a flower; the corolla is deeply cut. SOLANACEAE.

Schizoco'don m. Gr. *schizo*, to divide; *k;amodōn*, a bell; the segments of the bell-shaped corolla are elegantly cut into a deep fringe. DIAPENSIACEAE.

Schizolo'bium n. Gr. *schizo*, to divide; *lŏbŏs*, pod; the thin inner coat of the pod splits away from the firm outer coat. LEGUMINOSAE.

Schizope'talon n. Gr. *schizo*, to divide; *petalon*, a petal; the flowers have deeply cut petals. CRUCIFERAE.

schizope'talus, -a, -um With cut petals.

Schizophrag'ma n. Gr. *schizo*, to divide; *phragma*, a fence or screen, hence septum with reference to the inner wall of the capsule dividing into transverse horizontal fibres. HYDRANGEACEAE.

schizophyl'lus, -a, -um With cut leaves.

Schizosty'lis f. Kaffir lily. Gr. *schizo*, to divide; *stylis*, a column, style. The style is divided into three parts. IRIDACEAE.

schlippenba'chii In honour of Baron Alexander von Schlippenbach, Russian naval officer of Baltic German origin, who collected in Korea in 1854.

Schlumber'gera f. Named for Frederick Schlumberger, Belgian horticulturist, *c.* 1900. CACTACEAE.

schoenopra'sum Specific epithet for chives (*Allium schoenoprasum*) compounded from Gr. *schoinos*, rush; *prasŏn*, leek; with reference to its rush-like leaves.

schola'ris, -is, -e Relating to school. *Alstonia scholaris* was so named because children wrote their exercises on boards made from its wood.

Schomburg'kia f. Named for Sir Robert Hermann Schomburgk (1804–1865), German-born, naturalized British explorer who found the *Victoria amazonica* (*V. regia*) water-lily in Guyana in 1837. It had, however, been discovered in the Rio Mamore, a tributary of the Amazon, in 1801 by Thaddaeus Haenke, also German-born, then employed by the Spanish government. Moritz Richard Schomburgk (1811–1890), his brother, accompanied him to Guyana in 1840 and was later the director of the Adelaide botanic garden. ORCHIDACEAE.

Scho'tia f. Named for Richard van der Schot (*c.* 1730–1819), Dutch head gardener at the palace of Schönbrunn, near Vienna. LEGUMINOSAE.

Schran'kia f. Named for Franz von Paula von Schrank (1747–1835), German botanist, professor at Munich. LEGUMINOSAE.

Schuber'tia f. In honour of Gotthilf Heinrich von Schubert (1780–1860), German doctor and naturalist, 1827 onwards professor of natural history in Munich, who collected in 1837 in Egypt, Palestine and Greece. ASCLEPIADACEAE.

Sciado'pitys f. Umbrella pine. Gr. *skias*, *skiados*, umbel; *pitys*, a fir tree. The spreading whorls of needles resemble the ribs of an umbrella or parasol, for which the Greek name was *skiadĕiŏn*.

Scil'la f. Squill. The Greek name *skilla* for sea-squill (*Urginea maritima*). The English bluebell used to be named S. *nutans*, but is now placed in the genus *Hyacinthoides* as *H. nonscripta*. LILIACEAE (HYACINTHACEAE).

scilloi'des Resembling *Scilla*.

Scindap'sus m. Ivy arum. Gr. *skindapsos*, a word with varied applications, one being to an ivy-like tree. ARACEAE.

scintil'lans Gleaming.

Scir'pus m. Clubrush. The Latin name. Once called bulrush; the biblical bulrush was the related *Cyperus papyrus*, but Victorian illustrators portrayed the reedmace or cat-tail (*Typha*) instead. See *Typha*. CYPERACEAE.

scla'rea Italian plant name.

Scleran'thus m. From Gr. *sklēros*, hard, dry, harsh; *anthos*, flower; in allusion to the hardened flower. CARYOPHYLLACEAE.

Sclerocac'tus m. Gr. *sklērŏs*, hard, harsh, cruel, plus *Cactus*; in allusion to the sharp spines. CACTACEAE.

sclerocar'pus, -a, -um With hard fruits.

Scolio'pus m. From Gr. *skŏlios*, curved, bent; with reference to the twisting flower-stalks. LILIACEAE (TRILLIACEAE).

Scolopen'drium n. Gr. *skŏlŏpĕndrion*, from *skŏlŏpĕndra*, millipede; in allusion to the rows of sori. ASPLENIACEAE.

Sco'lymus m. Latin name for Spanish oyster-plant. This biennial has edible roots and is closely related to salsify. COMPOSITAE.

scopar'ius, -a, -um Broomlike.

Scopo'lia f. In honour of Giovanni Antonio Scopoli (1723–1788), Italian physician, botanist, zoologist, chemist and geologist, 1754–1767 state physician in Carniola, professor in Schemnitz, then Pavia. SOLANACEAE.

scopulo'rum Of cliffs, crags, projecting rocks.

scorodopra'sum Gr. *skŏrŏdŏprasŏn*, a kind of *Allium* between the garlic, *skŏrŏdŏn* or *skŏrdŏn*, and the leek, *prasŏn*.

scorpioi'des Resembling a scorpion, Gr. *skŏrpiŏs*.

Scorpiu'rus m. Gr. *skorpios*, a scorpion; *oura*, a tail; from the form of the pods. LEGUMINOSAE.

Scorzone'ra f. Black salsify. From Old French *scorzon*, Italian *scorzone*, viper. The root was once regarded as a cure for snakebite. COMPOSITAE.

sco'ticus, -a, -um Scottish.

scotophy'lus, -a, -um From Gr. *skotos*, dark; *phyllon*, leaf.

scottia'nus, -a, -um In honour of Munro B. Scott (1889–1917), Kew botanist killed in action in the First World War.

scrobicula'tus, -a, -um Pitted with minute depressions.

Scrophula'ria f. Fig-wort. L. *scrofule*, scrophula, which it was supposed to cure. SCROPHULARIACEAE.

scul'ptus, -a, -um Carved.

scuta'tus, -a, -um Armed with a shield, shield-shaped; the *scutum* was the long oblong shield of Roman legionaries.

Scutella'ria f. Skullcap. L. *scutella*, a small dish or saucer; from the pouch on the fruiting calyx. LABIATAE.

Scutica'ria. f. L. *scutica*, a whip, lash; from the shape of the long leaves of these orchids. ORCHIDACEAE.

scu'tum A shield.

Scyphan'thus m. From Gr. *skyphŏs*, cup, beaker; *anthos*, flower. LOASACEAE.

Scyphoste'gia f. From Gr. *skyphŏs*, beaker; *stĕgē*, covering, shelter. SCYPHOSTEGIACEAE.

se'bifer, sebi'fera, -um Tallow-bearing.

sebo'sus, -a, -um Full of grease or tallow.

Seca'le n. Rye. Latin name for some cereal grain, perhaps rye. GRAMINEAE.

secali'nus, -a, -um Resembling rye.

sechella'rum Of the Seychelle Islands in the Indian Ocean.

Se'chium n. From the West Indian name. CUCURBITACEAE.

seclu'sus, -a, -um Hidden.

secundiflo'rus, -a, -um; secun'dus, -a, -um One-sided, secund. Applied to leaves or flowers arranged on one side of a stalk only.

Securi'daca f. L. *securis*, an axe; from the shape of the wing at the end of the pods. POLYGALACEAE.

secu'riger, securi'gera, -um Axe-bearing.

Securi'gera f. Hatchet vetch. L. *securis*, an axe; *gero*, to bear; from the shape of the pods. LEGUMINOSAE.

Se'dum n. Stonecrop, wall-pepper. Latin name used by the ancients for various succulents, mostly species of *Sedum* or *Sempervivum*, from *sedo*, to sit; from the manner in which some species grow on rocks and walls. CRASSULACEAE.

segeta'lis, -is, -e; se'getum Of cornfields.

seinghkuen'sis, -is, -e From the Seinghku Valley, Upper Burma.

Selaginel'la f. Club-moss. Diminutive of *Selago*, the name of another moss-like plant (*Lycopodium selago*). SELAGINELLACEAE.

selaginoi'des Resembling club-moss.

Selenice'reus m. Night-blooming cereus. Gr. *sĕlēnē*, the moon, plus *Cereus*. The flowers are mostly nocturnal. CACTACEAE.

Selenipe'dium n. From Gr. *sĕlēnē*, the moon; *pĕdilŏn*, sandal; referring to the lip of the flowers. ORCHIDACEAE.

selen'sis, -is, -e Of the Sie-la pass, Yunnan, China.

Sellie'ra f. In honour of François Noël Sellier (1737–*c.* 1800) of Paris, who engraved botanical illustrations for Cavanilles and Desfontaines between 1780 and 1800. GOODENIACEAE.

sel'loi, sellovia'nus, -a, -um; sello'wii In honour of Friedrich Sellow (1789–1831), German traveller and naturalist, who made extensive collections in Brazil and Uruguay. The family name was 'Sello' but when in Brazil he altered it to 'Sellow', hence the varied spelling of epithets of South American plants commemorating him.

Se'mele f. A climbing shrub named for Sĕmĕle, in Greek mythology the daughter of Kadmos and mother of Dionysos. LILIACEAE (RUSCACEAE).

semi- In Latin compound words signifies half.

semperflo'rens Ever-blooming.

semper'virens Evergreen.

Sempervivel'la f. Diminutive of *Sempervivum*. CRASSULACEAE.

sempervivoi'des Resembling *Sempervivum*.

Sempervi'vum n. House-leek. L. *semper*, always; *vivus*, alive, living. Houseleek, *S. tectorum* (of roofs), was regarded as effective against lightning and for that reason this fleshy herb was planted on roofs. CRASSULACEAE.

Sene'cio m. Groundsel, ragwort. L. *senex*, an old man; from the hoary pappus of these plants. COMPOSITAE.

senecioi'des Resembling *Senecio*.

senes'cens Growing old.

seni'lis, -is, -e Aged, hence white-haired.

sensib'ilis, -is, -e; sensiti'vus, -a, -um Sensitive; responding quickly to touch, changes in light, etc.

sentico'sus, -a, -um Thorny, full of thorns.

sepiar'ius, -a, -um; se'pium Growing in hedges or used for hedging.

sept- In compound words signifying seven. Thus:
septangula'ris (seven-angled); **septem'fidus** (with seven cuts); **septem'lobus** (with seven lobes); **septempuncta'tus** (seven-spotted).

septentriona'lis, -is, -e Northern; from *septemtriones* and *septentriones*, the seven stars of the Plough or Wain.

sepulcra'lis, -is, -e Growing in burial places.

sepul'tus, -a, -um Buried.

Sequoi'a f. Redwood. The name commemorates the inventor of the Cherokee alphabet Sequoia or Sequoiah (1770–1843), otherwise George Gist, the son of a German-American merchant and an American Indian girl. The name *sequoia*, in Cherokee the name of the opossum, was used as a nickname for a half-breed. He was born at a village by the Cherokee (now Tennessee) River and brought up as a full-blooded Indian without knowledge of English. The adoption of his syllabary by the Cherokees in 1821 quickly made half the tribe literate, but did not save them from eviction from their territory in 1838 by the U.S. Army and the death of 4,000 Cherokees on behalf of white Americans greedy for their land. TAXODIACEAE.

Sequoiaden'dron n. Big tree. *Sequoia* plus Gr. *dendron*, tree. In England *Wellingtonia* is the commonly accepted synonym of *S. gigantea*. The first seeds seem to have been collected in the early 1850's by G. H. Woodruff, a down-and-out gold miner in California, who found himself in a shower of seeds dropped by squirrels feeding on the cones. He sent the seeds by pony express at a cost of $25 to the nursery of Ellwanger and Barry of Rochester, New York. They obtained about 4,000 seedlings, of which 400 went to England. Every great European house and botanic garden wanted an avenue. Woodruff's share of the profits came to $1,030.60. The tree was originally listed as '*Washingtonia gigantea*, . . . *Wellingtonia* of the English'. TAXODIACEAE.

Sera'pias f. From the Egyptian god Serapis. ORCHIDACEAE.

Sere'noa f. Saw palmetto. Named for Sereno Watson (1826–1892), of Harvard University, distinguished American botanist who named and described many of the new species found during the pioneer botanical exploration of western and middle North America. PALMAE.

serican'thus, -a, -um With silky flowers.

seri'ceus,-a, -um Silky.

seri'cifer, seri'cifera, -um; seri'cofer, seri'cofera, -um Silk-bearing.

Sericocar'pus f. Gr. *sērikon*, silk; *karpŏs*, fruit. The dry fruits (*cypsalae*) are covered with silky hairs. COMPOSITAE.

Seris'sa f. From the East Indian name for this shrub. RUBIACEAE.

Serja'nia f. Named in 1703 by Plumier in honour of the Rev. Father Philippe Sergeant of Caux, France, a monk skilled in botany and medicine, for many years domiciled in Rome. SAPINDACEAE.

sero'tinus, -a, -um Late in flowering or ripening.

ser'pens Creeping.

serpenti'nus, -a, -um Serpentine; relating to snakes, or to serpentine rocks.

serpyllifo'lius, -a, -um With leaves like *Thymus serpyllum*.

serratifo'lius, -a, -um With serrated or saw-toothed leaves.

Serra'tula f. L. *serrula*, a small saw; from the serrated leaves. COMPOSITAE.

serra'tus, -a, -um Saw-toothed.

serrula'tus, -a, -um With small saw-like teeth.

Se'samum n. Sesame. The Greek name *sēsamŏn*, from a Semitic name for this important oil plant. PEDALIACEAE.

Sesba'nia f. From the Arabic name. LEGUMINOSAE.

Se'seli n. Gr. *Sĕsĕli*, ancient name of an umbelliferous plant. UMBELLIFERAE.

Sesle'ria f. In honour of Leonardo Sesler (d. 1785), doctor at Venice, who had a private botanic garden there. GRAMINEAE.

sesquipeda'lis, -is, -e One and one-half feet in length or height, applied to a Madagascar species of *Angraecum* because of the length of the spur of this orchid, which led Darwin to predict that a moth with a tongue (proboscis) long enough to reach the nectar in it would later be discovered; such a moth was indeed later discovered and named *Xanthopan morgani* subsp. *praedicta*.

ses'silis, -is, -e Stalkless; sessile. In compound words *sessili-*.
sessili'florus, -a, -um With sessile flowers.
seta'ceus, -a, -um Bristled.
Seta'ria f. Foxtail millet. L. *seta*, a bristle; from the bristles on the spikelet. GRAMINEAE.
setchuenensis, -is, -e Of Sichuan (Szechwan) province, China.
seti- In compound words signifies bristled. Thus:
setifo'lius (with bristly or bristle-like leaves); **se'tifer** (bearing bristles); **se'tiger** (bearing bristles); **seto'sus** (full of bristles, bristly); **setulo'sus** (full of small bristles).
Severi'nia f. Named for Marco Aurelio Severino (1580–1656), professor of anatomy at Naples. RUTACEAE.
sexangula'ris, -is, -e With six angles.
shal'lon Rendering of a western American Indian (Chinook) name *Kikwu-salu* for *Gaultheria shallon*.
Shepher'dia f. Named for John Shepherd (1764–1836), curator of the Liverpool botanic garden. One of the first gardeners to raise ferns from spores. ELEAGNACEAE.
sherri'ffii Specific epithet of many Himalayan plants named after Major George Sheriff (1898–1967), who explored south-eastern Tibet in company with F. Ludlow.
Shibataea f. Japanese bamboo named for Keita Shibata (1877–1949), Japanese botanist, professor at Tokyo 1912–1938. GRAMINEAE.
Shor'tia Evergreen stemless herbs named for Dr. Charles W. Short (1794–1863), Kentucky botanist. The American *S. galacifolia*, sometimes called Oconee bells, came to the notice of Asa Gray in 1839 when he went to Paris to study André Michaux's North American herbarium made some fifty years before. Michaux's specimen lacked flowers but Gray was much impressed. Michaux had collected it somewhere in Carolina, in 1788, but it was not rediscovered until 1888. Sargent in 1888 found the place in the mountains of Carolina where Michaux had camped in December 1788 and there grew the *Shortia*. DIAPENSIACEAE.
shwelien'sis, -is, -e Of the Shweli river (Longchuan Jiang), western China.
sia'meus, -a, -um Of Siam (Thailand).
Sibbal'dia f. In honour of Sir Robert Sibbald (1641–1722), first professor of medicine in the University of Edinburgh and author of *Scotia Illustrata*. ROSACEAE.
Sibirae'a f. From the Siberian habitat of one species. ROSACEAE.
Sibthor'pia f. Named for Humphrey Sibthorp (1713–1797), English botanist and professor of botany at Oxford from 1747 to 1783, during which period he gave only one lecture; he was the father of John Sibthorp (1758–1796),

professor of botany at Oxford from 1783 to 1795, celebrated for his travels in Greece, Cyprus and Asia Minor, commemorated in plants named *sibthorpii*. SCROPHULARIACEAE.

Sica'na f. From the Peruvian name. CUCURBITACEAE.

siculifor'mis, -is, -e Dagger-shaped.

si'culus, -a, -um; sicilie'nsis, -is, -e Of Sicily; Sicilian.

Si'cyos m. Bur-cucumber. Greek for cucumber, to which this is closely related. CUCURBITACEAE.

Si'da f. Gr. name for a water plant, transferred to this genus. MALVACEAE.

Sidal'cea f. False mallow. From *Sida* and *Alcea*, both related genera. MALVACEAE.

Side'rasis f. Said by the author of the name, Rafinesque, to mean 'rusty fur', in allusion to the dense reddish hair-covering, presumably from Gr. *sīdērŏs*, iron; *sisura*, goat-skin. COMMELINACEAE.

Sideri'tis f. The Greek name *sidērītis* for plants used as wound dressings, etc., from *sīdērŏs*, iron; being supposed to heal wounds caused by swords and other iron weapons. LABIATAE.

siderophloi'us, -a, -um Iron-barked.

siderophyl'lus, -a, -um With rusty leaves.

Siderox'ylon n. Miraculous berry of West Africa. From Gr. *sīdērŏs*, iron; *xylŏn*, wood; from the hardness of the heartwood. SAPOTACEAE.

sie'beri In honour of Franz Wilhelm Sieber (1789–1844) of Prague, natural history collector and traveller, who himself visited Italy, Crete, Egypt, Palestine, Mauritius, Australia and South Africa and sent collectors elsewhere.

siebol'dii; sieboldi'anus, -a, -um In honour of Philipp Franz van Siebold (1796–1866), German doctor, in Japan from 1823 to 1830, during which period he introduced many Japanese plants into European gardens and collected material for his important publications on Japan; he was again in Japan from 1859 to 1863.

Sigesbe'ckia f. A weedy small-flowered plant named by Linnaeus after his bitter opponent Johann Georg Siegesbeck (1686–1755). COMPOSITAE.

signa'tus, -a, -um Well-marked.

sikkim'ensis -is, -e Of Sikkim, eastern Himalayan region between Nepal and Bhutan.

sikokia'nus, -a, -um Of the island of Shikoku, southern Japan.

Sile'ne f. Catchfly, campion. The Greek name for another plant (*Viscaria*), now applied to these common herbs. CARYOPHYLLACEAE.

sili'ceus, -a, -um Growing in sand.

sillamontanus, -a, -um Of Silla mountain, north Mexico.

Sil'phium n. Rosinweed. Perennial American herbs to which has been transferred the Greek name *silphiŏn* of an umbelliferous plant, generally considered to be a species of *Ferula* now extinct but possibly *Cachrys ferulacea*. It was in ancient times of great economic importance in Cyrenaica as the source of an aromatic gum and hence portrayed on coins. COMPOSITAE.

silva-taroucan'us, -a, -um In honour of Count Ernst Emmanuel Silva-Tarouca (1860–1936), Austrian botanist, horticulturist and dendrologist who formed a remarkable arboretum at Pruhonice Park near Prague, Czechoslovakia and produced authoritative handbooks in German on herbaceous perennials, conifers and deciduous trees.

silva'ticus, -a, -um; silves'tris, -is, -e Growing in woods or growing wild.

silvi'cola An inhabitant of woods.

Si'lybum n. Lady's thistle, milk thistle, blessed thistle. The Greek name, *silybŏn*, for some thistlelike plant. The English names apply to S. *marianum*, also called Our Lady's thistle from the white spots on the leaves supposed to have been the result of milk dropped on them by the Virgin Mary. COMPOSITAE.

Sime'this f. In Gr. mythology a nymph, mother of the Sicilian shepherd Acis. LILIACEAE (ASPHODELACEAE).

sim'ilis, -is, -e Similar; like.

sim'plex Simple, unbranched. Also:
simplicicau'lis with simple (unbranched) stems; **simplicifo'lius** with simple (entire) leaves.

sim'sii In honour of John Sims (1749–1831), medical man and botanist, editor of *Curtis's Botanical Magazine* from 1800 to 1826.

sim'ulans Resembling.

Sina'pis f. L. name, also spelled *sinapi* and *sinape*, for mustard plant. CRUCIFERAE.

Sinarundina'ria f. From *sino-*, Chinese, plus *Arundinaria*; with reference to the Chinese habitat of these bamboos. GRAMINEAE.

sinen'sis, -is, -e; si'nicus, -a, -um Chinese.

Sinnin'gia f. Tropical herbs named for Wilhelm Sinning (1794–1874), head gardener, University of Bonn. This is the gloxinia of the florists (*S. speciosa*). GESNERIACEAE.

sino- In compound words refers to China; the Chinese were vaguely known to Ancient Greek and Roman geographers as the *Sinai*, *Sinae*. Their land was known in Ancient India as *China* from the Ch'in dynasty (255–206 B.C.).

Sinome'nium n. Chinese moonseed. From Gr. *sinai*, the Chinese; *mēnē*, the moon.

Sinowilso'nia f. Named for Ernest Henry Wilson (1876–1930), of the Arnold Arboretum, Boston. Known as Chinese Wilson, he travelled widely in China and introduced more than a thousand plants, including *Lilium regale*, to the gardens of England and America. English-born and Kew-trained, he brought unusual imagination and perception to all he did, including a remarkable ability to get along with Chinese at all levels. He was sent out to China in 1899 by the firm James Veitch and Sons of Chelsea primarily to introduce into cultivation *Davidia involucrata*, but his expedition proved so successful in revealing the wealth of plants available only in China that he made further Chinese expeditions in 1903 to 1905, 1907 to 1908 and 1910 to 1911. In 1914 and 1918 he visited Japan. He escaped innumerable dangers in the Far East only to be killed in a motor accident in the United States. Among his books, *China, Mother of Gardens*, is well worth reading as a story of adventurous travel in distant parts of China. HAMAMELIDACEAE.

sinua'tus, -a, -um; sinuo'sus, -a, -um With a wavy margin.

Siphocam'pylus m. From Gr. *siphōn*, tube; *kampylŏs*, curve; with reference to the curved corolla. CAMPANULACEAE.

Siphonosman'thus m. From Gr. *siphōn*, tube plus *Osmanthus*; with reference to the long tube of the corolla which distinguishes this group from typical *Osmanthus*. OLEACEAE.

sisala'nus, -a, -um Pertaining to Sisal, on the coast of Yucatan, Mexico.

Sisymb'rium n. Old Greek name, *sisymbriŏn*, for various plants. CRUCIFERAE.

Sisyrin'chium n. The ancient Greek name for another plant. These plants are native to North and South America, and naturalized elsewhere. IRIDACEAE.

sitchen'sis Of Sitka, Alaska.

Si'um n. Ancient Greek name, *siŏn*, applied to some water plants. UMBELLIFERAE.

Skim'mia f. From the Japanese name *Shikimi*. RUTACEAE.

smarag'dinus, -a, -um Emerald-green.

Smilaci'na f. False Solomon's seal, false spikenard. Diminutive of *Smilax*. LILIACEAE (CONVALLARIACEAE).

smilaci'nus, -a, -um Relating to *Smilax*.

Smi'lax f. Greenbrier. The Greek name for these climbers although the smilax of florists is generally *Asparagus asparagoides*. SMILACACEAE.

Smithian'tha f. In honour of Matilda Smith (1854–1926), botanical artist at Kew, who provided drawings for *Curtis's Botanical Magazine* from 1878 to 1923. GESNERIACEAE.

smi'thii Of the many botanists with the surname Smith, the first to be commemorated was Sir James Edward Smith (1759–1828) of Norwich, England, purchaser of Linnaeus's collections and a founder and first president of the Linnean Society of London. Among others are Christen Smith (1785–1816), Norwegian botanist who died on a Congo expedition; John Smith (1798–1888), Scottish gardener and pteridologist at Kew; Joannes Jacobus Smith (1867–1947), Belgian botanist many years at Buitenzorg, and Karl A. Harald Smith (1889–1971), Swedish botanist often known as Harry Smith, who made extensive collections in China.

Smyr'nium n. Greek name, *smyrniŏn*, from *smyrna*, myrrh; with reference to its smell. UMBELLIFERAE.

sobo'lifer, soboli'fera, -um Having creeping rooting stems.

Sobra'lia f. Named for Francisco Martinez Sobral, Spanish physician, *c.* 1790. ORCHIDACEAE.

socia'lis, -is, -e Forming colonies.

socotra'nus, -a, -um Of the island of Socotra, Indian Ocean.

sodo'meus, -a, -um Of the Dead Sea area; from *Sŏdoma*, Sodom.

Solan'dra f. Named for Daniel Carlsson Solander (1736–1782), a pupil of Linnaeus, who emigrated to England in 1760, sailed with Joseph Banks as botanist on Captain Cook's first voyage of exploration in the *Endeavour* in 1768 to 1771 and became keeper of the Natural History Department of the British Museum, London, in 1773. He was a very able and industrious biologist as is shown by his numerous descriptions of animals as well as plants made on Cook's voyage. SOLANACEAE.

Sola'num n. Latin name for some plant, probably *Solanum nigrum*, assigned to this huge and varied genus which includes such plants as the potato, Jerusalem cherry and deadly nightshade. SOLANACEAE.

sola'ris, -is, -e Of the sun; growing in sunny situations.

Soldanel'la f. Apparently a diminutive of Italian *soldo*, a small coin; in allusion to the rounded leaves. PRIMULACEAE.

soldanelloi'des Resembling a *Soldanella*.

Soleiro'lia f. In honour of Joseph François Soleirol (d. 1863) who made vast collections of specimens of Corsican plants in the first half of the 19th century. *Soleirolia soleirolii* is the latest name for the little creeping herb usually known as *Helxine soleirolii* or ironically as 'Mind-your-own-business'. URTICACEAE.

Solida'go f. Goldenrod. From *solido*, to make whole; in allusion to the reputed healing qualities of these perennial herbs. COMPOSITAE.

Solidas'ter m. Hybrid between *Aster* and *Solidago*. COMPOSITAE.

so'lidus, -a, -um Solid; dense.

Soli'sia f. In honour of Octavio Solis of Mexico City, a cactus-specialist *c.* 1923. CACTACEAE.

Sol'lya f. Named for Richard Horsman Solly (1778–1858), English plant physiologist and anatomist. PITTOSPORACEAE.

solstitia'lis, -is, -e Relating to midsummer.

som'nifer, somni'fera, -um Sleep-producing.

Son'chus m. Sowthistle. The Greek name *sŏnchŏs*. COMPOSITAE.

Sone'rila f. From the native Malabar name, *soneri-ila*. MELASTOMATACEAE.

songa'ricus, -a, -um Of Dzungaria, eastern central Asia.

Sopho'ra f. From the Arabic name. LEGUMINOSAE.

Sophro'nitis f. From Gr. *sōphrōn*, modest; in allusion to the small but pretty flowers; *Sophronia* is a genus of orchids to which *Sophronitis* is related. ORCHIDACEAE.

Sorba'ria f. False *Spiraea*. From the Latin *sorbum*; meaning here resembling *Sorbus*. ROSACEAE.

sorbifo'lius, -a, -um With leaves like *Sorbus*.

Sor'bus f. Mountain ash, rowan. L. *sorbum*, the fruit of the service tree (*Sorbus domestica*). Hybrids with *Pyrus* are called *Sorbopyrus*. ROSACEAE.

sor'didus, -a, -um Dirty.

Sorg'hum n. Latinized from Italian *sorgo* but origin obscure. GRAMINEAE.

soro'rius, -a, -um Sisterly, i.e. very closely related.

soulangia'nus, -a, -um In honour of the chevalier Etienne Soulange-Bodin (1774–1846), horticulturist at Fromont near Paris, raiser of a celebrated hybrid Magnolia.

sou'liei In honour of Jean André Soulie (1858–1905), French missionary and plant collector in western China, murdered by lamas.

spachia'nus, -a, -um; spa'chii In honour of Edouard Spach (1801–1879), Alsatian botanist, from 1829 onwards at the Musée d'Histoire Naturelle, Paris, noted for careful and precise work on the flora of the Near East and other publications which somewhat paradoxically split genera and lump species.

spadi'ceus, -a, -um Date-brown, dark chestnut-brown.

Spara'xis f. From Gr. *sparasso*, to tear; the spathe is lacerate. IRIDACEAE.

Sparga'nium n. Bur-reed. Greek name *sparganion* used by Dioscorides. SPARGANIACEAE.

Sparman'nia f. Named for Dr. Andreas Sparrman (1748–1820), Swedish botanist who sailed with Captain Cook in the *Resolution* on his second voyage of exploration (1772–1775), and travelled with Thunberg in South Africa. TILIACEAE.

spar'sus, -a, -um Few; far-between.

spar'teus, -a, -um Pertaining to *Spartium*, i.e. like *Spartium* in growth.

Spar'tium n. Spanish broom. Gr. *spartion*, from *spartŏn*, i.e. esparto grass (*Stipa tenacissima*), a kind of grass used for weaving and cordage. LEGUMINOSAE.

Spathiphyl'lum n. Gr. *spathe; phyllon*, a leaf; from the leaflike spathe of these low herbs. ARACEAE.

Spatho'dea f. From Gr. *spathe*; *-ōdēs*, of the nature of; in allusion to the spathe- or boat-like calyx. The Scarlet Bell or Fountain Tree (S. *campanulata*) is among the showiest of tropical trees. BIGNONIACEAE.

Spathoglot'tis f. Gr. *spathe; glotta*, a tongue; in allusion to the mid-lobe of the lip of the flower. ORCHIDACEAE.

spathul'atus, -a, -um Like a spatula.

specio'sus, -a, -um Showy.

specta'bilis, -is, -e Spectacular; showy.

Specula'ria f. Venus' looking-glass. L. *speculum*, a mirror; from the old name for these annuals, *speculum Veneris*, literally, the looking glass of Venus. CAMPANULACEAE.

specula'tus, -a, -um Shining (as if with mirrors).

Spence'ria f. In honour of Spencer Le Marchant Moore (1850–1931), botanist first at Kew, then at the British Museum (Natural History). ROSACEAE.

Sper'gula f. Spurrey. Origin obscure. CARYOPHYLLACEAE.

-spermus, -a, -um In Gr. compounds means -seeded, e.g. *argyrospermus*, with silvery seeds; *megaspermus*, large-seeded.

sphacela'tus, -a, -um Withered as if dead.

Spha'cele f. Gr. *sphakos*, sage; from resemblance in foliage. LABIATAE.

Sphaeral'cea f. Globe-mallow. Gr. *sphaira*, a globe; *Alcea*, mallow; in allusion to the globose fruit. MALVACEAE.

sphae'ricus, -a, -um Spherical. Also:
sphaerocar'pus (with round fruits); **sphaeroce'phalus** (with a round head); **sphaeroi'des** (resembling a sphere).

spheco'des See *sphegodes*.

sphego'des Wasp-like. In medieval Latin *c* and *g* were sometimes interchanged; hence the old spelling *sphegodes* should not be altered to *sphecodes*, from *sphēx*, *sphēkŏs*, wasp.

sphenanthe'rus, -a, -um With wedge-shaped anthers.

sphenochi'lus, -a, -um With wedge-shaped lip.

Sphenogy'ne f. From Gr. *sphēn*, wedge; *gynē*, woman, hence ovary or pistil. COMPOSITAE.

Spheno'meris f. From Gr. *sphēn*, wedge; *mĕris*, part; in allusion to the wedge-shaped segments of the fronds. DENNSTAEDTIACEAE.

spi'cifer, spici'fera, -um Spike-bearing.

spicifor'mis, -is, -e Spike-shaped. Also:
spi'ciger (spike-bearing); **spiculiflo'rus** (with flowers in small spikes).

Spige'lia f. Indian pink, pink-root. Named for Adrian van der Spiegel (1578–1625), professor of anatomy at Padua and author of *Isagoges* (1606) on botany. LOGANIACEAE.

Spilan'thes f. From Gr. *spilŏs*, spot, fleck; *anthē*, flower; in allusion to the white flowers of the type-species being flecked with black pollen. COMPOSITAE.

Spina'cia f. Spinach. Origin obscure, Latinized from Italian *spinace* or Spanish *espinaca* and possibly from L. *spina*, a spine; in allusion to the spiny husk of the fruit. CHENOPODIACEAE.

spines'cens; spi'nifer, spini'fera, -um; spin'ifex; spino'sus, -a, -um Spiny.

spinu'lifer, spinuli'fera, -um Bearing little spines.
Spirae'a f. Bridal wreath, spiraea. Gr. *speiraira*, a plant used for garlands, from *speira*, spiral, anything twisted. ROSACEAE.
spira'lis, -is, -e Spiral.
Spiran'thes f. Ladies' tresses. Gr. *speira*, spiral, twisted object; *anthos*, a flower; from the spiral inflorescence. ORCHIDACEAE.
Spirone'ma n. From Gr. *speira*, spiral; *nēma*, a thread; in allusion to the spiral cells in the stamen filaments. COMMELINACEAE.
spis'sus, -a, -um Thick, crowded, dense.
splen'dens; splen'didus, -a, -um Splendid.
Spon'dias f. Greek for the plum. Assigned to this genus of tropical trees because of the fruit. ANACARDIACEAE.
spoo'neri In honour of Herman Spooner (1878–1975), botanist on staff of Messrs. Veitch at Chelsea, later of Imperial Institute, Kensington, London.
Spreke'lia f. Named for Johann Heinrich von Sprekelsen (1691–1764), a lawyer of Hamburg, who possessed a fine garden and botanical library visited by Linnaeus in 1735. AMARYLLIDACEAE.
spren'geri Named for Karl (Carlo) Sprenger (1846–1917), German nurseryman and introducer of new plants at Vomero, near Naples, Italy.
spuma'rius, -a, -um Frothing.
spu'rius, -a, -um False.
Spyri'dium n. Gr. *spyris*, a basket; *eidos*, like; from the form of the calyx. RHAMNACEAE.
squa'lens, -a, -um; squa'lidus, -a, -um Dirty.
squama'tus, -a, -um Squamate; with small scalelike leaves or bracts.
squamo'sus, -a, -um Full of scales.
squarro'sus, -a, -um With parts spreading or recurved at the ends.
stachyoi'des Resembling *Stachys*.
Sta'chys f. Woundwort. Gr. *stachys*, ear of corn, but by extension of application used by Dioscorides and Pliny for a Labiate presumably with a similar inflorescence. LABIATAE.
Stachytar'pheta f. Gr. *stachys*, ear of corn, hence a spike; *tarphys*, thick; from the thick flower spikes. VERBENACEAE.
Stachyu'rus m. Gr. *stachys*, ear of corn, hence a spike; *oura*, a tail; from the form of the racemes of these shrubs. STACHYURACEAE.
stah'lii In honour of Christian Ernst Stahl (1848–1919), German botanist of Alsatian origin, professor at Jena.
stami'neus, -a, -um With prominent stamens.
standi'shii In honour of John Standish (*c.* 1809–1875), English nurseryman, partner in Standish and Noble of Bagshot, Surrey, who raised the Chinese and Japanese plants introduced by Robert Fortune 1848 to 1860.

Stange'ria f. In honour of William Stanger (1812–1854), English naturalist, surveyor and medical man who discovered the remarkable cycad S. *paradoxa* in Natal and sent it to the Chelsea Physic Garden in 1851. STANGERIACEAE.

Stanho'pea f. Named for Philip Henry Stanhope (1781–1855), 4th Earl Stanhope, President of the Medico-Botanical Society of London, 1829 to 1837. ORCHIDACEAE.

stans Erect, upright.

Stape'lia f. Carrion-flower. *Cactus*-like plants named by Linnaeus for Johannes Bodaeus van Stapel (d. about 1636), of Amsterdam, editor of Theophrastus's major work on plants.

stap'fii; stapfianus, -a, -um In honour of Otto Stapf (1857–1933), British botanist of Austrian origin, from 1890 to 1908 Assistant in the Kew Herbarium, 1909 to 1921 Keeper of the Herbarium, editor of *Index Londinensis* and *Curtis's Botanical Magazine* (1922–1933).

Staphy'lea f. Bladdernut. Gr. *staphyle*, a cluster; from the arrangement of the flowers. STAPHYLEACEAE.

Sta'tice f. Marsh rosemary, sea lavender, sea pink, thrift. Greek name for thrift or sea pink, now abandoned owing to its application by some authors to the sea pinks (*Armeria*), by others to the sea lavenders (*Limonium*), thus creating confusion and ambiguity. Thrift or sea pink has been assigned to *Armeria* and sea lavender or marsh rosemary to *Limonium*. Gardeners, however, will probably long continue to call the latter *Statice*. PLUMBAGINACEAE.

Staunto'nia f. Named for Sir George Leonard Staunton (1737–1801), Irish medical man and naturalist, who accompanied Lord Macartney on his futile and costly mission to the Emperor of China in 1792. LABIATAE.

stauracan'thus, -a, -um With crosswise spines.

stearnii; stearnianus, -a, -um In honour of William Thomas Stearn (b. 1911), botanist, historian and bibliographer.

Steirone'ma n. Loosestrife. Gr. *steiros*, sterile; *nema*, a thread; referring to the staminodes (sterile stamens) alternating with the fertile stamens. PRIMULACEAE.

Ste'lis f. Gr. *stĕlis*, a name for the mistletoe (*Viscum*) transferred to these epiphytic orchids because they also grow on trees though not parasitic. ORCHIDACEAE.

Stella'ria f. Chickweed, stitchwort, starwort. From L. *stella*, a star; from the starry flowers. CARYOPHYLLACEAE.

stella'ris, -is, -e; stella'tus, -a, -um Starry, star-like, with spreading leaves or petals star-wise.

Stel'lera f. In honour of Georg Wilhelm Steller (1709–1746), German naturalist and traveller in Siberia, after whom the extinct Steller's sea-cow is also named. THYMELEACEAE.

stel'liger, stelli'gera, -um Star-bearing.

Stenan'drium n. From Gr. *stěnŏs*, narrow; *aner*, *andros*, man, hence stamen; with reference to the narrow stamens. ACANTHACEAE.

Stenan'thium n. From Gr. *stěnŏs*, narrow; *anthŏs*, a flower; in allusion to the narrow sepals and petals. LILIACEAE (MELANTHIACEAE).

ste'naulus, -a, -um With narrow furrows.

steno- In compound words signifies narrow. Thus:
stenocar'pus (with narrow fruits); **stenoce'phalus** (with narrow head); **stenope'talus** (with narrow petals); **stenophyl'lus** (narrow-leaved); **steno'pterus** (narrow-winged); **stenosta'chyus** (narrow-spiked).

Stenocar'pus m. From Gr. *stěnŏs*, narrow; *karpos*, fruit; from the flat, narrow fruits. PROTEACEAE.

Stenochlae'na f. Gr. *stěnŏs*, narrow; *chlaina*, a cloak. In these ferns there is no indusium covering the sori. BLECHNACEAE.

Stenoglot'tis f. From Gr. *stěnŏs*, narrow; *glōtta*, a tongue; in allusion to the narrow division of the lip of the flower. ORCHIDACEAE.

Stenosperma'tion n. Gr. *stěnŏs*, narrow; *spermation*, a little seed, diminutive of *sperma*, seed. ARACEAE.

Stenota'phrum n. Gr. *stěnŏs*, narrow; *taphros*, a trench; from the cavities on the stem of the inflorescence on which the flower spikelets are situated. A variegated form of *S. secundatum* is cultivated as a glasshouse basket-plant. GRAMINEAE.

Stephanan'dra f. Gr. *stephanŏs*, a crown; *aner*, *andros*, a man, hence stamen. The numerous short stamens persist as a rather obvious crown around the capsule. ROSACEAE.

stephan'ensis, -is, -e Epithet of a hybrid *Jasminum* raised at Saint-Etienne (Latinized as *S. Stephani fanum*).

Stephano'tis f. Wax-flower, Madagascar jasmine. Gr. *stěphanōtis*, a name of the myrtle which was used for making crowns (Gr. *stěphanŏs*, a crown) but applied to these shrubs also, in reference to the auricles in the staminal crown. ASCLEPIADACEAE.

Stercu'lia f. Trees named for Sterculius, Roman god of privies (L. *stercus*, dung). The flowers and wood of the type species, *S. foetida*, stink. STERCULIACEAE.

ste'rilis, -is, -e Infertile; sterile.

Sternber'gia f. Winter daffodil. Named for Count Kaspar M. von Sternberg (1761–1838), Austrian botanist, clergyman and palaeontologist, a founder of the Bohemian National Museum in Prague. AMARYLLIDACEAE.

sternii; sternia'nus, -a, -um In honour of Frederick Claude Stern (1884–1967), merchant banker interested in peonies, snowdrops and gardening on chalk.

Stetso'nia f. Named in honour of Francis Lynde Stetson of New York in 1920. CACTACEAE.

steve'nii In honour of Christian von Steven (1781–1863), Finnish botanist for many years in the Crimea.

Stevenso'nia f. Named for Sir William Stevenson, Governor of Mauritius, 1857–1863. PALMAE.

Ste'via f. Named for Pedro Jaime Esteve (d. 1566) of Valencia, Spanish botanist and physician. COMPOSITAE.

Stewar'tia f. See *Stuartia*. The name of the royal stewards, later kings, of Scotland was spelled Stewart until Mary Queen of Scots lived in France; there being no *w* in French, the royal name was henceforth spelled Stuart.

stewartia'nus, -a, -um In honour of Laurence Baxter Stewart (1876–1934), curator of Edinburgh Botanic Garden.

Stictocardia f. From Gr. *stiktos*, spotted; *kardia*, heart; in allusion to the heart-shaped leaves spotted beneath with numerous black minute glands. CONVOLVULACEAE.

stictocar'pus, -a, -um From Gr. *stiktos*, spotted; *karpos*, fruit.

Stigmaphyl'lon n. From Gr. *stigma*, stigma; *phyllon*, a leaf. The stigma is broad and somewhat leaf-like. MALPIGHIACEAE.

Sti'pa f. Feather-grass. Gr. *stuppē*, tow; from the feathery inflorescence on these perennial grasses. GRAMINEAE.

stipula'ceus, -a, -um; stipula'ris, -is, -e; stipula'tus, -a, -um Having stipules or well-developed stipules, the outgrowths found at the base of the leaf-stalks of many flowering plants.

Stizolo'bium n. Velvet-bean. Gr. *stizo*, to prick; *lŏbŏs*, a lobe, pod. The pods of some species have stinging hairs. See *Mucuna*. LEGUMINOSAE.

Stoke'sia f. Stokes' aster. Named for Dr. Jonathan Stokes (1755–1831) of Chesterfield, English botanical author. COMPOSITAE.

stolo'nifer, stoloni'fera, -um Having stolons or rooting runners.

Stoma'tium n. From Gr. *stŏma*, mouth, opening. AIZOACEAE.

stra'gulus, -a, -um Covering like a mat or carpet, hence mat-forming.

strami'neus Straw-coloured.

strangula'tus, -a, -um Constricted.

Stranvae'sia f. Named for William T. H. Fox-Strangways (1795–1865), 4th Earl of Ilchester, English diplomatist and botanist, also commemorated in the genera *Strungweja* and *Foxia*. ROSACEAE.

Stratio'tes m. Water soldier. Gr. *stratiōtēs*, name for the water-lettuce (*Pistia stratiotes*) from *stratiōtēs*, soldier, and hence transferred to this aquatic herb on account of its sword-shaped leaves. HYDROCHARITACEAE.

Strelit'zia f. Bird-of-paradise flower. Named for Charlotte Sophia of Mecklenburg-Strelitz (1744–1818), who in 1761 became Queen to George III and bore him 15 children. STRELITZIACEAE.

stre'pens Creaking, rattling, clattering; usually referring to seeds in a capsule.

strepto- In compound words signifies twisted. Thus: **streptocar'pus** (with twisted fruits); **streptope'talus** (with twisted petals); **streptophyl'lus** (with twisted leaves); **streptose'palus** (with twisted sepals).

Streptocar'pus m. Cape primrose. From Gr. *strĕptŏs*, twisted; *karpos*, fruit. The capsules are twisted in a spiral. GESNERIACEAE.

Strep'topus m. Twisted stalk. Gr. *strĕptŏs*, twisted; *pous*, a foot; with reference to the twisted flower-stalks. LILIACEAE (CONVALLARIACEAE).

Streptoso'len m. From Gr. *strĕptŏs*, twisted; *sōlēn*, pipe, tube; referring to the twisted corolla-tube. SOLANACEAE.

stria'tus, -a, -um Striped.

stric'tus, -a, -um Erect, upright.

strigillo'sus, -a, -um With short appressed bristles.

strigo'sus, -a, -um With stiff bristles.

striola'tus, -a, -um Faintly striped or with fine lines.

Strobilan'thes f. Gr. *strŏbilŏs*, a cone; *anthos*, a flower; from the form of the flower-head in some species. ACANTHACEAE.

strobi'lifer, strobili'fera, -um Cone-bearing.

Stroman'the f. From Gr. *strōma*, a bed; *anthos*, a flower; from the form of the inflorescence. MARANTACEAE.

Strombocac'tus m. Gr. *strŏmbŏs*, a spinning-top; plus *Cactus*; from the shape of the plants. CACTACEAE.

Strongy'lodon m. From Gr. *strŏngylŏs*, round; *ŏdŏus*, *ŏdŏntŏs*, tooth; in allusion to the rounded teeth of the calyx. LEGUMINOSAE.

strongylophyl'lus, -a, -um Round-leaved.

Strophan'thus m. From Gr. *strŏphŏs*, twisted band or cord; *anthos*, flower; in allusion to the long appendages of the corolla. APOLCYNACEAE.

struma'rius, -a, -um; strumo'sus, -a, -um Having cushionlike swellings.

Strych'nos f. Greek name for various poisonous plants and applied here by Linnaeus because so many of these are poisonous. The genus includes *S. nux-vomica* or strychnine. A few species are the source of some of the most lethal arrow poisons known. LOGANIACEAE.

Stuartia f. Named by Linnaeus in honour of John Stuart (1713–1792), 3rd Earl of Bute, British Prime Minister 1762 to 1763, a zealous patron of botany and horticulture; for whom *Butea* was also named. Linnaeus erroneously believed his family name to be Stewart and, owing to misinformation, named the genus *Stewartia*, later corrected to *Stuartia*. See *Stewartia*. THEACEAE.

Styli'dium n. Trigger-plant. Gr. *stylos*, a column. The stamens are united with the style to form a column in these plants. STYLIDIACEAE.

Stylo'phorum n. Celandine poppy. Gr. *stylos*, style; *phoros*, bearing; with reference to the long columnar style. PAPAVERACEAE.

stylo'sus, -a, -um With a prominent or well-developed style.
styracif'luus, -a, -um Flowing with gum.
Sty'rax m. Snowbell, storax-tree. The classical Greek name derived from a Semitic name for these resin-producing plants. STYRACACEAE.
suave'olens Sweet-scented.
sua'vis, -is, -e Sweet.
sub- In compound words possibly the most used prefix, signifying somewhat; almost; rather; slightly; partially; under; etc.
subacau'lis, -is, -e Without much of a stem.
subalpi'nus, -a, -um Growing in the lower mountain ranges.
subauricula'tus, -a, -um Somewhat eared.
subcaeru'leus, -a, -um Slightly blue.
subca'nus, -a, -um Greying.
subcarno'sus, -a, -um Rather fleshy.
subcorda'tus, -a, -um Rather heart-shaped.
subdenta'tus, -a, -um Nearly toothless.
subdivarica'tus, -a, -um Somewhat spreading.
subelonga'tus, -a, -um Somewhat elongated.
suberec'tus, -a, -um Almost upright.
subero'sus, -a, -um Cork-barked.
subfalca'tus -a, -um Somewhat curved or hooked.
subglau'cus, -a, -um Somewhat glaucous, i.e. with a somewhat blue-green or grey-green appearance.
subhirtel'lus, -a, -um Somewhat hairy.
subluna'tus, -a, -um Rather crescent-shaped.
submer'sus, -a, -um Submerged.
subpetiola'tus, -a, -um With a very short leaf-stalk.
subscan'dens Tending to climb.
subterra'neus, -a, -um Underground.
subula'tus, -a, -um Awl-shaped.
subvillo'sus, -a, -um With rather soft hairs, or somewhat villous.
Succi'sa f. Devil's bit. From L. *succido*, cut off below; with reference to the truncate end of the rhizome, looking as if bitten off by the Devil. DIPSACACEAE.
succi'sus, -a, -um Cut off below.
succotri'nus, -a, -um Of the island of Socotra, Indian Ocean.
succulen'tus, -a, -um Fleshy; juicy.
sude'ticus, -a, -um Referring to the Sudetenland of Czechoslovakia and Poland.
sue'cicus, -a, -um Swedish.
suenderma'nnii In honour of Franz Sündermann (1864–1946), German gardener specializing in rock garden plants.

suffrutes'cens; suffrutico'sus, -a, -um Somewhat shrubby.
suio'num Of the Swedes.
sulca'tus, -a, -um Furrowed.
sulfu'reus, -a, -um; sulphu'reus, -a, -um Sulphur-yellow.
sulta'ni Of the Sultan, referring in *Impatiens sultani* to the Sultan of Zanzibar.
sumatra'nus, -a, -um Of Sumatra, Indonesia.
sunten'sis, -is, -e Epithet of a hybrid *Hibiscus* raised at Sunte House, Haywards Heath, England.
super'bus, -a, -um Superb.
supercilia'ris, -is, -e Eyebrowlike.
supi'nus, -a, -um Prostrate.
supranu'bius, -a, -um Above the clouds, i.e. growing high in mountains.
surculo'sus, -a, -um Producing suckers or shoots.
susia'nus, -a, -um From Susa in western Iran (Persia).
suspen'sus, -a, -um Hanging.
sutchuenen'sis, -a, -um Of Sichuan (Szechwan) province, China.
Sutherlan'dia f. Shrubs named for James Sutherland (*c.* 1639–1719), superintendent, botanic garden, Edinburgh, and professor of botany, Edinburgh, 1676 to 1715. LEGUMINOSAE.
Sutto'nia f. Named for the Reverend Charles Sutton (1756–1846), clergyman at Norwich, England, a keen-eyed amateur botanist. This genus is sometimes included in *Myrsine*. MYRSINACEAE.
Swain'sona f. Darling River pea. Named for Isaac Swainson (1746–1812), London physician who had a private botanic garden at Twickenham. LEGUMINOSAE.
Swer'tia f. Herbaceous plants commemorating Emanuel Sweert (1552–1612), Dutch florist, author and artist of *Florilegium* (1612–1614). GENTIANACEAE.
Swi'da f. Czech name for dogwood; see *Cornus*.
Swiete'nia f. Mahogany tree. Named for Gerard van Swieten (1700–1772), Dutch botanist and physician who settled in Vienna in 1745 and became physician to the Empress Maria Theresa, reforming medical education in Vienna. This is the original mahogany of Central America and the West Indies. African and Philippine mahogany, commercially so called, are members of several other genera although the wood, in most cases, bears some resemblance to that of *Swietenia*. MELIACEAE.
Syco'psis f. From Gr. *sykōn*, fig; *ŏpsis*, appearance. HAMAMELIDACEAE.
sylva'ticus, -a, -um; sylves'ter; sylves'tris, -is, -e Growing in woods, forest-loving, wild.
sylvi'cola An inhabitant of woods.
Sym'pagis f. From Gr. *sympagēs*, joined together; in allusion to the basal membrane joining the stamens. ACANTHACEAE.

Symphoricar'pos m. Snowberry. From Gr. *symphorein*, bear together; *karpŏs*, fruit. The berries are borne in clusters. CAPRIFOLIACEAE.

Symphyan'dra f. Pendulous bellflower. Gr. *symphyio*, to grow together; *aner*, *andros*, man. The anthers are joined. CAMPANULACEAE.

Sym'phytum n. Comfrey. Gr. name of two herbs which were reputed to heal wounds, one *S. bulbosum*, from *symphyo*, make to grow together; *phyton*, plant. BORAGINACEAE.

Symplocar'pus m. Skunk-cabbage. Gr. *symplŏkē*, combination, connection, embrace; *karpŏs*, fruit. The ovaries of this swamp-loving perennial herb grow together to make one fruit. ARACEAE.

Sym'plocos f. Gr. *symplŏkē*, combination. The stamens are variously united. SYMPLOCACEAE.

Synade'nium n. Gr. *syn*, with, together; *adēn*, a gland; from the united glands of the inflorescence. EUPHORBIACEAE.

Synechan'thus m. Gr. *synĕchēs*, holding together, continuous; *anthos*, a flower; from the form of the flower heads. PALMAE.

Syngo'nium n. From Gr. *syn*, with, together; *gŏne*, reproductive organs, womb; in allusion to the united ovaries. ARACEAE.

Synne'ma n. From *syn*, together; *nēma*, thread; in allusion to the united stamens. ACANTHACEAE.

Syn'thyris f. Gr. *syn*, together; *thyris*, a small door; in allusion to a characteristic of the fruits. SCROPHULARIACEAE.

syria'cus, -a, -um Syrian.

Syrin'ga f. Lilac. Originally also applied to *Philadelphus*. The name is still very commonly applied to mock-orange as a vernacular name. The name derives from Gr. *syrinx*, a pipe, and refers to the hollow stems. OLEACEAE.

syringan'thus, -a, -um With flowers like lilac.

syringifo'lius, -a, -um With leaves like lilac.

Syzy'gium n. From Gr. *syzygos*, joined; with reference to the paired leaves and branchlets of a Jamaican species (*Calyptranthes suzygium*) for which the name was used originally. MYRTACEAE.

szovitsia'nus, -a, -um In honour of Johann Nepomuk Szovits (d. 1830), Hungarian apothecary who collected in the Caucasus.

T

Tabebu'ia f. From the native Brazilian name *tabebuia* or *taiaveruia*. BIGNONIACEAE.

Tabernaemonta'na f. Named for Jakob Theodor von Bergzabern (d. 1590), personal physician to the Count of the Palatine at Heidelberg, West Germany.

He Latinized his name as Tabernaemontanus and is also commemorated by species named for him in *Amsonia*, *Potentilla* and *Scirpus*. He was the author of a celebrated herbal *Neuw Kreuterbuch* (1588–1591), of which the illustrations were issued separately at Frankfurt-am-Main in 1590 under the title *Eicones Plantarum*. The woodcuts were mostly copied from those in other herbals but make an attractive book. The London printer, John Norton, acquired them from the Frankfurt printer Nicolaus Bassaeus and used them in 1597 to illustrate Garard's *Herball*. APOCYNACEAE.

tabula'ris, -is, -e Flat, like a table or board; growing on Table Mountain, South Africa.

tabulifor'mis, -is, -e Flat, like a table or board.

tacamahac'ca American vernacular name said to be of Aztec origin converted to Hackmatack and used for both *Populus balsamifera* and *Larix laricia*.

Tac'ca f. From the Indonesian name, *taka*. TACCACEAE.

tae'diger, taedi'gera, -um Torch-bearing.

Tage'tes f. Marigold. African and French. Named for an Etruscan deity, Tages, said to have sprung from the earth as it was being ploughed and to have taught the Etruscans the art of divination. An extract from the roots is inimical to soil nematodes. COMPOSITAE.

taggia'nus, -a, -um In honour of Harry Frank Tagg (1874–1933), botanist at the Edinburgh botanic garden, authority on *Rhododendron*.

tagliabua'nus, -a, -um In honour of the brothers Alberto Linneo Tagliabue and Carlo Tagliabue, Italian nurserymen at Lainate near Milan in 1858. The hybrid between *Campsis grandiflora* and *C. radicans* is named after them.

taiwanen'sis, -is, -e Of Taiwan (Formosa).

Taiwa'nia f. From Taiwan (Formosa) where the one species of this genus, *T. cryptomerioides*, was originally found. TAXODIACEAE.

talien'sis, -is, -e Of the Tali (Dali) Range, Yunnan, west China.

Tali'num n. Derivation of the name is obscure. PORTULACACEAE.

Tamarin'dus m. Tamarind. From the Arabic name *tamar*, date; *hindi*, Indian. LEGUMINOSAE.

tamariscifo'lius, -a, -um With leaves like tamarisk.

Ta'marix f. Tamarisk. The Latin name, of which *tamariscus* is another form. TAMARICACEAE.

Ta'mus f. Black bryony. From the Latin name, *tamnus*, for another climbing plant. DIOSCOREACEAE.

tanacetifo'lius, -a, -um Tansy-leaved.

Tanace'tum n. Tansy, pyrethrum. From the medieval Latin name *tanazita*, still used in some European places, ultimately derived from Gr. *athanasia*, immortality. Once regarded as a specific for intestinal worms, it was formerly used in Europe as well as in some rural areas of New England in funeral winding sheets, to discourage worms; see p. 318.

Tanakae'a f. In honour of a pioneer Japanese botanist Yoshio Tanaka (1838–1916).

tanasty'lus, -a, -um Long-styled.

tancarvil'leae, tankervil'leae In honour of Lady Emma Tankerville (d. 1836), wife of Charles, Earl of Tankerville. The title is derived from the earldom of Tancarville in Normandy, created in 1419 for John Grey.

tangu'ticus, -a, -um From Kansu (Gansu), north-west China; named for the Tangut people of this region.

Tapeinochi'los m. From Gr. *tapeinos*, low; *cheilos*, lip; in allusion to the short labellum. ZINGIBERACEAE.

tapetifor'mis, -is, -e Carpet-like.

taraxacifo'lius, -a, -um With leaves like a dandelion.

Tara'xacum n. Dandelion. Medieval name traceable through Arabic to Persian *talkh chakok*, meaning bitter herb. COMPOSITAE.

tardiflor'us, -a, -um Late-flowering.

tardi'vus, -a, -um; tar'dus, -a, -um Late.

taronen'sis, -is, -e Of the Taron Gorge, Yunnan, China.

tarta'reus, -a, -um With a rough surface.

tarta'ricus, -a, -um Of Central Asia, formerly called Tartary.

tasma'nicus, -a, -um Of Tasmania, named in honour of the Dutch navigator Abel Janszoon Tasman (1603–1659), who discovered it in 1642 or 1643.

tatsienen'sis, -is, -e Of Tatsienlu (K'angding), western China.

tau'ricus, -a, -um Of the Crimea, in ancient geography Taurica Chersonesus.

tauri'nus, -a, -um Of the neighbourhood of Turin, Italy.

taxifo'lius, -a, -um With leaves like yew.

Taxo'dium n. Swamp cypress. L. *Taxus*, yew; Gr. *eidos*, resemblance; from a similarity of leaf shape. TAXODIACEAE.

Ta'xus f. Yew. The Latin name. TAXACEAE.

tazet'ta Italian vernacular name used for a *Narcissus*, meaning a small cup (*tazza*), from the form of the corona.

tchonos'kii In honour of a Japanese botanical collector, Tchonoski, who between 1862 and 1867 made extensive collections of Japanese plants for Carl Maximowicz, which included many new species.

tech'nicus, -a, -um Used for a special purpose.

Teco'ma f. Abbreviated from the Mexican name *tecomaxochitl.* BIGNONIACEAE.

Tecoma'ria f. Cape honeysuckle. From *Tecoma* which it closely resembles. BIGNONIACEAE.

Tecophilae'a f. In honour of an Italian botanical artist Tecophila Billotti, apparently the daughter of the Italian botanist Luigi Colla (1766–1848), of

Turin, whose works she illustrated. *T. cyanocrocus* has a crocus-like flower of an intense and wonderful blue. LILIACEAE (TECOPHILAEACEAE).

Tec'tona f. From the Tamil name for the teak tree (*T. grandis*). This magnificent forest tree, growing up to 150 feet, supplies the teak of commerce. The wood is extremely durable and proof against the ravages of white ants. VERBENACEAE.

tecto'rum Of the roofs of houses. Specific epithet of *Sempervivum* (house-leek) which grows freely on stone and slate roofs and which, in some places, is thought to avert lightning by acting as a conductor of electricity. Various plants growing on thatched roofs in Sweden were named *tectorum* by Linnaeus.

tec'tus, -a, -um Concealed; covered.

Tele'kia f. In honour of a Hungarian nobleman, Samuel Teleki de Szék, patron of the botanist J. C. Baumgarten (1765–1843). COMPOSITAE.

tele'phium Specific epithet of *Sedum*, of uncertain meaning, possibly connected with Telephus, a king of Mysia in Asia Minor.

Tel'lima f. Anagram of *Mitella* to which these herbaceous perennials are closely related. SAXIFRAGACEAE.

telmate'ius, -a, -um Of marshes.

telonen'sis, -is, -e Of Toulon (ancient Telo or Telonium), France.

Telo'pea f. Waratah. Gr. *tēlōpŏs*, seen from afar. The crimson flowers of these tall shrubs are conspicuous at a distance. PROTEACEAE.

Telos'ma f. From Gr. *tēlĕ*, far; *osmē*, fragrance, smell; in allusion to the great distance over which the scent carries. ASCLEPIADACEAE.

teme'nius, -a, -um Gr. *tĕmĕniŏs*, belonging to a sacred place.

Templeto'nia f. In honour of John Templeton (1766–1825) of Belfast, Irish botanist who gave special attention to mosses, fungi, lichens and algae. LEGUMINOSAE.

temulen'tus, -a, -um; tem'ulus, -a, -um Drunken.

te'nax Strong; tough; matted.

tenebro'sus, -a, -um Of shaded places.

tenel'lulus, -a, -um; tenel'lus, -a, -um Tender; delicate.

te'nens Holding, persistent, enduring.

ten'uis, -is, -e Slender; thin. Also:
tenuicau'lis (slender-stemmed); **tenuiflo'rus** (with slender flowers);
tenuifo'lius (slender-leaved); **tenuipe'talus** (with slender petals).

Tephro'sia f. Gr. *tĕphrŏs*, ash-coloured; from the grey appearance of the leaves. LEGUMINOSAE.

terebinthina'ceus, -a, -um; terebinthi'nus, -a, -um Pertaining to turpentine, from Gr. *tĕrĕbinthŏs*, itself probably of pre-Greek origin, the name of *Pistacia terebinthus* and its resin.

te'res Cylindrical; circular in section.

Termina'lia f. Indian almond. L. *terminus*, end. The leaves are borne at the ends of the shoots. COMBRETACEAE.

termina'lis, -is, -e Terminal; relating to boundaries; coloured forms of *Cordyline terminalis* are sometimes planted as boundary markers.

ter'minans Ending.

terna'teus, -a, -um Of the island of Ternate in the Moluccas.

terna'tus, -a, -um In clusters of three.

Ternstroe'mia f. Named for Christopher Tärnström (1703–1746), Swedish student of Linnaeus, who went out in 1745 to investigate the natural history of China but died at Pulo Condor off the coast of Indochina. THEACEAE.

terres'tris, -is, -e Of the ground; growing in the ground as opposed to growing on trees or in water.

tessella'tus, -a, -um Checkered in squares.

testa'ceus, -a, -um Brick-coloured.

testicula'ris, -is, -e; testicula'tus, -a, -um Like testicles.

testudina'rius, -a, -um Like, for example, tortoise shell with the curved surface divided into areas.

testu'do A tortoise.

tetra- In compound words signifies four. Thus:
tetracan'thus (four-spined); **tetrago'nus** (with four angles); **tetran'drus** (with four anthers); **tetran'thus** (four-flowered); **tetraphyl'lus** (with four leaves or leaflets); **tetra'pterus** (four-winged).

Tetracen'tron n. Gr. *tĕtra*, four; *kĕntrŏn*, a spur. The fruit is four-spurred. TETRACENTRACEAE.

Tetracli'nis f. Gr. *tĕtra*, four; *kline*, a bed. The leaves of this evergreen tree are grouped in fours. CUPRESSACEAE.

Tetrago'nia f. New Zealand spinach. Gr. *tĕtra*, four; *gōnia*, an angle; from the form of the fruits. AIZOACEAE.

Tetragono'lobus m. From Gr. *tetra*, four; *gōnia*, an angle; *lŏbŏs*, pod; in allusion to the four-angled or -winged pods. LEGUMINOSAE.

tetra'lix Old Greek name, *tĕtralix*, transferred to this heath having leaves in fours.

Tetrane'ma n. From Gr. *tetra*, four; *nēma*, thread; in allusion to the four stamens. SCROPHULARIACEAE.

Tetra'panax m. Rice paper tree. Gr. *tetra*, four, plus *Panax*. The flowers are in fours and have some resemblance to *Panax*. ARALIACEAE.

Tetrathe'ca f. Gr. *tetra*, four; *thēkē*, case. The anthers of these heath-like shrubs are often four-lobed. TREMANDRACEAE.

teucrioi'des Resembling *Teucrium*.

Teu'crium n. Germander. The Greek name, possibly named for Teucer, first king of Troy. LABIATAE.

texa'nus, -a, -um; texen'sis, -is, -e Of Texas, U.S.A.

tex'tilis, -is, -e Used in weaving, hence interwoven.

Tha'lia f. Named for Johannes Thal (1542–1583), German physician at Nordhausen whose Flora of the Harz region, *Sylva Hercynia*, was posthumously published in 1588. MARANTACEAE.

thalictroi'des Like *Thalictrum.*

Thalic'trum Meadow rue. The Greek name for a plant which may have been of this genus. RANUNCULACEAE.

The'a Tea. From the Dutch rendering *thee* of the Chinese (Amoy) word for tea, *t'e*, which in Mandarin is *ch'a*. *Thea sinensis* is now placed in the genus *Camellia.* The cultivation of tea was virtually a Chinese monopoly until in the 1840's Robert Fortune exported both tea plants and skilled labour to India to establish the industry there. THEACEAE.

theba'icus, -a, -um Of Thebes (*Thebae Aegypti*).

the'ifer, thei'fera, -um Tea-bearing.

Thelesper'ma n. From Gr. *thēlē*, a nipple; *sperma*, a seed; in allusion to the form of the seeds (cypselae) in some species. COMPOSITAE.

Thelocac'tus m. From Gr. *thēlē*, a nipple, plus *Cactus*. CACTACEAE.

Theobro'ma n. Cacao. A name coined by Linnaeus, from *thĕŏs*, a god; *brōma*, food; to replace the name *Cacao*, from the Mexican *cacahoaquahuitl* and its variants, which Linnaeus rejected as a barbarous generic name although he retained it as a specific epithet. The source of both cocoa and chocolate, and a highly profitable crop both on a large and on a small scale, cacao (*Theobroma cacao*) is grown throughout the tropics wherever the right climatic and other conditions prevail. The cacao tree requires protection of shade from other trees, hence a plantation may often take on a rather unkempt look. STERCULIACEAE.

Theophras'ta f. In honour of the Ancient Greek philosopher and botanist Theophrastos (371–*c*.287 B.C.), disciple and successor of Aristotle at Athens. THEOPHRASTACEAE.

therma'lis, -is, -e Of warm springs.

Thermop'sis f. Gr. *thermos*, lupin; *opsis*, like. The flower heads resemble yellow lupins. LEGUMINOSAE.

Thespe'sia f. Bhendi tree. Gr. *thĕspĕsiŏs*, divine; with reference to *Thespesia populnea* 'being regarded as a sacred plant in Tahiti at the time of Captain James Cook's visit in 1769 and being grown around places used for worship'. MALVACEAE.

Theve'tia f. Named for André Thevet (1502–1592), French monk, who travelled in Brazil and Guiana. APOCYNACEAE.

Thladian'tha f. From Gr. *thladias*, eunuch (from *thlao*, crush, bruise); *anthos*, flower; the suppression of stamens and their conversion into staminodes in the female flowers suggesting that they had been castrated. CUCURBITACEAE.

Thlas'pi f. Penny-cress. The Greek name for a cress. CRUCIFERAE.

Thoma'sia f. Named in 1821 for a Swiss family of plant-collectors, Peter Thomas and his brother Abraham Thomas and the latter's sons Philip (d. 1831), Louis (d. 1823) and Emanuel. STERCULIACEAE.

thomso'nii Of Thomas Thomson (1817–1878), a Scots doctor who became a surgeon in the Bengal Army and was superintendent of the Calcutta Botanic Garden, 1854–1861. Many Himalayan plants commemorate him.

Thri'nax f. Thatch palm. Fan palm. Gr. *thrinax*, a trident. PALMAE.

Thu'ja (also **Thuya**) f. Arbor-vitae, Northern white cedar. The Greek name *thuia*, for a kind of Juniper. CUPRESSACEAE.

Thujo'psis f. *Thuja; opsis*, like, from a similarity to *Thuja*. CUPRESSACEAE.

Thunber'gia f. Named for Carl Peter Thunberg (1743–1828), Swedish botanist, a student of Linnaeus who was persuaded by Dutch lovers of new plants to enter the service of the Dutch East India Company as a doctor and send back plants from Japan to Europe. He travelled in South Africa and Japan and became professor of botany at Uppsala. ACANTHACEAE.

thu'rifer, thuri'fera, -um Incense-bearing.

thuyoi'des Resembling *Thuja*.

thymifo'lius, -a, -um Thyme-leaved.

thymoi'des Like Thyme.

Thy'mus m. The ancient Greek name. LABIATAE.

thyrsiflo'rus, -a, -um With flowers in a thyrse, a many-flowered kind of inflorescence.

Thysano'tus m. Gr. *thysanōtŏs*, fringed, with reference to the three inner perianth segments fringed with long hairs. LILIACEAE (ANTHERICACEAE).

Tiarel'la f. Foamflower. Diminutive of Gr. *tiara*, a small crown; in reference to the form of the fruit. SAXIFRAGACEAE.

tibe'ticus, -a, -um Of Tibet.

tibici'nus, -a, -um Of a flute-player; used in *Schomburgkia* to refer to the long narrow pseudobulbs.

Tibouchi'na f. Spider-flower. From the native name in Guiana. MELASTOMATACEAE.

Tigrid'ia f. Tiger-flower. L. *tigris*, a tiger; because of the spots on the flowers (see *tigrinus*). IRIDACEAE.

tigri'nus, -a, -um Either striped like the Asiatic tiger or spotted like the jaguar, known as a tiger in South America. Also, tiger-toothed.

Ti'lia f. The Latin name for the linden or lime tree, known in southern Sweden (Småland) as linn and the origin of the name Linnaeus. TILIACEAE.

tilia'ceus, -a, -um Linden-like.

tiliifo'lius, -a, -um Linden-leaved.

Tillan'dsia f. Spanish moss. American herbs, mostly epiphytic, named for Elias

Til-Landz (d. 1693), a Swedish botanist and professor of medicine at Abo (Turku), Finland, who to avoid sea-sickness travelled by land between Finland and Sweden. BROMELIACEAE.

Tinan'tia f. Named for François A. Tinant (1803–1858), forester in Luxemburg, author of *Flore Luxembourgeoise* (1836 and 1855). COMMELINACEAE.

tinctor'ius, -a, -um Used in dyeing.

tinctor'um Of dyers.

tinc'tus, -a, -um; tin'gens Coloured.

tingita'nus, -a, -um Of Tangiers, ancient Tingis, North Africa.

Tin'nea f. A genus commemorating a scientific expedition up the Nile in 1861 by three courageous Dutch ladies, Madame Henrietta L. M. Tinne, her daughter Alexandria Tinne, and her sister Adrienne van Capellen. LABIATAE.

ti'nus Latin name for the laurustinus, *Viburnum tinus*.

Tipua'na f. From the South American name. LEGUMINOSAE.

tipulifor'mis, -is, -e Shaped like a daddy-long-legs insect, which belongs to the genus *Tipula*.

tirolen'sis, -is, -e Of the Tyrol.

titanius, -a, -um Very large.

Titano'psis f. L. *Tītān*, the sun god; Gr. *ŏpsis*, like; from the resemblance of the flower to the sun. AIZOACEAE.

Titho'nia f. Annual herbs named for Tithonus, a young man much loved by Aurora, the dawn-goddess. COMPOSITAE.

Toco'ca f. From the native name in Guyana. MELASTOMATACEAE.

To'dea f. In honour of Heinrich Julius Tode (1733–1797), German botanist specially interested in the fungi of Mecklenburg, north Germany. OSMUNDACEAE.

toku'dana Japanese name for a species of *Hosta*.

Tolmie'a f. Pickaback plant. Named for Dr. William Fraser Tolmie (1830–1886), Scottish physician, fur-trader, ethnologist and botanist in north-west America. SAXIFRAGACEAE.

Tol'pis f. Yellow hawkweed. Name coined by Adanson, who gave no derivation. COMPOSITAE.

tomento'sus, -a, -um Densely woolly; with matted hairs.

tommasinia'nus, -a, -um In honour of Muzio Giuseppe Spirito de Tommasini (1794–1879), magistrate and botanist at Trieste, celebrated for his work on the Dalmatian flora.

tonduzi In honour of J. F. Adolphe Tonduz (1862–1921), Swiss botanist from 1888 to 1919 in Costa Rica.

ton'sus, -a, -um Sheared; smooth-shaved.

Tore'nia f. Named for the Reverend Olof Torén (1718–1753), chaplain to the

Swedish East India Company at Surat, India, and in China, whose account of his travels was published by Linnaeus in 1759. SCROPHULARIACEAE.

tormina'lis, -is, -e Relating to or effective against colic.

toro'sus, -a, -um Cylindrical, with contractions at intervals.

Torrey'a f. Named for Dr. John Torrey (1796–1873), one of the giants of North American botany and co-author with Asa Gray of *The Flora of North America*. Regarded as an ultimate authority, he described hundreds of plants brought back by such explorers as Fremont and Pickering. TAXACEAE.

tor'ridus, -a, -um Growing in hot, dry places.

tor'tilis, -is, -e; tor'tus, -a, -um Twisted.

Tournefor'tia f. In honour of Joseph Pitton de Tournefort (1656–1708), celebrated French botanist who travelled with Aubriet and Gundelsheimer in the Levant and was the first botanist properly to define genera. He was an important fore-runner of Linnaeus. BORAGINACEAE.

Townsen'dia f. Rocky Mountain herbs named for David Townsend (1787–1858) of Pennsylvania. COMPOSITAE.

toxica'rius, -a, -um; to'xicus, -a, -um; to'xifer, toxi'fera, -um Poisonous.

Trache'lium n. Throatwort. Gr. *trachēlos*, a neck; supposed to be good for afflictions of the throat (trachea). CAMPANULACEAE.

Trachelosper'mum n. From Gr. *trachēlos*, a neck; *sperma*, a seed. APOCYNACEAE.

Trachycar'pus m. From Gr. *trachys*, rough; *karpŏs*, a fruit; in allusion to the fruit of some species. PALMAE.

Trachyme'ne f. From Gr. *trachys*, rough; *mēninx*, a membrane; in reference to the fruits of some species. UMBELLIFERAE.

Trachyste'mon m. Gr. *trachys*, rough; *stēmōn*, a stamen. BORAGINACEAE.

Tradescan'tia f. Widow's-tears, wandering-Jew, spiderwort. Named for John Tradescant senior (d. 1638), English gardener who went to north Russia in 1618 and Algiers in 1620 and received the type species, *T. virginiana*, from Virginia before 1629, and his son John Tradescant junior (1608–1662) who visited Virginia in 1654; both were successively gardeners to King Charles I. COMMELINACEAE.

Tragopo'gon m. Goatsbeard. Gr. *tragŏpōgōn* from *tragŏs*, goat; *pōgōn*, beard; possibly in allusion to the silky pappus. COMPOSITAE.

translu'cens Translucent.

transpa'rens Transparent.

transylva'nicus, -a, -um From Transylvania, Romania.

Tra'pa f. Water-chestnut. Aquatic floating herbs with edible fruit. A contraction of L. *calcitrappa*, called a caltrop or crow's foot—a weapon of defensive war with four sharp iron points which, thrown on the ground, has one point always pointing upward particularly to pierce the hooves of cavalry horses. The reference here is to the similarly four-pointed fruit. TRAPACEAE.

trapezifor'mis, -is, -e With four unequal sides.

Trautvette'ria f. Named for Ernst Rudolf von Trautvetter (1809–1889), Russian botanist of Baltic German origin, for many years associated with the St. Petersburg botanic garden. RANUNCULACEAE.

tremuloi'des Resembling the quivering poplar.

tre'mulus, -a, -um Quivering; trembling.

Treve'sia f. Named for Enrichetta Treves de Bonfigli of Padua, 19th-century supporter of botanical research. ARALIACEAE.

tri- In compound words signifies three. Thus:
triacan'thus (three-spined); **trian'drus** (with three stamens); **triangula'ris/triangula'tus** (with three angles); **tricauda'tus** (three-tailed); **trice'phalus** (three-headed); **tricho'tomus** (three-branched); **tricoc'cus** (with deeply three-lobed fruit, each division usually one-seeded); **tri'color** (three-coloured); **tricor'nis** (with three horns); **tricuspida'tus** (three pointed); **tri'dens/tridenta'tus** (three-toothed); **tri'fidus** (cut in three); **triflo'rus** (three-flowered); **trifolia'tus/trifo'lius** (with three leaves or leaflets); **trifurca'tus/trifur'cus** (three-forked); **trigonophyl'lus** (with three-cornered leaves); **triloba'tus** (three-lobed); **trimes'tris** (of three months); **triner'vis** (three-nerved); **trinota'tus** (three-spotted or-marked); **tripe'talus** (three-petalled); **triphyl'lus** (three-leaved); **tri'pterus** (three-winged); **trisper'mus** (with three seeds); **tri'stachyus** (three-spiked); **triterna'tus** (thrice in threes).

Trichan'tha f. From Gr. *thrix*, *trichos*, hair; *anthos*, flower, in allusion to the profuse long hairs on the corolla. GESNERIACEAE.

Trichi'lia f. Gr. *tricha*, three parts; the ovary and capsule are usually three-celled. MELIACEAE.

tricho- In compound words signifies hairy or hair-like, from Gr. *thrix*, *trichos*, hair; *trichion*, small hair. Thus:
tricho'calyx (with a hairy calyx); **tricho'gynus** (with a hairy ovary); **tricho'phorus** (bearing hairs); **trichophyl'lus** (with hairy leaves); **trichosper'mus** (hairy-seeded); **trichosto'mus** (with a hairy mouth).

Trichoce'reus m. From Gr. *thrix*, a hair, plus *Cereus*. The flowering parts are hairy. CACTACEAE.

Trichodiade'ma n. From Gr. *thrix*, hair; *diadēma*, band around the head, diadem; in allusion to the crown of bristles at the tips of leaves and sepals of these succulent plants formerly placed in *Mesembryanthemum*. AIZOACEAE.

Tricholae'na f. From Gr. *thrix*, a hair; *chlaina*, a cloak. The spikelets have a hairy coating. GRAMINEAE.

Trichoma'nes f. Gr. *trichŏmanēs*, name of a fern mentioned by Theophrastus and Dioscorides. HYMENOPHYLLACEAE.

Trichopi'lia f. From Gr. *thrix*, a hair; *pilion*, a cap. The anthers are concealed under a cap. ORCHIDACEAE.

Trichosan'thes f. Snakegourd. Gr. *thrix*, hair; *anthos*, a flower; in allusion to the conspicuously fringed corolla. CUCURBITACEAE.

Trichoste'ma n. Bluecurls, bastard pennyroyal. From Gr. *thrix*, hair; *stēma*, penis, stamen; the stamens have hair-like very long and slender filaments. LABIATAE.

Tricyr'tis f. Japanese toad-lily. Gr. *tri-*, three; *kyrtos*, humped; the bases of the three outer petals swollen and sacklike. LILIACEAE (CONVALLARIACEAE).

Tri'dax f. Greek name for another plant used as a vegetable, not these perennial herbs. COMPOSITAE.

Trienta'lis f. Star flower. Latin word meaning one-third of a foot in height, approximately that of these small perennials. PRIMULACEAE.

Trifo'lium n. Clover. The Latin name, from *tri-*, three; *folium*, a leaf; because of the trifoliolate leaves. LEGUMINOSAE.

Trigonel'la f. Fenugreek. Diminutive of L. *trigōnus*, three-cornered, triangular, from Gr. *tri-*, three; *gōnia*, angle. The flowers appear triangular. LEGUMINOSAE.

Tri'lisa f. Carolina vanilla. Anagram of *Liatris*, indicating a close relationship to that genus. COMPOSITAE.

Tril'lium n. Wake-robin, birthroot. From L. *tri-*, three, triple. Leaves and other parts are in threes. LILIACEAE.

Trime'za f. From Gr. *tri-*, three; *mĕizōn*, greater (comparative of *mĕgas*, great). The three outer flower segments are bigger than the inner. IRIDACEAE.

Trios'teum n. Horse-gentian. From Gr. *tri-*, three; *ŏstĕŏn*, a bone; from the three very hard seeds. CAPRIFOLIACEAE.

Tripha'sia f. Limeberry. Gr. *triphasios*, triple; the parts of the flowers are in threes. RUTACEAE.

Trip'laris f. Gr. *triplex*, triple; The parts of the flowers are in threes. POLYGONACEAE.

Tripleurosper'mum n. From Gr. *tri-*, three; *pleuron*, rib; *sperma*, seed; in allusion to the three-ribbed achenes (cypselae). COMPOSITAE.

Tripogan'dra f. From Gr. *tri-*, three; *pōgōn*, beard; *anēr*, *andrŏs*, man, hence stamen; in allusion to the three longer stamens being hairy, but the three shorter ones glabrous. COMMELINACEAE.

Trip'sacum n. Origin obscure. GRAMINEAE.

Triptery'gium n. Gr. *tri-*, three; *pteryx*, a wing. The fruits are equipped with three membranous wings. CELASTRACEAE.

Trise'tum n. From L. *tri-*, three; *seta* (*saeta*) bristle; in allusion to the three awns of the lemnas. GRAMINEAE.

Tristagma n. Gr. *tri-*, three; *stagma*, that which drips; in allusion to the three nectar-producing pores on the ovary. LILIACEAE (ALLIACEAE).

Trista'nia f. Named for Jules Tristan (1776–1861), French botanist. MYRTACEAE.
tri'stis, -is, -e Dull; sad.
Trithri'nax f. Gr. *tri-*, three; *thrinax*, trident; from the form of the leaves. PALMAE.
Tri'ticum n. Wheat. The classical Latin name. Next to rice, wheat is the most important cereal crop. GRAMINEAE.
tritifo'lius, -a, -um With polished leaves.
Trito'nia f. According to Gawler the name is derived from *Triton*, in the signification of a weather-cock and alluding to the variable direction of the stamens of the different species. IRIDACEAE.
trium'phans Splendid; triumphant.
trivia'lis, -is, -e Common; ordinary.
Trochoden'dron n. Gr. *trŏchōs*, a wheel; *dĕndrŏn*, a tree; in allusion to the spreading stamens. TROCHODENDRACEAE.
trochopteran'thus, -a. -um Epithet for a Cyclamen with somewhat propeller-like flowers, from Gr. *trŏchŏs*, wheel; *pteron*, wing; *anthos*, flower.
troja'nus, -a, -um Of Troy, western Asia Minor.
trolliifo'lius With leaves like *Trollius*.
Trol'lius m. Globe flower. From the Swiss German name, *Trollblume*, Latinized by Conrad Gessner in 1555 as *Trollius flos*, rounded flower, cognate with Middle English *troll*, to trundle or roll. RANUNCULACEAE.
Tropa'eolum n. The nasturtium of gardeners, though not that of botanists. Named by Linnaeus, from Gr. *tropaion*, trophy; L. *tropaeum*. The plant growing up a support reminded Linnaeus of a classical trophy, with round shields and golden helmets hung on a pillar, which was a sign of victory set upon a battlefield. The word *nasturtium* is Latin for a pungent-tasting plant. TROPAEOLACEAE.
tro'picus, -a, -um Of the tropics; tropical.
trullipe'talus, -a, -um With petals shaped somewhat like a bricklayer's trowel.
trunca'tus, -a, -um Cut off square.
tsarongen'sis, -is, -e From Tsarong province, Mekong-Salwin divide, south-eastern Tibet.
Tsu'ga f. Hemlock. From the Japanese name. PINACEAE.
tubaefor'mis, -is, -e Trumpet-shaped.
tuba'tus, -a, -um Trumpet-shaped.
Tuberar'ia f. From the thickened rootstock of the type species. CISTACEAE.
tubercula'tus, -a, -um Tuberculate, i.e. covered with wart-like excrescences.
tuberculo'sus, -a, -um Covered with wart-like excrescences; tubercled.
tuberge'nii, tubergenianus, -a, -um In honour of Messrs. C. G. van Tubergen, bulb-growers of Haarlem, Holland, founded in 1868 by Cornelis

Gerrit van Tubergen (1844–1919), continued by his two nephews Th. M. Hoog and J. M. C. Hoog.

tubero'sus, -a, -um Tuberous.

tu'bifer, tubi'fera, -um; tubulosus, -a, -um Tubular; pipelike.

tubiflo'rus, -a, -um Tubular-flowered.

tugurio'rum Of native or peasant huts, i.e. growing over or around them.

Tulba'ghia f. Named for Rijk Tulbagh (1699–1771), Dutch Governor of the Cape of Good Hope. LILIACEAE (ALLIACEAE).

Tu'lipa f. Tulip. Latinized version of the Turkish *tulbend*, a turban. With many species native to the Caucasus, Anatolia, and the Near East generally, tulips have been grown in Turkish gardens for centuries and were widely used as a decorative motif in their ceramics. They seem to have been introduced to the West by Ogier Ghiselin de Busbecq, Ambassador of the Holy Roman Empire to Suleiman the Magnificent, who saw them on his way to Constantinople in 1554. Conrad Gessner described tulips he saw growing in Augsburg in 1559. It is likely that Clusius sent tulips to England around 1578. He took bulbs to Holland in 1593 when he became professor of botany at Leyden where tulips must already have been known and appreciated, for he soon lost his bulbs by theft. A few years later there developed in Holland and, afterwards, in Turkey the extraordinary financial hysteria known as the 'Tulipomania'. It reached its most extravagant heights and sudden collapse in Holland in 1634–1637. Fabulous prices included up to 100,000 florins for a single bulb. LILIACEAE.

tuli'pifer, tulipi'fera, -um Tulip-bearing.

tu'midus, -a, -um Swollen.

Tu'nica f. A synonym of *Petrorhagia* . CARYOPHYLLACEAE.

tuolumnen'sis From Tuolumne county, California, U.S.A., so named from the Talmalamne Indians who once lived there.

Tupidan'thus m. Gr. *tupis*, a mallet; *anthos*, a flower; from the shape of the flower buds. ARALIACEAE.

turbina'tus, -a, -um Shaped like a spinning top.

tur'gidus, -a, -um Inflated; full.

Turnera f. In honour of William Turner (*c.* 1508–1568), the 'Father of English Botany', clergyman, physician and herbalist, who spent several years in exile on account of his ardent Protestant views and published the first botanical works in English with any claim to originality. TURNERACEAE.

Turra'ea f. Trees or shrubs named for Giorgia della Turre (1607–1688), professor of botany at Padua. MELIACEAE.

Tussa'cia f. Tropical American plants named for Richard de Tussac (1751–1837), French botanist and author of a book on the flora of the Antilles. GESNERIACEAE.

Tussila'go f. Coltsfoot. L. *tussis*, a cough: *-ago*, action; with reference to the use of flowers and leaves in cough remedies, including herbal tobacco. COMPOSITAE.

Ty'pha f. Cat-tail, reed-mace. The Greek name. In England *Typha* is commonly called bulrush which is properly a *Scirpus*. The biblical bulrushes were almost certainly *Cyperus papyrus*, the source of papyrus and the material used for making small rafts. The infant Moses is reported to have been found by Pharaoh's daughter in an ark of bulrushes laid in the flags by the river's brink. See *Scirpus*. TYPHACEAE.

typhi'nus, -a, -um Resembling *Typha*.

ty'picus, -a, -um Typical; agreeing with the type of a group.

tytthocar'pus, -a, -um Gr. *tytthos*, little; *karpos* fruit; small-fruited.

U

uitewaalia'nus, -a, -um In honour of Adriaan Joseph Antoon Uitewaal (1889–1963), Dutch photographer, editor of *Succulenta* from 1947 to 1956 and specialist on the genus *Haworthia*.

U'lex m. Gorse. The ancient Latin name. LEGUMINOSAE.

ulici'nus, -a, -um Resembling *Ulex*.

uligino'sus, -a, -um Of swamps and wet places.

Ul'lucus m. From the native Peruvian name. BASELLACEAE.

ulma'ria f. From L. *ulmus*, elm; referring to the leaflets of the meadow-sweet (*Filipendula ulmaria*).

ulmifo'lius, -a, -um With leaves like elm.

ulmo'ides Resembling elm.

Ul'mus f. Elm. The Latin name. The Romans used pollarded elms in vineyards over which to grow their vines. Trees are still so used in parts of Italy and Portugal. ULMACEAE.

ulva'ceus, -a, -um Resembling the seaweed *Ulva*.

umbella'tus, -a, -um Furnished with an umbel, the flower-stalks all arising from the same place.

Umbellula'ria f. California laurel. L. *umbella*, an umbel; from the shape of the inflorescence. LAURACEAE.

Umbili'cus m. Navelwort. L. for navel; in allusion to the sunken centre of the leaves of these succulents. CRASSULACEAE.

umbona'tus, -a, -um Having a stout projection at the centre.

umbracu'lifer, umbraculi'fera, -um Umbrella-bearing.

umbrel'la Of umbrella-like shape.

umbro'sus, -a, -um Shade-loving.

uncina'tus, -a, -um Hooked at the end.

unda'tus, -a, -um; undula'tus, -a, -um Waved, wavy, as in *undulifolius*, wavy-leaved.

u'nedo Latin name for the strawberry tree (*Arbutus*) and its attractive-looking fruit.

unger'nii In honour of Baron Franz Ungern-Sternberg (1808–1885), Baltic-German botanist and physician, monographer of *Salicornia*.

Ungna'dia f. Mexican buckeye. Named for Baron David von Ungnad, Austrian Ambassador at Constantinople (1576–1582), who sent horse-chestnut and other seeds to Clusius. SAPINDACEAE.

unguicular'is, -is, -e; unguicula'tus, -a, -um Furnished with a claw, i.e. contracted into a long narrow base.

unguipe'talus, -a, -um With claw-shaped petals.

unguispi'nus, -a, -um With hooked spines.

uni- In compound words signifies one. Thus:

uni'color (of one colour); **unicor'nis** (with one horn); **unidenta'tus** (with a single tooth); **uniflo'rus** (one-flowered); **unifo'lius** (with one leaf); **unilatera'lis** (one-sided); **uniner'vis** (one-nerved); **uniseria'tus** (in one row); **univitta'tus** (with one stripe).

Uni'ola f. Latin name for an unidentifiable plant, not these perennial grasses. GRAMINEAE.

uplan'dicus Of Uppland, Sweden.

urba'nii In honour of Ignatz Urban (1848–1931), German botanist in Berlin, an outstanding taxonomist most renowned for his work on plants of the West Indies.

urba'nus, -a, -um; ur'bicus, -a, -um Belonging to towns.

urceola'tus, -a, -um Urn-shaped.

Urceo'lina f. From L. *urceolus*, a little pitcher; in allusion to the shape of the flowers. AMARYLLIDACEAE.

u'rens Stinging; burning.

U'rera f. Cow-itch. L. *ure*, burn or sting. One member (*U. baccifera*) of this genus of trees and shrubs is also called chichaste and described as 'one of the most dangerous plants of Central America… . When one is struck by the coarse hairs, the effect is almost like that of an electric shock and there often follows the most intense pain… . The plant is often used for hedges, which few larger animals care to penetrate.' (Standley, *Flora of Costa Rica.*) URTICACEAE.

Urgi'nea f. From the name of an Arab tribe in Algeria, the Beni Urgin. These bulbous herbs, closely related to *Scilla*, are the source of the commercial 'squill'. Sometimes included in *Drimia*. LILIACEAE (HYACINTHACEAE).

ur'niger, urni'gera, -um Urn-bearing, for example with urn-shaped fruits.

urophyl'lus, -a, -um With leaves having a long tail-like tip.

Ursi'nia f. Named for Johannes Heinrich Ursinus (1608–1667) of Regensburg, a German botanical author. COMPOSITAE.

ursi'nus, -a, -um Like a bear, in shagginess or other respects. Also, sometimes, Northern (from the Great Bear constellation).

Urti'ca f. Stinging nettle. The Latin name. All species sting but some are occasionally so virulent (e.g. *U. urentissima*) as to cause death. The power to sting lasts even in the dried plant. For instance, when the great Linnaean herbarium was being photographed, specimen by specimen, in 1941 for record in case of destruction during the Second World War, the photographer, Miss Gladys Brown, was stung on the arm, raising a blister, by one of the specimens of *Urtica* dried and mounted approaching two hundred years earlier. URTICACEAE.

urticifo'lius, -a, -um Nettle-leaved.

urticoi'des Resembling nettle.

usitatis'simus, -a, -um Most useful.

ustula'tus, -a, -um Burned, scorched.

u'tilis, -is, -e Useful.

Utricula'ria f. Bladderwort. L. *utriculus*, a small bottle; in allusion to insect-trapping bladders borne on the leaves and runners of these aquatic and terrestrial herbs. LENTIBULARIACEAE.

utricula'tus, -a, -um; utriculo'sus, -a, -um Bladderlike.

Uva'ria f. From L. *uva*, bunch of grapes; in allusion to the somewhat grape-like fruits. ANNONACEAE.

uva-ur'si Bear's grape. Specific epithet of *Arctostaphylos*.

uvariifo'lius, -a, -um With leaves like *Uvaria*.

u'vifer, uvi'fera, -um Bearing grapes.

Uvula'ria f. Bellwort. From the anatomical term *uvula*, the lobe hanging from the back of the soft palate in man; in reference to the hanging blossoms of this plant. LILIACEAE (CONVALLARIACEAE).

V

Vacca'ria f. Possibly from L. *vacca*, cow. CARYOPHYLLACEAE.

vaccinifo'lius, -a, -um With leaves like *Vaccinium*.

vaccinioi'des Resembling *Vaccinium*.

Vacci'nium n. Blueberry. Bilberry. Cranberry. A Latin name apparently derived from the same prehistoric Mediterranean language as the Gr. *hyakinthos* and transferred to these berry-bearing shrubs. ERICACEAE.

vacil'lans Variable.

va'gans Of wide distribution; wandering.

vagina'lis, -is, -e; vagina'tus, -a, -um; vagin'ifer, vaginifera, -um Sheathed; having a sheath.

valdivia'nus, -a, -um Of Valdivia, Chile.

valenti'nus, -a, -um Of Valentia in Spain.

Valeria'na f. Valerian. The medieval Latin name, possibly derived from L. *valere*, to be healthy; in allusion to the plant's medicinal uses in nervousness and hysteria. VALERIANACEAE.

Valerianel'la f. Corn-salad. Lamb's lettuce. Diminutive of *Valeriana*, from the resemblance of these herbs to *Valeriana*. VALERIANACEAE.

va'lidus, -a, -um Strong; well-developed.

Valla'ris f. L. *vallus*, a stake in a palisade. These twining shrubs grow over fences. APOCYNACEAE.

valli'cola. A dweller in valleys.

Vallisne'ria f. Eelgrass. Named for Antonio Vallisnieri de Vallisnera (1661–1730), professor at Padua. These are fresh-water plants. Salt-water eelgrass so much appreciated by wild ducks is *Zostera marina*.

Vallo'ta f. Scarborough-lily. Named for Pierre Vallot (1594–1671), French physician and botanical author. AMARYLLIDACEAE.

Vancouve'ria f. Named for Captain George Vancouver, Royal Navy (1758–1798), British explorer. As a young man he sailed twice with Captain Cook. His outstanding feat was his four-year voyage (1791–1795) during which he carried out the first detailed surveys of the Pacific coast of North America as far north as Cook Inlet. Vancouver Island is named for him. The botanical results were due to surgeon Archibald Menzies (1754–1842). See *Menziesia*. BERBERIDACEAE.

Van'da f. From a Sanskrit name for an epiphytic orchid. ORCHIDACEAE.

Vando'psis f. From *Vanda*, plus Gr. *opsis*, similar to (see *Vanda*). ORCHIDACEAE.

vanhout'tei In honour of Louis Benoît Van Houtte (1810–1876), celebrated Belgian nurseryman, editor of *Flore des Serres*.

Vanil'la f. Vanilla. Spanish *vainilla*, a small pod; with reference to the shape of the fruit. *Vanilla* is the only genus of orchids from which a commercial product is derived. ORCHIDACEAE.

varia'bilis, -is, -e; va'rians; varia'tus, -a, -um Variable; varying.

variega'tus, -a, -um Irregularly coloured; variegated.

va'rius, -a, -um Differing; diverse.

vase'yii In honour of George Vasey (1822–1893), medical man, later botanist of the U.S. Department of Agriculture.

vayre'dae Named for Estanislao Vayreda y Vila (1848–1901), Spanish pharmacist and botanist in Catalonia.

vedrarien'sis, -is, -e From Verrières near Paris, nursery of Messrs. Vilmorin-Andrieux.

ve'getus, -a, -um Vigorous.

Veit'chia f. Named for James Veitch (1815–1869) and his son John Gould Veitch (1839–1870), of Exeter and Chelsea, England. The Veitchs were the leading nurserymen of their day and their family remained in business continuously for 106 years until the Chelsea firm was brought to an end in 1914 on the retirement of Sir Harry James Veitch (1840–1924). They introduced many good plants as shown by the many specific names commemorating them (*veitchianus, veitchii, veitchiorum*, also *harryanus*). They sent their own collectors abroad and trained a legion of excellent gardeners. Dr. E. H. Wilson was among their collectors, see *Sinowilsonia*. PALMAE.

vela'ris, -is, -e Pertaining to a veil.

velle'reus, -a, -um Fleecy.

ve'lox Swift, quick-growing.

Velthei'mia f. Named for August Ferdinand von Veltheim (1741–1801) of Brunswick, German patron of botany. LILIACEAE (HYACINTHACEAE).

veluti'nus, -a, -um Velvety.

vena'tor Hunter; in allusion to scarlet ('pink') coats of British fox-hunters.

venena'tus, -a, -um Poisonous. Also, *venenosus*, very poisonous.

vene'tus, -a, -um Of Venice, Italy.

Veni'dium n. From L. *vena*, a vein; because of the ribbed fruits. COMPOSITAE.

veno'sus, -a, -um Full of veins, prominently veined.

ventrico'sus, -a, -um Having a swelling on one side (L. *ventricosus*, big-bellied, pot-bellied).

venus'tus, -a, -um Handsome; charming.

Vera'trum n. False hellebore. The Latin name (*vērātrum*). LILIACEAE (MELANTHIACEAE).

verbascifo'lius, -a, -um With leaves like mullein.

Verbas'cum n. Mullein. The ancient Latin name. SCROPHULARIACEAE.

Verbe'na f. Vervain. The Latin name for leaves and shoots of laurel, myrtle, etc., used in religious ceremonies and also in medicine. VERBENACEAE.

Verbesi'na f. From the resemblance of leaves to *Verbena*. COMPOSITAE.

verecun'dus, -a, -um Modest.

ve'ris Of the spring; spring-flowering. Specific epithet of the cowslip (*Primula veris*).

vermicula'ris, -is, -e; vermicula'tus, -a, -um Wormlike.

verna'lis, -is, -e Of spring; spring-flowering.

verni'cifer, vernici'fera, -um Producing varnish.

vernico'sus, -a, -um Varnished.

ver'nix Varnish.

Verno'nia f. Ironweed. Named for William Vernon (d. *c.* 1711), English botanist who collected in Maryland in 1698. COMPOSITAE.

ver'nus, -a, -um Of the spring.

Vero'nica f. Speedwell. Reputedly named for St. Veronica (see *Hebe*). SCROPHULARIACEAE.

Ve'ronicastrum n. From *Veronica* plus *-astrum*, indicating incomplete resemblance. SCROPHULARIACEAE.

verruco'sus, -a, -um Warty.

verruculo'sus, -a, -um With small warts.

Verschaffel'tia f. Named for Ambrose Colletto Alexandre Verschaffelt (1825–1886) of Ghent, Belgian nurseryman and author of a book on camellias. PALMAE.

versi'color Variously coloured.

verticilla'ris, -is, -e; verticilla'tus, -a, -um Having whorls; forming a ring around an axis.

Verticor'dia f. Juniper-myrtle. The name is a compliment to Venus to whom myrtle was sacred (L. *verto*, I turn; *cor*, the heart). MYRTACEAE.

ve'rus, -a, -um True to type; standard.

ves'cus, -a, -um Thin; feeble; edible.

Vesica'ria f. L. *vesica*, a bladder; from the inflated pods. CRUCIFERAE.

vesica'rius, -a, -um Bladder-like.

vesiculo'sus, -a, -um Furnished with small bladders or vesicles.

vespertilio'nis, -is, -e Of a bat, applied to plants with leaves suggestive of a bat's wings.

vesperti'nus, -a, -um Of the evening; evening-blooming.

Vestia f. In honour of Lorenz Chrysanth von Vest (1776–1840), Austrian botanist and physician, professor of medicine at Klugenfurt, then of medicine, botany and chemistry at Graz. SOLANACEAE.

vesti'tus, -a, -um Covered; clothed, usually with hairs.

Vetive'ria f. Knus-Knus. From Southern Indian vernacular name *vettiveri*, derived like *Zingiber* from Dravidian *ver*, root. GRAMINEAE.

vex'ans Annoying; wounding.

vexilla'ris, -is, -e Having a standard (flag). Also *vexillarius*, a standard-bearer.

via'lii. Named for Paul Vial (1855–1917), French missionary and ethnologist in China.

via'rum Of waysides.

viburnifo'lius, -a, -um With leaves like *Viburnum*.

Vibur'num n. Arrow-wood, wayfaring tree. Latin name of one species of this genus, possibly *V. lantano*. CAPRIFOLIACEAE.

Vi'cia f. Vetch. The Latin name for these herbs. LEGUMINOSAE.

viciifo'lius, -a, -um Vetch-leaved.

Victo'ria f. Royal water-lily, Queen Victoria water-lily. Named for Queen Victoria (1819–1901). The great water-lily, *V. amazonica* (*V. regia*), was found in Guiana by Schomburgk, having earlier been discovered by Haenke, but introduced into cultivation from Bolivia in 1846 by Thomas Bridges (1807–1865), who brought seeds to Kew. The first detailed account of it appeared in the *Bot. Mag.* t. 4275 (1847). A seedling was obtained from Kew by Joseph Paxton, head gardener to the Duke of Devonshire at Chatsworth. Paxton's feat of bringing the plant to bloom caused a sensation and a special telegram was dispatched to announce the fact to Her Majesty. The leaves, which may be up to six feet across, are reputed to be able to support the weight of a small child—provided he does not fidget and his weight is distributed correctly. NYMPHAEACEAE.

victoria'lis, -is, -e Victorious. Bulbs of *Allium victorialis* were once worn by German miners for protection against evil spirits and for victory over them.

Vi'gna f. Named for Dominico Vigna (d. 1647), professor of botany at Pisa, author in 1625 of a commentary on Theophrastus. LEGUMINOSAE.

Villare'sia f. Named for Mattias Villares, a monk who had a botanic garden in the monastery of Santa Espina, Chile, *c.* 1790. ICACINACEAE.

Villar'sia f. In honour of Dominique Villars (1745–1814), French botanist, chiefly noted for his work on the flora of Dauphiné. He also spelled his name as Villar, which is the name of his birth-place. MENYANTHACEAE.

villo'sus, -a, -um Covered with soft hairs.

vilmorinia'nus, -a, -um; vilmorinii; vilmorinae In honour of the firm of Vilmorin-Andrieux of Paris, celebrated French nurserymen and seedsmen, notably Maurice de Vilmorin and his wife.

vimina'lis, -is, -e; vimi'neus, -a, -um With long slender shoots, like osiers.

Vin'ca f. Periwinkle. Contracted from the Latin name *vinca pervinca* or *vincapervinca*, apparently from *vincio*, bind, wind around; in allusion to the use of the long flexible shoots in making wreaths. APOCYNACEAE.

Vinceto'xicum n. L. *vinco*, to conquer; *toxicum*, poison. This genus was supposed to contain an antidote to snake-bite. ASCLEPIADACEAE.

vinea'lis, -is, -e Belonging to vines, growing in vineyards.

vi'nifer, vini'fera, -um Wine-bearing.

vino'sus, -a, -um Wine-red.

Vi'ola f. Pansy, heart's-ease, violet. The Latin name for various sweet-scented flowers, such as violets, stocks, wallflowers, and derived from the same source as the Gr. *ion* (which in its earlier form had an initial letter corresponding to *v* or *w*, the digamma, which was later lost); see p. 315. VIOLACEAE.

viola'ceus, -a, -um Violet-coloured.

violas'cens Becoming violet-coloured.

virens Green. Also **virescens**, becoming green.

virga'tus, -a, -um Twiggy.

Virgi'lia f. Named for the Latin poet Virgil (70–19 B.C.). LEGUMINOSAE.

virgina'lis, -is, -e; virgi'neus, -a, -um White; virginal.

virginia'nus, -a, -um; virgi'nicus, -a, -um; virginien'sis, -is, -e Of Virginia, U.S.A., named for Queen Elizabeth I, England's 'Virgin Queen'.
virgulto'rum Of thickets.
vi'ridis, -is, -e Green. Also in compound words *viridi.* Thus: **virides'cens** (becoming green); **viridiflo'rus** (with green flowers); **viridifo'lius** (green-leaved); **viridifus'cus** (green-brown); **viridis'simus** (very green); **viri'dulus** (somewhat green).
viscidifo'lius, -a, -um With sticky leaves.
vis'cidus, -a, -um; visco'sus, -a, -um Sticky; clammy, like birdlime obtained from mistletoe berries.
Vis'cum n. Latin name for mistletoe. VISCACEAE.
vita'ceus, -a, -um Vinelike.
vitelli'nus, -a, -um Like the yolk of an egg; egg-yellow.
Vi'tex f. Latin name for *V. agnus-castus* or chaste tree. VERBENACEAE.
vitifo'lius, -a, -um With leaves like the grapevine.
Vi'tis f. Grape. The Latin name for the grapevine. V. *vinifera* is the European grape. It has been used by man since prehistoric times but no one knows when the secret of wine-making was discovered. VITACEAE.
Vittadi'nia f. Named for Carlo Vittadini (1800–1865), Italian mycologist and physician. COMPOSITAE.
vitta'tus, -a, -um Striped lengthwise. L. *vitta,* a band.
vit'tiger, vitti'gera, -um Marked with stripes.
volga'ricus, -a, -um Of the Volga River region.
volu'bilis, -is, -e Twining.
volu'tus, -a, -um With rolled leaves.
vomito'rius, -a, -um Emetic; causing vomiting.
Vrie'sia f. Named for Willem Hendrik de Vriese (1806–1862), Dutch botanist, professor of botany first at Amsterdam, then at Leiden. BROMELIACEAE.
vulca'nicus, -a, -um Growing on a volcano.
vulga'ris, -is, -e; vulgatus, -a, -um Common.
vulpi'nus, -a, -um Relating to foxes; used for species with an inferior sort of fruit.

W

Wachendor'fia f. In honour of Everardus Jacobus van Wachendorff (1702–1758), Dutch botanist professor at Utrecht. HAEMODORACEAE.

Wahlenber'gia f. Named for Georg Wahlenberg (1780–1851), Swedish botanist, professor at Uppsala, noted for his studies of European plant geography. CAMPANULACEAE.

Waldstei'nia f. Named for Count Franz Adam Waldstein-Wartenburg (1759–1823), Austrian botanist, part-author of a finely illustrated book on the plants of Hungary. ROSACEAE.

Wallich'ia f. Named for Nathaniel Wallich (1786–1854), originally Nathan Wolff, Danish botanist, who went out to India in 1807 as surgeon to the Danish settlement at Serampore near Calcutta, entered the service of the Hon. East India Company in 1813 and was superintendent of the Calcutta botanic garden from 1814 to 1841, during which period he made immense contributions to the knowledge of the Indian flora, many species of which bear the epithet *wallichii*. PALMAE.

Wardaster m. See *wardii*.

war'dii Specific epithet of many plants from eastern Asiatic mountains and highlands where for many years Frank Kingdon-Ward (1885–1958), English botanist and geographer, carried out intensive botanical exploration, bringing back many plants, some of which are now widely cultivated in temperate gardens.

warleyen'sis, -is, -e After Warley Place, Brentwood, Essex, England, the garden of Miss E. A. Willmott. See *willmottiae*.

Warscewiczel'la f. Named for Joseph Warsczewicz (1812–1866), who escaped from Poland after taking an active part in the 1830 rebellion. He travelled widely in South America collecting orchids for Messrs. van Houtte, Belgian nurserymen. Eventually he was able to return home to become Inspector of the botanic gardens in Krakow. (Name pronounced 'var-shev-i-chel-la'.) ORCHIDACEAE.

Warzewiczia See *Warscewiczella*.

Washingto'nia Named for George Washington (1732–1799), first President of the United States. PALMAE.

waso'nii In honour of Rear-Admiral Cathcart Romer Wason (1874–1941) of the Royal Navy.

Watso'nia f. Bugle lily. Named for Sir William Watson (1715–1787), English scientist and physician. IRIDACEAE.

watso'nii This epithet is likely to commemorate Sereno Watson (1826–1892) of Harvard University (see under *Serenoa*) if a North American plant, but it has also been used to commemorate Hewett Cottrell Watson (1804–1881), pioneer British plant-geographer, and William Watson (1858–1925), curator of the Royal Botanic Gardens, Kew.

Wattaka'ka f. From a Malabar (Malayalam) name *watta kakadodi* recorded by van Rheede tot Draakenstein in 1689. ASCLEPIADACEAE.

wat'tii In honour of Sir George Watt (1851–1930), Scottish medical man and economic botanist, author of *Dictionary of Economic Products of India*.

Wede'lia f. Named for Georg Wolfgang Wedel (1645–1721), professor of botany at Jena. COMPOSITAE.

Wei'gela f. Named for Christian Ehrenfried Weigel (1748–1831), German botanist, professor at Greifswald. CAPRIFOLIACEAE.

Weinman'nia f. Named for Johann Wilhelm Weinmann (1737–1745) of Regensburg, German apothecary, who published between 1737 and 1745 a big illustrated work on plants entitled *Phytanthozaiconographia*. CUNONIACEAE.

Welde'nia f. In honour of Ludwig van Welden (1780–1853), master of the ordnance in the Austrian army. COMMELINACEAE.

Welwitschia f. One of the world's most remarkable plants, named for Friedrich Martin Joseph Welwitsch (1806–1872), Austrian botanist and explorer in Angola where he made immense plant collections. WELWITSCHIACEAE.

werck'lei In honour of Karl (Carlos) Wercklé (1866–1924), investigator of the flora of Costa Rica.

Westrin'gia f. Australian rosemary. Named for Johan Peter Westring (1753–1833), physician to the King of Sweden and a keen lichenologist. LABIATAE.

weyri'chii In honour of Heinrich Weyrich (1828–1868), Russian naval surgeon of German extraction who botanized in eastern Asia.

wher'ryi In honour of Edgar Theodor Wherry (1885–1982), American chemist, geologist, crystallographer and botanical professor in Philadelphia, U.S.A.

Wigan'dia f. Named for Johannes Wigand (1523–1587), Bishop of Pomerania, who wrote on Prussian plants. HYDROPHYLLACEAE.

wi'ghtii In honour of Robert Wight (1796–1872), Superintendent of Madras Botanic Garden and author of important works on Indian plants.

Wilco'xia f. Named for General Timothy E. Wilcox, U.S. Army, a keen student of plants, *c.* 1900. CACTACEAE.

wilkesia'nus, -a, -um In honour of Charles Wilkes (1798–1877), American naval officer, leader of the United States Exploring Expedition of 1838–1842 in the Pacific Ocean.

william'sii In honour of Williams, the family name of at least seven distinguished horticulturists or botanists. Nepal plants with the epithet *williamsii* commemorate L. H. John Williams (1915–1991), English botanist, leader of British Museum Natural History expedition to Nepal in 1954.

willmot'tiae, willmottia'nus, -a, -um In honour of Miss Ellen Ann Willmott (1860–1934), celebrated English amateur gardener, who created a richly stocked and beautiful garden at Warley Place, Essex, employing 100 gardeners before 1914, produced *The Genus Rosa* (1910–1914), introduced new plants and, through her reckless extravagance, ended her days an impoverished woman employing two gardeners and the garden becoming derelict.

wilso'nii See *Sinowilsonia*.

Wistaria See *Wisteria*.

Wiste'ria f. Deciduous climbing flowering shrubs named for Caspar Wistar (1761–1818), professor of anatomy at the University of Pennsylvania and one of the early owners of what is now Vernon Park, Philadelphia. Despite the apparent anomaly Wisteria is the correct spelling according to the international rules of botanical nomenclature. LEGUMINOSAE.

wolga'ricus, -a, -um Of the Volga River region.

Wood'sia f. Ferns named for Joseph Woods (1776–1864), English architect and botanical author, best known for his *Tourist's Flora* (1852). ASPLENIACEAE.

Woodwar'dia f. Chain-fern. Named for Thomas Jenkinson Woodward (1745–1820), English botanist. BLECHNACEAE.

Wulfe'nia f. Named for Franz Xavier von Wulfen (1728–1805), Austrian botanical author, teacher and Jesuit abbot. SCROPHULARIACEAE.

Worsley'a f. Blue Amaryllis. In honour of Arthington Worsley (1861–1943), English civil engineer and gardener. AMARYLLIDACEAE.

Wye'thia f. Mule-ears. Named for Nathaniel J. Wyeth (1802–1856), American botanist who discovered these plants. COMPOSITAE.

X

Xantheran'themum n. From Gr. *xanthos*, yellow; *Eranthemum* which is a related genus. ACANTHACEAE.

xan'thinus, -a, -um Yellow. In compound words *xanth-*. Thus:
xanthacan'thus (yellow-spined); **xanthoca'lyx** (yellow calyx);
xanthocar'pus (yellow-fruited); **xanthochlo'rus** (yellowish green);
xanthoco'don (yellow bell); **xantholeu'cus** (yellowish white);
xanthoner'vis (yellow-nerved); **xanthophyl'lus** (with yellow leaves);
xanthorrhi'zus (yellow-rooted); **xanthoxy'lon** (with yellow heartwood).

Xanthis'ma f. Gr. *xanthos*, yellow; alluding to the colour of the flowers. COMPOSITAE.

Xantho'ceras n. From Gr. *xanthos*, yellow; *keras*, a horn; from the yellow hornlike growths between the petals. SAPINDACEAE.

Xanthorhi'za f. From Gr. *xanthos*, yellow; *rhiza*, root. RANUNCULACEAE.

Xanthorrho'ea f. Gr. *xanthos*, yellow; *rheo*, to flow. A yellow resinous gum is extracted from these perennials which are known as Botany Bay gums or blackboys. XANTHORRHOEACEAE.

Xanthoso'ma n. Gr. *xanthos*, yellow; *sōma*, a body; with reference to the yellow inner tissues of some species. ARACEAE.

xanthostepha'nus, -a, -um With a yellow crown or garland.

Xeran'themum n. Immortelle. From Gr. *xēros*, dry; *anthos*, a flower. The dry flower heads retain their form and colour for years. COMPOSITAE.

Xerophyl'lum n. Turkey's beard. From Gr. *xeros*, dry; *phyllon*, a leaf. The leaves are dry and grass-like. LILIACEAE (MELANTHIACEAE).

Xi'phium n. Greek name for a *Gladiolus* referring to the sword-like leaves, from *xiphŏs*, sword. IRIDACEAE.

Xylo'bium Gr. *xylon*, wood; *bios*, life. They grow on trees. ORCHIDACEAE.

xylocan'thus, -a, -um With woody spines.

Xylos'teum n. From Gr. *xylon*, wood; *ŏstĕŏn*, bone.

Y

yakuinsula'ris, -is,-e; yakusima'nus, -a, -um Of the island of Yakushima, southern Japan.

yedoen'sis, -is,-e Of Yedo (now Tokyo), Japan.

Yuc'ca f. From the Carib name for manihot or cassava (a genus belonging to the *Euphorbiaceae*), unfortunately applied to these liliaceous evergreen shrubs or small trees with rosettes of sword-shaped leaves. Adam's needle and Spanish bayonet are familiar names of some species. AGAVACEAE.

yungningen'sis, -is, -e From Yungning, Sichuan (Szechwan), China.

yunnanen'sis, -is, -e; yunna'nicus, -a, -um From Yunnan, west China.

Yushan'ia f. Of Yushan (Mount Morrison), Taiwan, where this bamboo was first found. GRAMINEAE.

Z

zabelii In honour of Hermann Zabel (1832–1912), German forester and dendrologist.

zaleu'cus, -a, -um Very white.

za'lil Afghan vernacular name for a kind of Delphinium.

Zaluzian'skya f. Annual or perennial herbs named for Adam Zaluziansky von Zaluzian (1558–1613), botanist at Prague. SCROPHULARIACEAE.

Za'mia f. Name derived from *zamiae*, a false rendering in some texts of Pliny for *azaniae*, referring to pine-cones. The roots of the coontie (*Z. floridana*) were used by the Indians to make soap. ZAMIACEAE.

Zantedes'chia f. Calla-lily (U.S.), arum-lily (U.K.). Named for Francesco Zantedeschi (b. 1797), Italian botanist. ARACEAE.

Zanthox'ylum n. Prickly ash, Hercules' club, tooth-ache tree. Gr. *xanthos*, yellow; *xylon*, wood; from the colour of the heartwood of some species. RUTACEAE.

Zauschne'ria f. In honour of Johann Baptist Zauschner (1737–1799), professor at Prague. ONAGRACEAE.

zawad'zkii. In honour of Alexander Zawadzki (1798–1868), Austrian botanist, mathematician and physicist at Lemberg and Brno.

Ze'a f. Corn (U.S.), maize, Indian corn (U.K.). From the Greek name for another cereal. The plant is widely cultivated in the tropical and temperate zones from sea level to about 12,000 feet. It is the staple grain in Mexico, Central America, and southern Africa. In the United States, about 90 per cent of the crop is fed to animals and goes to market as meat, milk, and eggs. GRAMINEAE.

Zebri'na f. From Portuguese *zebra*; in allusion to the striped leaves of one species. COMMELINACEAE.

zebri'nus, -a, -um Zebra-striped.

Zel'kova f. From the Caucasian name. ULMACEAE.

Zeno'bia f. Named for Zenobia, Queen of Palmyra *c.* A.D. 266. ERICACEAE.

Zephyran'thes f. Zephyr-lily. Gr. *zĕphyrŏs*, the west wind; *anthos*, a flower; because native of the Western hemisphere. AMARYLLIDACEAE.

zeylan'icus, -a, -um Of Ceylon (Sri Lanka).

zibethi'nus, -a, -um Ill-smelling like a civet cat. From the Italian *zibetto* (derived from the Arabian *zabad* or *zubad*), a civet. Specific epithet of the durian (*Durio*) whose fruit contains a pith which can smell horribly.

Zigade'nus m. Gr. *zygŏs*, a yoke; *adēn*, a gland. The floral glands are in pairs at the base of the perigon. LILIACEAE (MELANTHIACEAE).

Zin'giber n. Ginger. From the Greek name which, in turn, is said to derive from an East Indian word cognate with Pali *singivera*. The root of *Z. officinale* is the source of commercial ginger; see p. 318. ZINGIBERACEAE.

Zin'nia f. Named for Johann Gottfried Zinn (1727–1759), professor of botany, Göttingen. COMPOSITAE.

Ziza'nia f. Wild rice, Canadian wild rice. Greek name for another wild grain, not this. *Z. aquatica* is eagerly eaten by wild ducks. GRAMINEAE.

zizanioi'des Resembling *Zizania*.

Zi'zyphus f. Jujube. From the Persian name *zīzfūm* or *zīzafūn*. RHAMNACEAE.

zona'lis, -is,-e; zona'tus, -a, -um Banded or with a girdle usually of a distinct colour.

Zoy'sia f. Named for Karl von Zoys (1756–1800), Austrian botanist. GRAMINEAE.

Zygade'nus See *Zigadenus*.

Zygocac'tus m. Christmas cactus. From Gr. *zygŏs*, a yoke, plus *Cactus*; from the way in which the stems are jointed. CACTACEAE.

Zygope'talum n. Gr. *zygŏs*, a yoke; *pĕtalŏn*, a petal; from the swelling at the base of the lip which seems to yoke together the lateral segments of the flowers. ORCHIDACEAE.

Final Definition

Lexico'grapher ... A writer of dictionaries; an harmless drudge, that busies himself in tracing the original, and detailing the figuration of words.

Samuel Johnson, *A Dictionary of the English Language* (1755)

An Introduction to Vernacular Names

All the world over and from remote times onwards men have devised names for plants of special interest to them and in their naming have been motivated by much the same ways of thought because, as fishermen, hunters, primitive farmers and herdsmen, they have had much the same ways of life. Other plants were left nameless until in comparatively recent times, that is after A.D. 1500, the urge to describe and record in books the diversity of the whole natural world led to the development of botany as a special field of enquiry distinct from herbalism and medicine and thereby to the study of these hitherto unimportant plants. Hence the greater part of the world's plants lack common or vernacular names; they have been recorded only under botanical names, usually of Latin form, since 1753. Modern botanical nomenclature originated, however, in the vernacular names of plants used for economic, medicinal or decorative purposes by the Ancient Greeks and Romans, some of which had descended to them out of languages dead and gone long before the invention of writing. Thus the related Greek *rhŏdŏn* (ʽροδον) and Latin *rosa*, Greek *lĕiriŏn* (λειριον) and Latin *lilium*, Greek *iŏn* (ʼιον), earlier *wiŏn* (Fιον), and Latin *viola*, Greek *sukŏn* (συκον) and Latin *ficus*, Greek *ĕlaia* (ἐλαια), earlier *ĕlaiwa* (ἐλαιFα), and Latin *oliva*, Greek *kuparissŏs* (κυπαρισσος) and Latin *cupressus* were not derived the one from the other. They must have come independently from common sources, among them a supplanted Mediterranean language sometimes called Thraco-pelasgian spoken maybe four thousand years ago. Plant names like *Hyacinthus* and *Terebinthus* are also considered to be relicts of a pre-Greek language. The origins of such English plant names as those of the rose, violet, fig, cypress, mint, hyacinth, ginger, lily and crocus go back to very remote antiquity. As Walter de la Mare wrote:

> ...No man knows
> Through what wild centuries
> Roves back the rose.

The connexion between the botanical and the vernacular nomenclature of plants is much closer, both as regards method and the names themselves, than is commonly noticed. Indeed during the past three hundred years many vernacular names out of many diverse languages have been taken into botanical nomenclature, with little or no alteration, for use as generic names (see p. 9). Even Linnaeus, despite his puristic ruling in his *Critica botanica* (1737) that 'generic names which have not a root derived from Greek and Latin are to be rejected', in fact accepted names having quite other origins. Thus the generic name *Trollius*, adopted by Linnaeus in 1753 for the globeflowers, originated as a contraction of the name *Trollius flos* used by Conrad Gessner in 1555 instead of the Swiss-German name *Trollblume* meaning 'a rounded flower'. Vernacular names adopted as specific epithets are even more numerous; some examples illustrating their diversity are *Abies pinsapo*, *Actinidia kolomikta*, *Albizia lebbek*, *Brassica kaber*, *Cajanus cajan*, *Camellia sasanqua*, *Diospyros kaki*, *Gentiana kurroo*, *Krigia dandelion*, *Malus zumi*, *Myrica gale*, *Narcissus tazetta*, *Ophiopogon jaburan*, *Picea omorika*, *Pinus mugo*, *Pittosporum tobira*, *Prunus mume*, *Schinus molle*, *Silene schafta*, *Spondias mombin*, *Taraxacum kok-saghyz*, *Terminalia catappa*, *Veronica beccabunga*, *Zea mays*.

Clearly then vernacular names should not be despised and disregarded. When only a few plants need to be distinguished within a limited area, vernacular names can be just as useful, precise and stable for local use as their scientific equivalents intended for international use. It is important, however, to ascertain their application by examining the plants designated and so to equate them with standard scientific names, for the same vernacular name may be given to different plants in different places within the same country.

In 1892 Nathaniel Colgan of Dublin set out to ascertain the botanical identity of **'true shamrock'**. That year and the next he collected and received about St. Patrick's Day living specimens certified as 'true shamrock' from twenty Irish counties, grew them and identified them when in flower. Of these 19 were white clover (*Trifolium repens*), 12 were lesser yellow trefoil (*Trifolium dubium*, syn. *T. minus*), 2 red clover (*Trifolium pratense*) and 2 spotted medick (*Medicago arabica*); not one was wood-sorrel (*Oxalis acetosella*) sometimes called 'shamrock' in England. A summary of Colgan's investigation into the history of the shamrock (Irish *seamróg*, a diminutive of *seamar*, clover), originally published in the *Journal of the Royal Society of Antiquaries of Ireland* 26: 211–226, 349–361 (1896), will be found in Lloyd Praeger's *The Way That I Went* (1937). A similar enquiry by E. Charles Nelson in 1988 for 'true shamrock' resulted in the receipt of 221 plants from thirty Irish counties. They included the same four species received by Colgan and in much the same proportions, as related in Nelson's comprehensive book *Shamrock* (1991).

Unfortunately all the early lives of St. Patrick are completely silent about shamrock, apparently an Irish famine-food down to about 1682, and Colgan traced the legend of it as exemplifying the Trinity no further back than 1727. This example will suffice to indicate the need for caution, indeed for painstaking enquiry, when accepting vernacular names. Their interpretation links botany, etymology, and folklore. Collections of such names, notably English ones assembled by Britten and Holland and commented upon by Prior and by Grigson, Danish ones by Lange, French ones by Rolland, German ones by Marzell, as well as those in the multilingual work of Gerth van Wijk, reveal how many variants may arise from a single basic concept or motive in the course of centuries of use as part of common speech over a wide area. They also illustrate, from the wealth of local names applied to a single species, how bewildering would be reliance for communication upon these alone. For example, the marsh-marigold or kingcup (*Caltha palustris*) has over 80 local names in Britain, about 60 in France and at least 140 in Germany, Austria and Switzerland, some of which may also be applied to other species.

English plant names can be divided into several groups according to their origin. Out of the linguistic heritage of the Teutonic peoples the Anglo-Saxons brought to England their names for trees or shrubs such as ash (Old English *aesc*, Old High German *ask*, Modern German *Esche*, Dutch *es*, Swedish *ask*), beech (Old English *bece*, Old High German *buohha*, Modern German *Buche*, Dutch *beuk*, Swedish *bok*), birch or birk (Old English *berc*, *birce*, Modern German *Birke*, Dutch *berk*, Swedish *björk*), hawthorn (Old English *haguthorn*, *hawethorn*, Modern German *Hagedorn*, Dutch *haagdoorn*, Swedish *hagtorn*), hazel (Old English *haesel*, Modern German *Hasel*, Dutch *hazelaar*, Swedish *hassel*), linden or lime (Old English *lind*, Modern German *Linde*, Dutch *linde*, Swedish *lind* with a Småland variant *linn*, whence the family name of Linnaeus), oak (Old English *ac*, Old High German *eih*, Modern German *Eiche*, Dutch *eik*, Swedish *ek*), and willow (Old English *welig*, *withig*, Dutch *wilg*, Modern German *Weide*). These were important woody plants in their forested Continental homeland. The name 'rowan' (Norwegian *rogn*, *raun*, Swedish *ronn*) for *Sorbus aucuparia* is a Scandinavian legacy to northern Britain.

Names of Latin or Greek origin attest indebtedness to southern Europe for such cultivated decorative or culinary plants as the lily (Latin *lilium*), fennel (*foeniculum*), peony (*paeonia*), rose (*rosa*), rue (*ruta*), sage (*salvia*) and thyme (*thymus*), which were probably medieval monastic introductions into Britain. Their short names have undergone little change during the past two thousand years. People receiving a plant from elsewhere usually retain or try to retain its original name if this is short, distinctive and memorable. Such names may accordingly indicate the origin and the transfer from people to people of cultivated plants. Thus the sweet-scented *Jasminum officinale* has the name 'jasmine' or 'jessamine' in England, *ye-si-min* in China and *dzasasimi* in Madagascar, all independently derived from the Persian *yasamin* or *yasmin*. The English word 'ginger', investigated in detail by Alan S. C. Ross, goes back through a complicated linguistic history to a plant name coined in remote antiquity somewhere in south-eastern Asia, probably southern India, and apparently alluding to the horny appearance of the rhizome of this ancient spice plant (*Zingiber officinale*); one element is the Sanskrit *śrnga*, 'horn', another the Dravidian *vēr*, 'root', giving the Pali *singivera*. Traders brought the pungent rhizomes from India to the Ancient Greeks, who rendered its name as *ziggibĕris*, and to the Romans, who rendered it as *zingiber*, a word giving rise later to Old French *gimibre*, Old High German *gingibero* and Middle English 'ginger', and so to the name 'wild ginger' for *Asarum*, and 'ginger-lily' for *Hedychium* or *Zingiberaceae* for its botanical family. Long names transferred into another language, with their original meaning no longer obvious, tend to be corrupted or changed into forms more easily memorized. Thus the Greek *glycorrhiza*, referring to the sweetish root of the liquorice plant, became *liquorita* in Late Latin, *licorys* in Middle English and thus 'licorice' or 'liquorice' in Modern English. The bitter strong-smelling leaves of tansy (*Tanacetum vulgare*) were formerly placed in coffins or rubbed on corpses, as mentioned by Linnaeus in his Swedish Flora (*Flora Suecica*, 2nd ed. 284; 1755); this was to preserve them against worms (cf. the Dutch name 'boerenwormkruid') and the plant accordingly acquired the Greek name *athanasia*, 'immortality' (from *a*, without, not; *thanatos* death) which became changed into Late Latin *tanasia*, Old French *tanesie* and English 'tansey' or 'tansy'.

Such names exemplify the **tendency to simplify unfamiliar words**. Associated with this, as Cassidy has noted in a study of Jamaican names and folklore, is the replacement of difficult or foreign words by easier native ones; even though the result may be nonsensical or comical, it may not seem more so than the original name to the uninitiated. Thus *Carpobrotus edulis*, now naturalized on the south-western coast of England and long included in the genus *Mesembryanthemum*, has become colloquially known as 'Sally-my-handsome'; *Galinsoga* has 'gallant soldier' as a vernacular equivalent, although there is nothing gallant or soldierly about this diffuse weed; *Antholyza*, in which *Curtonus* was formerly included, has been converted into 'Aunt Eliza'. Similarly, in Jamaica, *Delonix regia*, long included in *Poinciana*, has acquired, by way of the intermediate 'Panchilana', the name 'Fancy Anna' fitting its flamboyance. *Aloe*, at one time known as a *Sempervivum* or 'sempervive', is now sometimes called 'single bible' in Jamaica. The English name 'bullace' (earlier 'bolays') for *Prunus insititia* applied by English colonists to Jamaican species of *Bumelia* (*Dipholis*) with likewise small dark fruits had by 1657 been converted into 'bully' or 'bully tree', which a century or so later became 'bullet' or 'bullet tree', now the most commonly used name. The quaint Jamaican name 'Panchalum' for a species of *Cordia* would be inexplicable but for still used vernacular intermediates between this and the name 'Spanish Elm'. These are well-documented quite recent changes; if the original forms of such names were unknown, it would be difficult or impossible to deduce or reconstruct them: thus it should not be surprising that many plant names of great antiquity, which may have been similarly transformed, have no plausible meaning; possibly some names associated with mythical figures, e.g. *Hyacinthus* and *Narcissus*, were modified loan-words with originally no such association.

The value of such apparently facetious names as aids to memory, which has led to their retention and spread, used to be illustrated by Sir Herbert Maxwell (1845–1937) with tales about the *memoria technica* of Wyber, an old Scottish head gardener at Cardross, whose ingenuity and quiet sense of humour enabled him to triumph over the perplexities of Greek and Latin botanical names. Asked how his *memoria technica* worked, Wyber replied, 'Weel enough, Sir, weel enough. There's a tree now, the *Cryptomeria japonica*. When I'm teaching the lads I tell them when they look at that tree to say "Creep to the mear (mare) and jump on to her". Ane o' the lads was gaily surprised when I was giving him a lesson in the greenhouse ane day. I was showing him the *Selaginella stolonifera*, a kind o' moss, ye ken. If ye want to mind the name, say I, think of "Silly Jean and Nellie stole a neive-fu (a handful)". "Losh! Mr. Wyber", quo' he, "them's my twa sisters".' The Greek name *anemone* possibly originated in some such manner from a Semitic name connected with Na'aman.

Many **English names for common plants**, for example bindweed, daffodil, daisy, dandelion, henbane, hollyhock, honeysuckle, horehound, mugwort, poppy, primrose, wormwood and yarrow, originated during the Middle Ages. They first found their way into print in the anonymous *Herball* printed by Rycharde Banckes in 1525, the *Grete Herball* printed by Peter Treveris in 1526 and the more original works of the clergyman, physician and naturalist, William Turner (*c.* 1508–1568), whose little books *Libellus de Re herbaria* (1538) and *Names of Herbes* (1548) had become among the rarest of botanical publications until reprinted in 1877 and 1882 and again in 1965. In the first of these, Turner recorded the English names known to him as used in Northumberland and East Anglia. They include the curious name 'Alleluya' or 'Allelua' which, also in the form 'Hallelujah', still persists locally in Britain for wood sorrel (*Oxalis acetosella*). The earliest mention of this traced by Sprague is to be found in a thirteenth-century manuscript Latin herbal (first printed in 1946) by an Italian monk, Rufinus of Genoa, who stated that the herb-gatherers supplied it under that name to the druggists of Bologna for making a cooling syrup. It also occurs in a Latin-Flemish manuscript plant glossary of about 1350 (first printed by Vandewiele and Braekman in 1968 in *Scient. Historia* 10: 114–144). It was so called, according to Turner's *New Herbal* (1568), 'because it appeareth about Easter when Alleluya is sung again'. As stated by Sprague, 'the wood sorrel flowers in April and May, and a liturgical hymn, each verse of which ends in Alleluia, is sung daily during the Office of Compline from Easter Eve to the Saturday in Whitsun week, that is during April and May. Hence the name

Alleluia.' There are other plant names as indirectly connected through religious service with their time of flowering. A particularly notable one is 'St. John's wort' (*Hypericums, perforatum*), called by Turner 'Saynte Johans grasse', a conspicuous plant with gland-dotted leaves which is in flower on St. John's Day (24 June). Apparently its magical associations make it particularly obnoxious to evil spirits and a young woman fearing rape by the Devil can effectively safeguard herself by sitting on it or keeping it under her dress. Thus frustrated long ago the Devil in revenge pricked the leaves all over with his needle; these needle marks can be seen by holding a leaf up against the light and provide evidence as convincing as other evidence for the inheritance of acquired characteristics.

The **Devil's dealings with the plant world** are commemorated in a diversity of names. The devil's bit scabious (*Succisa pratensis*) has a rootstock apparently bitten off at the end which formerly possessed medicinal properties so valuable that, according to the *Grete Herball*, 'the devyll had envy at the vertue thereof and bete the rute so for to have destroyed it'. The name 'devil's bit' was taken by early colonists to North America and there transferred to native plants with an abruptly terminated rootstock such as *Chamaelirium luteum*. The American botanist, Willard Clute, counting plant names in Europe and America associated with saints and demons, found that Europeans 'have more saints than devils in their flora while our situation is quite the reverse; for we have twice as many devils and witches as saints', a matter worthy of consideration by politicians and sociologists, for the European names connected with saints and saints' days originated before the Reformation, the American names connected with the Devil arose later in Protestant America. Plants associated with the Devil usually possess some noxious or unpleasant characteristic; thus 'Devil's guts' for dodder (*Cuscuta*), on account of its parasitic stems like a small mammal's intestines, 'Devil's milk' for *Euphorbia*, on account of its poisonous milky juice, or bindweed (*Calystegia* and *Convolvulus*) so difficult to eradicate on account of their long creeping roots, 'Devil's stinkpot' for the stinkhorn (*Phallus impudicus*), 'Devil's club' for the terribly prickly *Oplopanax horridus*. On the other hand there are also many old plant names associated with the Virgin Mary, among them 'Virgin Mary's milk-drops' for *Pulmonaria officinalis* and 'Lady's thistle' for *Silybum marianum*, both with spotted leaves.

A very interesting series of **European plant names connected with keys** show how names now meaningless were at one time quite apt. These have been the subject of a detailed and very scholarly paper in Norwegian by Rolf Nordhagen entitled 'Marilykjel, springstrå och jennurt'. In England among the many local names of the cowslip (*Primula veris*; Fig. *a*) are 'culverkeys' (also applied to the keys or fruits of ash), 'Herb Peter', 'Lady Keys', 'Peter', 'St. Peterwort'. Similar names for it have been used in Germany, such as

Schlüsselblume and *Peters-schlüssel*, in the Netherlands and Scandinavia, such as in Norway *marilykjel*, *himelsnykkel* and *nökleblom*. The German name *Himmelschlüsseli* has also been used for the spring gentian (*Gentiana verna*; Fig. *b*) which has solitary flowers, not bunched flowers as in *Primula veris*. The names for the cowslip have been explained as indicating a resemblance between its clustered flowers and a bunch of keys, which at first seems a far-fetched and implausible idea, for the individual flowers of the cowslip, as also of the gentian, in no way resemble keys now used. Nordhagen showed, however, in 1948, that these flowers with their long fluted calyx and smaller corolla-lobes somewhat resemble a type of key (Fig. *c*) belonging to a very old lock mechanism known

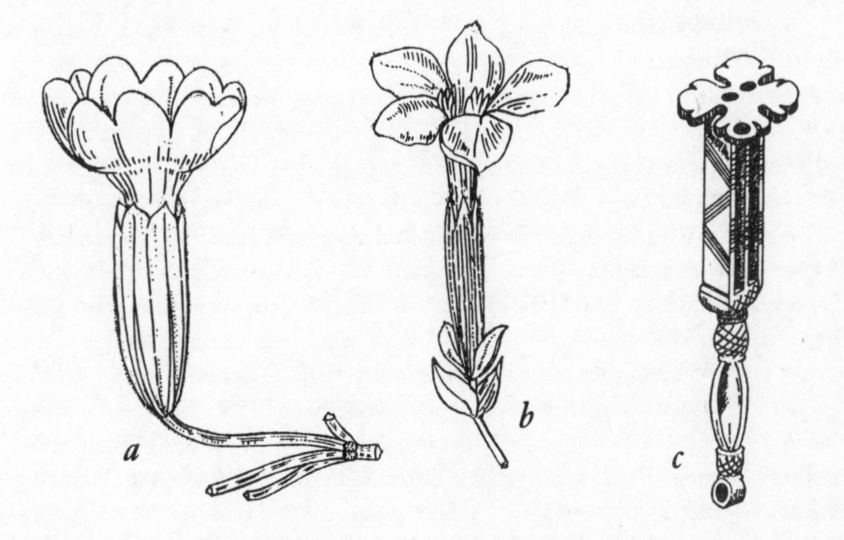

(*a*) flower of cowslip ('Herb Peter', 'Schlüsselblume'; *Primula veris*)
(*b*) flower of spring gentian ('Himmelschlüsseli'; *Gentiana verna*)
(*c*) key of 'Chinese padlock' from Sweden (after Nordhagen, 1948)
(*Drawings by Roberta E. Young*)

in ancient Rome, probably introduced into Scandinavia during the Viking period, evidently once widespread in Europe, and probably of Asiatic invention, sometimes called a 'Chinese padlock'. This key has a slender handle corresponding to the pedicel, a straight broader shank corresponding to the calyx and a spreading lobed limb with holes in the middle corresponding to the limb of the corolla.

The chain of associations by which an originally sensible or at least understandable plant name may give rise to a nonsensical one can be illustrated by some European vernacular names for moonwort (*Botrychium lunaria*; Fig. *d*), a small fern with 4 to 7 pairs of fan-shaped leaflets (pinnae), crudely horseshoe-like in outline, along the midrib (rachis) of the leaf (frond). One of these is 'Unshoe-the-horse' as recorded by Culpeper in 1653: 'Moonwort is an herb which they say will open locks and unshoo such horses as tread upon it . . . country people that I know, cal it *Unshoo the Horse*.' A variant of this is 'shoeless horse'. The superstition that moonwort possesses magical powers connected with locks and iron has been recorded in several European countries and gave rise to plant names such as German *Teufelsschlüssel* (Latinized as *clavis diaboli*) and *Walpurgiskraut*, Swedish *låsort* (lock-herb) and Norwegian *murulykjel* (i.e mare's key), which connects with the Anglo-Saxon *mare*, Norse *mara*, for a nocturnal horse-riding female demon, preserved in 'nightmare'. They evidently derive from the unusual shape of the leaf, but the association seemed far-fetched until elucidated by Nordhagen. Searching for old locks in Norway in 1944 and 1945, he found three which had to be opened by keys having 3 to 5 pairs of symmetrical projections (Fig. *e*), thus resembling in plan the *Botrychium* leaf. This was evidently the magical herb used by a burglar tried for witchcraft at Valdres, Norway, in 1937–8. It is easy to see how, from being associated with iron keys and the opening of locks as well as with the moon, the moonwort, on account of its leaflet shape, ultimately became also associated with horseshoes and with loosening them. Vestiges of long-vanished beliefs founded on sympathetic magic and the doctrine of signatures may thus linger in local vernacular names for plants.

The **deliberate coining of English plant names** began with William Turner, mentioned above as the pioneer sixteenth-century recorder of English plant names. He could find none for several plants: 'I have not hearde that it is named in englishe', 'it hathe no name in englishe that I knowe', 'howe that it is named in englishe, as yet I can not tel', he had several times to confess. He remedied this lack by introducing names of his own, mostly translated or modified from Greek, German or French. Thus of *Larix*, then unknown in Britain, he wrote 'the duch men [i.e. Germans] cal Laricem ein larchen baume. ... It may be called in englishe a Larche tree', and 'larch' in English it has remained. Of *Osyris* (i.e. *Linaria vulgaris*, toadflax) he stated that it 'groweth plenteously in Englande, but I do not remember what name it hath. If it have no name it may be called in englishe Lynary or todes flax, for the Poticares cal it Linariam and the duch [Germans] cal it Krotenflaks.' Similarly 'Barba hirci named in greeke Tragopogon ... may be called in englishe gotes bearde'. The enrichment of the English language by the deliberate coinage of new names or the adaptation of foreign names thus initiated by Turner has continued to the present time. The herbalist John Gerard (1545–1612) was an early coiner of such names, including 'traveller's joy' (*Clematis*). 'Shrimp plant' for *Beloperone* (or *Drejerella*) *guttata* alias *Justicia brandegeana* and 'Jade vine' for *Strongylodon macrobotrys*, for example, are among the welcome newcomers.

Since names are a means of communication and vernacular names belong to long oral tradition, passing from mouth to mouth without the intermediary of writing, they should be easy to hold in the mind and accordingly must have had originally some **apt or reasonable association with the plant** concerned, although this association may later be obscured or eliminated by transfer to some other plant or by corruption of the word itself as part of general linguistic change. Such associations may be very diverse. These include evident characters of the plant itself, e.g. 'arrowhead' or *Sagittaria* referring to leaf shape, 'reed-mace' or 'cat's-tail' (*Typha*), and 'mouse-tail' or *Myosurus* referring to inflorescence, 'tway-blade' (*Listera*) referring to paired leaves, 'woman's

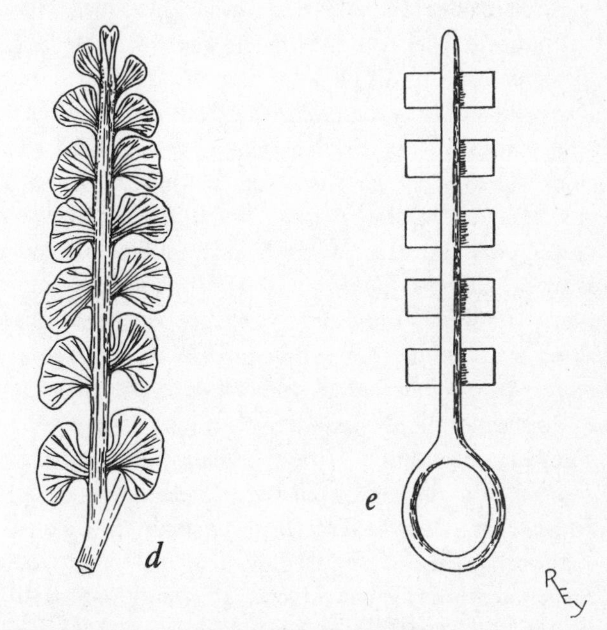

(*d*) leaf of moonwort ('Teufelsschlüssel; *Botrychium lunaria*)
(*e*) key of old Norwegian lock (after Nordhagen, 1947)
(*Drawings by Roberta E. Young*)

tongue' (*Albizia lebbek*) referring to long continually clattering pods, 'bindweed' or *Convolvulus* referring to habit of growth, 'rattle-box' or *Crotalaria* referring to rattling pods, 'buttercup' (*Ranunculus*) referring to flowers, 'ice-plant' (*Cryophytum*) from a frosted appearance, 'naked ladies' (*Colchicum*) referring to slender pink flowers bare of leaves; special properties including taste or smell, for example 'stinkhorn' (*Phallus*), 'soapwort' (*Saponaria*), 'skunk-cabbage' (*Symplocarpos*); 'bittersweet' (*Solanum dulcamara*), 'piss-a-bed' (*Taraxacum officinale*); place of growth, for example 'bog asphodel' (*Narthecium ossifragum*), 'cornflower' (*Centaurea cyanus*), 'wallflower' (*Cheiranthus cheiri*), 'mountain laurel' (*Kalmia*), 'marsh mallow' (*Althaea officinalis*), from the mucilaginous roots of which housewives formerly made their marshmallow sweetmeats; time of flowering, for example, pasque-flower (*Pulsatilla*). Others may refer to personages, real or mythical, for example 'herb-robert', 'timothy', 'joe-pye-weed' and 'geiger tree'. Real or supposed medicinal virtues, often linked with the doctrine of signatures, for example lungwort or *Pulmonaria*, form the basis of some names. The prevalence of medicinal plant names in North America, where grow asthma-weed, bellyache-root, Indian physic, boneset, ague-grass, kill-wart, fever-twig, colic root, cramp bark, mouth-root, consumption weed, toothache tree and others, testify to the pioneer period of self-medication with wild herbs, some of course derived from Indian herb-lore. Names referring to places often indicate whence a given plant was obtained, for example 'alsike' (*Trifolium hybridum*) from Alsike in Sweden, 'peach' (Latin *persica*) from Iran (Persia), into which in pre-Roman times it had been introduced from China, sisal (*Agave sisalana*) from Sisal, Mexico, and 'Madagascar periwinkle' (*Catharanthus roseus*): they can, however, be quite misleading, for example 'Jerusalem artichoke', 'Jerusalem cowslip' and 'Mexican bamboo', or become almost unintelligible, for example the West Indian 'dasheen' supposed to be contracted from '*madere de Chine*'.

Many plant names refer to animals. These are often derogatory and attached to the inferior of two kinds, for example toadflax (*Linaria*) contrasted with flax itself (*Linum*), horsechestnut (*Aesculus*) contrasted with sweet chestnut (*Castanea*), dog violet (*Viola canina*) as contrasted with sweet violet (*Viola odorata*). They may also refer to some character recalling that of an animal, for example 'hound's tongue' or *Cynoglossum*, 'adder's tongue' or *Ophioglossum*, 'cat's-tail' (*Typha*), 'rattlesnake fern' (*Botrychium virginianum*), 'tickseed' (*Coreopsis*), or to a common habitat, for example 'frog lily' (*Nuphar advena*), 'buffalo-berry' (*Shepherdia argentea*). Yet others may simply indicate the liking of an animal for that kind of plant, for example 'hogweed' (*Heracleum spondylium*), 'catmint' (*Nepeta cataria*), 'partridge-berry' (*Mitchella repens*), 'butterfly-bush' (*Buddleja*), 'bee-balm' (*Melissa* and *Monarda*).

On the contrary some, such as 'lamb-kill' (*Kalmia*) and 'wolf's bane' or *Lycoctonum* (*Aconitum*), refer to their poisonous nature. 'Dogwood', however, has nothing to do with dogs and refers to its use as a skewer wood; it is cognate with 'dagger' from French *dague*.

Some names of course are fanciful or playful, such as 'love-in-a-mist' for *Nigella damascena*, 'love-in-idleness', 'jump-up-and-kiss-me', 'johnny-jump-up' and 'kiss-me-at-the-garden-gate' for *Viola tricolor*; 'welcome-home-husband-though-never-so-drunk' for *Euphorbia cyparissias* and *Sedum acre*, 'love-lies-bleeding' for *Amaranthus caudatus*.

Others may preserve obsolete or obsolescent words. Thus the word 'loblolly' formerly referred to a kind of thick gruel or porridge and was also applied to a loutish person who could be described as 'soft and useless'; hence in the West Indies a tree with rather soft and useless wood as *Cupania* received in the seventeenth and eighteenth centuries the uncomplimentary name 'loblolly tree' or 'loblolly wood' and in the Southern United States a few notable species often growing in soft damp places, such as the loblolly bay (*Gordonia lasianthus*) and the loblolly pine (*Pinus taeda*), both soft-wooded, as well as the loblolly magnolia (*Magnolia grandiflora*), were likewise designated.

Since it is usually easier to adopt or adapt existing names than to coin new ones, settlers in a new area often apply the names of plants of their homeland to others which in some way resemble them although possibly not closely related. Thus the American evening-primrose (*Oenothera*) has nothing but a yellow colour in common with the British primrose (*Primula*). The mountain laurels (*Kalmia* and *Rhododendron*) of North America agree with the European laurels only in their foliage. The Moreton Bay chestnut (*Castanospermum*) has large brown seeds in general appearance somewhat like those of chestnut (*Castanea*) or horsechestnut, but is not otherwise like either. The white beech (*Gmelina*) of eastern Australia is botanically far removed from the European beech (*Fagus*) but has a straight grey-barked trunk recalling that of beech. Only taste and smell connect the *Eucalyptus* species called 'peppermint' in Australia with the herbaceous peppermint (*Mentha*). Only in name does *Eucalyptus regnans*, called 'mountain ash' in Australia, resemble the mountain ash or rowan (*Sorbus aucuparia*) of Britain. Usually plants, such as species of *Eucalyptus*, having no counterparts elsewhere, receive new names or have their aboriginal names retained or modified; to the one class belong such names as 'stringybark', 'woollybutt', 'ghost gum', 'blackwood', 'bloodwood', 'blackbutt', 'scribbly gum', 'messmate', 'tallowwood'; to the other, 'marri', 'karri', 'bangaley', 'coolabah', 'jarrah', 'wandoo'.

A single name suffices when only one of a kind is known but has to be qualified by adding a distinguishing word or words when rather similar but distinct other ones come to notice. Thus the names 'ash' and *fraxinus* were adequate and precise when *Fraxinus excelsior* was the only recognized species and served for many centuries as combined generic and specific names (mononomials) for this species. The discovery of other species has led to *Fraxinus* becoming a scientific generic name covering these also and to the use of the name 'European ash' (*Fraxinus excelsior*) for this and of 'white ash' (*F. americana*), 'red ash' (*F. pennsylvanica*), 'water ash' (*F. caroliniana*) and 'black ash' (*F. nigra*) for other species. The significant fact about these vernacular specific names and their scientific equivalents is that they are two-word names or binomials consisting of a one-word generic name, 'ash' or *Fraxinus*, associated with a one-word specific epithet such as 'white' or *americana*, 'black' or *nigra*;

moreover, they are simple designations, not mutually exclusive descriptive definitions. This binomial system of naming must go back to the very early stage of human speech and thought when men first found it necessary to distinguish several rather similar kinds of objects. It occurs, for example, in languages as unrelated as Greek, with species of poppy (*mēkōn*), for example, distinguished by Theophrastus as *mēkōn ē mĕlaina*, *mēkōn ē kĕratitis* and *mēkōn ē rhŏias*, and Malay, with species of *Eugenia* (*kelat*) distinguished as *kelat hitam*, *kelat jambu*, *kelat jangkang*, *kelat layu* and *kelat merah*; other languages likewise provide examples. Linnaeus coined Swedish binomial vernacular names for species of grasses in his *Flora Suecica* of 1745 before he applied binomial scientific names to them in his *Species Plantarum* of 1753. The two sets of names are more or less parallel, for example: 95. *Får-swingel* (that is sheep-fescue), *Festuca ovina*; 93. *Rōd-swingel* (red-fescue), *Festuca rubra*; 92. *Ång-swingel* (meadow fescue), *Festuca elatior*; 90. *Swin-swingel* (swine-fescue), *Festuca fluitans*; 84. *Rag-losta* (rye-brome), *Bromus secalinus*; 85. *Ren-losta* (fieldside brome), *Bromus arvensis*; 86. *Taklosta* (roof-brome), *Bromus tectorum*; 88. *Lang-losta* (tall-brome), *Bromus giganteus*; 89. *Sparr-losta* (ratchet-brome), *Bromus pinnatus*.

Thus, as has been said elsewhere, Linnaeus by his introduction in 1753 of consistent two-word names of Latin form for species thereby brought the scientific nomenclature of plants back to the simplicity and brevity of the vernacular nomenclature out of which it had grown, back to the kinds of names used from time immemorial by peasants, woodmen, herdsmen, hunters and herb-gatherers in many lands. Scientific and vernacular names should be regarded as complementary; neither can supersede the other for all purposes.

Experience has everywhere made evident the need for vernacular names adapted to everyday speech and their association with botanical names serving as reference points. Such a need develops with increasing knowledge of more and more kinds of plants and the desire to communicate and perpetuate this; it reflects a relatively high stage of advancement in the popularisation of learning. This explains the independent development of **parallel systems of plant nomenclature in Europe and Japan**. Down to the middle of the eighteenth century, as Philipp Franz von Siebold noted in 1825 when resident in Japan, empirical botany had reached as high a level there as in Europe. Siebold found that the Japanese possessed a remarkably detailed knowledge of their wild and cultivated plants, had produced much herbalistic and botanical literature about them and had given them precise and stable names in accordance with a dual system of nomenclature. They had coined vernacular Japanese names for everyday use comparable to European vernacular names and associated with these names they had learned names given a so-called Chinese reading and written in Chinese characters which were comparable in their use and status to the Latin-based botanical names of Europe. Siebold recorded, for example, the Japanese name of the *Hydrangea* now called *H. macrophylla* as *azisai* (now rendered *ajisai*) and its Chinese equivalent as *zu-hats-sen* (now *shū hassen*) with Chinese idiographs now romanized as *chü pa hsien*. The Japanese thus found it easy to correlate their own vernacular names with a third set of names, that is the European botanical names made known to them by Thunberg, Siebold, Maximowicz, Savatier and Franchet, since they could substitute these for their Chinese learned names as international standards of reference and link them with excellent woodcut illustrations in such works as the *Kwa-wi*, *Honzo dzufu* and *Somoku dzusetsu*, while retaining and indeed adding to their own Japanese vernacular names. In consequence they made more rapid progress in the study of their native plants and the adoption of Western systems of classification and nomenclature than any other Oriental nation. Romanized versions of these Japanese names are given along with botanical names throughout Ohwi's *Flora of Japan* (1965).

Because the diversity and abundance of local names and their varied applications necessarily restrict their use for general purposes, it has now become usual to select from among these one or two vernacular names as standard equivalents for the scientific names of conspicuous, interesting or economically important species. Such names are those adopted in Bailey's *Manual of Cultivated Plants*, 2nd ed. (1949), Fernald's *Gray's Manual of Botany*, 8th ed. (1950), Clapham, Tutin & Warburg's *Flora of the British Isles*, 2nd ed. (1962), Correll & Johnston's *Manual of the Vascular Plants of Texas* (1970), Doney, Jury & Perring, *English Names of Wild Flowers*, 2nd ed. (1986) and like comprehensive works. Their acceptance, which is undoubtedly desirable for convenience and precision, must lead sooner or later to the disappearance of the others; since these may be just as rich in linguistic and folklore associations and apt for the plants themselves, they should not be regarded with contempt and allowed to pass away unrecorded and unexplained.

Some Sources of Further Information

Bartlett, H. H. 1940. History of the generic concept in botany. *Bull. Torrey Bot. Club* 67:349–362.

Bartlett, H. H. 1953. English names of some East Indian plants and plant products. *Asa Gray Bull.* 2:149–176.

Britten, J. & Holland, R. 1878–86. *A Dictionary of English Plant-Names*. 2 vols. London.

Brφndegaard, V. J. 1979. Ein angelsächsischer Pflanzename: openars(e). *Sudhoffs Archiv* 63: 190–193.

Cameron, J. 1883. *Gaelic Names of Plants, Scottish and Irish*. Edinburgh.

Cassidy, F. G. 1953. Language and folklore. *Caribbean Quarterly* 3:4–12.

Cassidy, F. G. & Le Page, R. B. 1967. *Dictionary of Jamaican English*. Cambridge (Cambridge University Press).

Clute, W. N. 1923. *A Dictionary of American Plant Names*. Joliet, Illinois.

Clute, W. N. 1939. *A Second Book of Plant Names and their Meanings*. Indianapolis.

Clute, W. N. 1942. *The Common Names of Plants and their Meanings*. 2nd ed. Indianapolis.

Comber, T. 1876. The etymology of plant names. *Trans. Hist. Soc. Lancash. & Cheshire* 28: 15–71 (1876), 29: 43–83 (1877), 30: 1–38 (1878).

Dawkins, R. M. 1936. The semantics of Greek names for plants. *Journal of Hellenic Studies* 56: 1–11.

Earle, J. 1880. *English Plant Names from the tenth to the fifteenth Centuries*. Oxford.

Fernald, M. L. 1942. Some historical aspects of plant taxonomy. *Rhodora* 44: 21–43.

Fisher, R. 1932–4. *The English Names of our commonest wild Flowers*. Arbroath.

Friend, H. 1884. *Flowers and Flower Lore*. 2nd ed. London.

Egli, M. 1930. *Benennungsmotive bei Pflanzen an schweizerdeutschen Pflanzennamen untersucht*. Bulach.

Gerth van Wijk, H. L. 1911. *Dictionary of Plant-Names*. 2 vols. The Hague. Reprinted 1971. Reprinted in 1987 with introductions by Jane Grigson & W. T. Stearn.

Grigson, G. 1958. *The Englishman's Flora*. London (Phoenix House). Reprinted 1987 with introductions by Jane Grigson & W. T. Stearn.

Grigson, G. 1973. *A Dictionary of English Plant Names*. London (Allen Lane).

Guyot, L. & Gibassier, P. 1960. *Les Noms des Plantes (Que sais-je 856)*. Paris (Presses Universitaires de France).

Guyot, L. & Gibassier, P. 1960. *Les Noms des Arbres (Que sais-je 861)*. Paris (Presses Universitaires de France).

Hunt, T. 1989. *Plant Names of Medieval England.* Cambridge (D. S. Brewer).

Kartesz, J. T. & Thieret, J. W. 1991. Common names for vascular plants: guidelines for use and application. *Sida* 14: 421–434.

Lange, J. 1959-61. *Ordbog over Danmarks Plantenavne.* 3 vols. Copenhagen (Munksgaard).

Laurell, F. 1904. *Svenska Växtnamn och binä Nomenklatur.* Uppsala.

Leighton, A. (Mrs. I. L. L. Smith). 1970. *Early English Gardens in New England.* London (Cassell).

McClintock, D. 1966. *Companion to Flowers.* London (G. Bell).

Marzell, H. 1937 et seq. *Wörterbuch der Deutschen Pflanzennamen.* Leipzig (Hirzel).

Maxwell, H. 1923. *Flowers, a Garden Notebook.* Glasgow.

Nelson, E. C. 1990. Shamrock 1988. *Ulster Folklife* 36: 32–42.

Nelson, E. C. 1991. *Shamrock. Botany and History of an Irish Myth.* Aberystwyth & Kilkenny (Boethius Press).

Nordhagen, R. 1948. Studien over gamle plantenavn: marilykjel, springstrå og jennurt. *Bergens Mus. Årb.* 1946–47, Hist. Antikv. no. 3.

Prior, R. C. A. 1879. *On the Popular Names of British Plants.* 3rd ed. London.

Rolland, E. 1896–1914. *Flore populaire.* 11 vols. Paris.

Ross, A. S. C. 1952. *Ginger; a loan-word Study.* London (Philological Society).

Ross, A. S. C. 1958. *Etymology with especial Reference to English.* London (André Deutsch; 1965, Methuen).

Rydén, M. 1978. *Shakespearean Plant Names, Identifications and Interpretations (Stockholm Studies in English 43)*. Stockholm (Almqvist & Wiksell).

Rydén, M. 1978. The English plant names in Gerard's Herball (1597). *Studies presented to R. Rynell*, 142–150.

Rydén, M. 1980. Vaxtnamnsforskning–något om dess historia, problem och metoder, *Svensk Botanisk Tidskrift* 74: 425–436.

Rydén, M. 1984. The contextual significance of Shakespeare's plant names. *Studia Neophilologica* 56: 155–162.

Rydén, M. 1984. *The English Plant Names in the Grete Herball* (1526). *(Stockholm Studies in English 61)*. Stockholm (Almqvist & Wiksell).

Rydén, M. 1987. English names for *Convallaria majalis* L. *Stockholm Studies in Modern Philology* N. S. 8: 61–70.

Siebold, Ph. F. von. 1828. Einige Worte über den Zustand der Botanik auf Japan. *Nova Acta Acad. Nat. Curios. Caes-Leopold-Carol.* 14: 671–693.

Smith, P. G. & Welch, J. E. 1964. Nomenclature of vegetables and condiment herbs grown in the United States. *Proc. Amer. Soc. Hort. Sci.* 84: 535–548.

Smith, T. A. 1966. Common Names of South African Plants. *Republic S. Afr. Bot. Survey Mem.* 35. Pretoria.

Sprague, T. A. 1957. English plant-names. *Proc. Cotteswold Nat. Field Club* 32, ii: 22–27.

Stromberg, R. 1940. *Griechische Pflanzennamen (Göteborgs Högsk. Årsskr.* 46). Göteborg.

Turner, W. 1965. *Libellus de Re herbaria, 1538: The Names of Herbes, 1548 Facsimiles with introductory matter by J. Britten, B. D. Jackson and W. T. Stearn.* London (Ray Society).

Dictionary of Vernacular Names

absinthium *Artemisia absinthium*
acacia *Acacia*
acacia, false *Robinia pseudacacia*
acacia, rose *Robinia hispida*
aconite *Aconitum*
aconite, winter *Eranthis*
Adam's needle *Yucca smalliana*
adder's mouth *Pogonia*
adder's tongue *Erythronium*
adder's tongue fern *Ophioglossum*
African daisy *Arctotis; Lonas*
African corn-lily *Ixia*
African cypress *Widdringtonia*
African lily *Agapanthus*
African marigold *Tagetes*
African peach *Nauclea latifolia*
African ragwort *Othonna*
African tuliptree *Spathodea campanulata*
African valerian *Fedia*
African violet *Saintpaulia*
agrimony *Agrimonia*
ague-weed *Eupatorium*
air-plant *Aerides; Bryophyllum pinnatum*
akee tree *Blighia sapida*
alder *Alnus*
alderbuckthorn *Frangula alnus*
alder, white *Clethra*
alexanders *Smyrnium olusatrum*
Alexandrian laurel *Danaë racemosa*
alfalfa *Medicago sativa*
Alice, sweet *Pimpinella anisum*
alkanet *Alkanna tinctoria*
Allegheny vine *Adlumia fungosa*
alligator pear *Persea*
allspice *Calycanthus floridus*; *Pimenta dioica*
allspice tree *Pimenta diocia*
almond *Prunus dulcis (Amygdalus communis)*
almond, dwarf *Prunus tenella*
alsike clover *Trifolium hybridum*
alum-root *Heuchera*
alyssum, sweet *Lobularia maritima*
amaranth, Chinese *Amaranthus tricolor*
amaranth, globe *Gomphrena globosa*
amaryllis *Amaryllis*; *Hippeastrum*
amaryllis, blue *Worsleya procera*
Amazon lily *Eucharis grandiflora*
American cowslip *Dodecatheon*
American cranberry *Vaccinium macrocarpon*
American cress *Barbarea*
American feverfew *Chrysanthemum parthenium*
American pennyroyal *Hedeoma*
American wayfaring-tree *Viburnum alnifolium*
anchor-plant *Colletia cruciata*
Andromeda *Pieris*
anemone *Anemone*
anemone, Japanese *Anemone hybrida*
anemone, poppy *Anemone coronaria*
anemone, snowdrop *Anemone sylvestris*
anemone, wood *Anemone nemorosa*
angel's trumpet *Brugsmania* (or *Datura*) *suaveolens*
angelica *Angelica archangelica*
angelica-tree *Aralia spinosa*
angelica-tree, Chinese *Aralia chinenis*
angelica-tree, Japanese *Aralia japonica*
angelica, wild *Angelica sylvestris*
anise *Pimpinella anisum*
anise, sweet *Foeniculum vulgare*
aniseed tree *Illicium*
annatto *Bixa orellana*
Antarctic forget-me-not *Myosotidium hortensia*
apple *Malus pumila*
apple berry *Billardiera*

apple-of-Peru *Nicandra physalodes*
apple, Otaheite *Spondias dulcis*
apricot *Prunus armeniaca*
apricot, Japanese *Prunus mume*
aralia, false *Dizygotheca*
arbor-vitae *Thuja*
arbutus, trailing *Epigaea repens*
arrow-arum *Peltandra*
arrowhead *Sagittaria*
arrowroot *Maranta arundinacea*
arrow-wood *Viburnum dentatum*
artichoke, Chinese *Stachys affinis*
artichoke, globe *Cynara scolymus*
artichoke, Japanese *Stachys affinis*
artichoke, Jerusalem *Helianthus tuberosus*
artillery plant *Pilea microphylla*
arum-lily *Zantedeschia*
ash *Fraxinus*
asoka tree *Saraca indica*
asparagus *Asparagus officinalis*
asparagus fern *Asparagus plumosus*
asparagus pea *Tetragonolobus purpureus*
aspen *Populus tremula*
asphodel *Asphodelus*
asphodel, yellow *Asphodeline*
aster, annual } *Callistephus*
aster, China } *chinenis*
aster, Stokes' *Stokesia laevis*
aster, tree *Olearia*
astilbe *Astilbe*
Aunt Eliza *Curtonus* (formerly *Antholyza*)
Australian currant *Leucopogon*
Australian Desert kumquat *Eremocitrus*
Australian honeysuckle *Banksia*
Australian laurel *Pittosporum*
Australian rosemary *Westringia*
Australian sarsaparilla *Hardenbergia violacea*
Australian sword lily *Anigozanthos*
Austrian briar *Rosa foetida*
Austrian copper briar *Rosa foetida bicolor*
autumn crocus *Colchicum; Crocus*
avens *Geum urbanum*
avens, mountain *Dryas octopetala*
avocado *Persea americana*
azalea *Rhododendron* (deciduous species)

baboon-root *Babiana*
baby blue-eyes *Nemophila menziesii*
baby's breath *Gypsophila paniculata*
baby's tears *Soleirolia soleirolii*
bachelor's button Name applied to double-flowered forms of *Ranunculus* and *Achillea*, also to *Centaurea* and *Craspedia*
bael fruit *Aegle marmelos*
bald cypress *Taxodium distichum*
baldmoney *Meum athamanticum*
balloon flower *Platycodon*
balloon vine *Cardiospermum halicacabum*
balm *Melissa officinalis*; *Ocimum basilicum*
balm, bastard *Melittis melissophyllum*
balm, bee *Melissa officinalis*
balm, lemon *Melissa officinalis*
balm-of-Gilead *Cedronella triphylla*; *Populus gileadensis*
balsam *Impatiens*
balsam-apple *Momordica balsamina*
balsam-pear *Momordica charantia*
balsam poplar *Populus balsamifera*
bamboo *Bambusa*, *Sasa*, *Thamnocalamus*, etc.
bamboo, heavenly *Nandina domestica*
banana *Musa*
baneberry *Actaea*
banyan *Ficus benghalensis*
baobab tree *Adansonia digitata*
Barbados-cherry *Malpighia*
Barbados-gooseberry *Pereskia aculeata*
Barbados-pride *Caesalpinia pulcherrima*
barberry *Berberis*
bark, paper *Melaleuca*
barley *Hordeum*
barna, sacred *Crateva nurvala*
barren strawberry *Waldsteinia*
barrenwort *Epimedium*
basket willow *Salix viminalis*
basil *Ocimum basilicum*
bass-wood *Tilia*
bastard balm *Melittis melissophyllum*
bastard jasmine *Cestrum*
bastard pennyroyal *Trichostema*
bats-in-the-belfry *Campanula trachelium*
bay *Laurus*
bay, sweet *Magnolia virginiana*
bayberry *Myrica pensylvanica*; *Pimenta racemosa*
bay-tree *Laurelia*; *Laurus*
bay willow *Salix pentandra*

beach grass *Ammophila*
beach heather *Hudsonia*
beach pea *Lathyrus japonicus*
beach plum *Prunus maritima*
bead-plant *Nertera granadensis*
bead-tree *Melia azedarach*
bean, asparagus *Vigna sinensis sesquipedalis*
bean, broad *Vicia faba*
bean, Jack *Canavalia ensiformis*
bean, kidney *Phaseolus vulgaris*
bean, Lima *Phaseolus limensis*
bean, runner *Phaseolus coccineus*
bean, soya *Glycine max*
bean, yam *Pachyrhizus*
bearberry *Arctostaphylos*
bear-grass *Camassia; Nolina*
bear-tongue *Clintonia*
beard grass *Andropogon*
beard-tongue *Penstemon*
bear's breech *Acanthus*
beauty-berry *Callicarpa*
beauty bush *Kolkwitzia amabilis*
bedstraw *Galium*
bee-balm *Monarda*
beech *Fagus*
beech fern *Thelypteris phegopteris*
beech, southern *Nothofagus*
beefsteak plant *Perilla*
beefwood *Casuarina*
beet *Beta vulgaris*
beggar's ticks *Bidens*
bell, Canterbury *Campanula medium*
belladonna *Atropa belladonna*
bellbine *Calystegia*
bellflower *Campanula*; *Wahlenbergia*
bellflower, Chilean *Nolana*
bellflower, Chinese *Platycodon*
bellflower, giant *Ostrowskia magnifica*
bellflower, gland *Adenophora*
bellflower, pendulous *Symphyandra*
bell-heather *Erica cinerea*
bellwort *Uvularia*
belvedere *Kochia*
Benjamin bush *Lindera benzoin*
bent grass *Agrostis*
benzoin *Lindera benzoin*
bergamot *Monarda didyma*
Bermuda buttercup *Oxalis pes-caprae*
Bermuda grass *Cynodon dactylon*
betelnut *Areca catechu*
Bethlehem sage *Pulmonaria*
betonica *Stachys officinalis*
betony *Stachys officinalis*
betony, wood *Pedicularis*
bhendi tree *Thespesia populnea*
bidgee-widgee *Acaena anserinifolia*
bididy-bid *Acaena*
big tree *Sequoiadendron giganteum*
bilberry *Vaccinium myrtillus* (Britain)
Vaccinium uliginosum (U.S.A.)
bilimbi *Averrhoa bilimbi*
bindweed *Convolvulus, Calystegia*
birch *Betula*
birch, canoe *Betula papyrifera*
birch, cherry *Betula lenta*
birch, dwarf *Betula nana*
birch, grey *Betula populifolia*
birch, paper *Betula papyrifera*
birch, silver *Betula pendula*
birch, yellow *Betula lutea*
bird cherry *Prunus padus*
bird-of-paradise flower *Strelitzia*
bird, pepper *Capsicum annuum*
bird's foot *Ornithopus perpusillus*
bird's foot trefoil *Lotus corniculata*
birthroot *Trillium*
birthwort *Aristolochia*
bishopsweed *Aegopodium podagraria*
bistort *Persicaria bistorta*
bitterbush *Picramnia pentandra*
bitternut *Carya cordiformis*
bittersweet *Celastrus scandens*; *Solanum dulcamara*
bitterwort *Lewisia*
blackberry *Rubus*
blackberry lily *Belamcanda chinensis*
blackboy *Xanthorrhoea*
black bryony *Tamus communis*
black cohosh *Cimicifuga*
black-eyed Susan *Rudbeckia hirta*; *Rudbeckia serotina*; *Thunbergia alata*
black horehound *Ballota*
black lily *Fritillaria camschatcensis*
black medick *Medicago lupulina*
black mulberry *Morus nigra*
black oak *Quercus velutina*
black poplar *Populus nigra*
blackthorn *Prunus spinosa*

black wattle *Acacia decurrens mollis*
black willow *Salix nigra*
bladder campion *Silene vulgaris*
bladder-cherry *Physalis*
bladder fern *Cystopteris*
bladdernut *Staphylea*
bladder senna *Colutea*
bladderwort *Utricularia*
blanket-flower *Gaillardia*
blazing star *Chamaelirium luteum*; *Liatris*; *Tritonia*
bleeding-heart *Dicentra*
blessed thistle *Cnicus benedictus*; *Silybum marianum*
blistercress *Erysimum*
blood-leaf *Iresine*
blood-lily *Haemanthus; Scadoxus*
bloodroot *Sanguinaria canadensis*
bloodwood tree *Haematoxylum campechianum*
bluebead-lily *Clintonia borealis*
bluebell *Campanula* (Scotland); *Mertensia* (Virginia); *Hyacinthoides (England)*; *Wahlenbergia* (Australia)
bluebell, California *Phacelia*
blueberry *Vaccinium*
bluebonnet, Texas *Lupinus subcarnosus*
bluebush *Kochia*
blue cohosh *Caulophyllum thalictroides*
bluecurls *Trichostema*
blue daisy *Felicia amelloides*
blue-eyed grass *Sisyrinchium*
blue-eyed Mary *Collinsia*
blue gum *Eucalyptus*
blue lace-flower *Trachymene caerulea*
blue oxalis *Parochetus*
blue palmetto *Rhapidophyllum*
blue poppy *Meconopsis*
blue sowthistle *Cicerbita*
blue spiraea *Caryopteris*
blue wattle *Acacia cyanophylla*
bluet *Houstonia*
bogbean *Menyanthes trifoliata*
bog heather *Erica tetralix*
bog myrtle *Myrica*
bog-orchid *Arethusa bulbosa*; *Calypso bulbosa*
bog rosemary *Andromeda glaucophylla*
boneset *Eupatorium perfoliatum*
borage *Borago officinalis*
Boston ivy *Parthenocissus tricuspidata*
Botany Bay gum *Xanthorrhoea*
bottlebrush tree *Callistemon*; *Melaleuca*
bottle-gourd *Lagenaria*
bottle-tree *Brachychiton*
bouncing Bet *Saponaria officinalis*
bowstring hemp *Sansevieria*
box *Buxus; Eucalyptus* (Australia)
boxberry *Gaultheria procumbens*
box-elder *Acer negundo*
box, sweet *Sarcococca*
box-thorn *Lycium*
bracken; brake *Pteridium aquilinum*; *Pteris*
bramble *Rubus*
brasiletto *Caesalpina vesicaria*
Brazil-nut *Bertholletia*
breadfruit *Artocarpus altilis*
bread-nut *Brosimum alicastrum*
bridal wreath *Francoa; Spiraea*
bristlecone pine *Pinus aristata*
broad bean *Vicia faba*
brome grass *Bromus*
Brompton stock *Matthiola incana*
brooklime *Veronica beccabunga*
brookweed *Samolus valerandi*
broom *Cytisus*; *Genista*; *Spartium*
broom, butcher's *Ruscus aculeatus*
broom, hedgehog *Erinacea anthyllis*
broom, Spanish *Spartium junceum*
broom, sweet *Scoparia dulcis*
bryony *Bryonia*
bryony, black *Tamus communis*
bryony, white *Bryonia dioica*
buckbean *Menyanthes trifoliata*
bucket orchid *Coryanthes*
buckeye *Aesculus*
buckthorn *Rhamnus*
buckthorn, alder *Frangula alnus*
buckthorn, sea *Hippophae rhamnoides*
buckwheat *Fagopyrum esculentum*
buffalo-berry *Shepherdia argentea*
bugbane *Cimicifuga*
bugle *Ajuga*
bugle lily *Watsonia*
bugleweed *Ajuga*
bugloss *Anchusa*
bugloss, viper's *Echium*

bullace *Prunus spinosa insititia*
bulbous chervil *Chaerophyllum*
bullock's heart *Annona reticulata*
bulrush *Scirpus; Typha* (Britain)
bunchberry *Cornus*
bunchflower *Melanthium*
bur, New Zealand *Acaena*
bur-cucumber *Sicyos angulatus*
burdock *Arctium*
bur-marigold *Bidens*
burnet *Poterium*; *Sanguisorba minor*
burnet rose *Rosa spinosissima*
burning bush *Dictamnus albus*; *Euonymus atropurpureus* (U.S.A)
burstwort *Herniaria*
bush-clover *Lespedeza*
bush-honeysuckle *Diervilla*
bush-pea *Pultenaea*
butcher's broom *Ruscus aculeatus*
butter-and-eggs *Linaria vulgaris*
butter-bur *Petasites*
buttercup *Ranunculus*
buttercup, Bermuda *Oxalis pes-caprae*
butterfly bush *Buddleja davidii*
butterfly flower *Bauhinia monandra* (tropics); *Schizanthus*
butterfly iris *Moraea*
butterfly pea *Clitoria ternatea*; *Centrosema*
butterfly-weed *Asclepias tuberosa*
butternut *Juglans cinerea*
butterwort *Pinguicula*
button-bush *Cephalanthus*
buttonwood *Platanus occidentalis*
Buxbaum's speedwell *Veronica persica*

cabbage *Brassica oleracea capitata*
cabbage rose *Rosa centifolia*
cabbage tree *Andira; Cordyline*
cacao *Theobroma cacao*
calaba tree *Calophyllum*
calabash gourd *Lagenaria ciceraria*
calabash tree *Crescentia cujete*
calamint *Calamintha*
calico-bush *Kalmia*
California bluebell *Phacelia*
California golden bells *Emmenanthe*
California laurel *Umbellularia californica*
California poppy *Eschscholzia*
calla-lily *Zantedeschia*
calvary-clover *Medicago echinus*
camas *Camassia*
camomile (see chamomile)
campeachy-wood *Haematoxylum campechianum*
campion *Silene*
campion, bladder *Silene vulgaris*
campion, evening *Lychnis*
campion, moss *Silene acaulis*
campion, red *Silene dioica*
campion, rose *Lychnis coronaria*
campion, sea *Silene maritima*
campion, white *Silene alba*
Canada Ted *Gaultheria*
Canadian wild rice *Zizania aquatica*
canary creeper *Tropaeolum peregrinum*
canary-grass *Phalaris canariensis*
candle-plant *Kleinia articulata*
candle-tree *Parmentiera cereifera*
candlewick *Verbascum thapsus*
candlewood *Fouquiera*
candy carrot *Athamanta*
candy mustard *Aethionema*
candytuft *Iberis*
cane, dumb *Dieffenbachia seguine*
cane palm *Calamus*
cane, southern *Arundinaria*
cannon-ball tree *Couroupita guianensis*
canoe birch *Betula papyrifera*
cantaloupe *Cucumis melo*
Canterbury-bell *Campanula medium*
Cape chestnut *Calodendrum capense*
Cape cowslip *Lachenalia*
Cape figwort *Phygelius capensis*
Cape gooseberry *Physalis*
Cape honeysuckle *Tecomaria capensis*
Cape jasmine *Gardenia jasminoides*
Cape marigold *Dimorphotheca*
Cape pondweed *Aponogeton*
Cape primrose *Streptocarpus*
Cape stock *Heliophila*
caper *Capparis spinosa*
caper spurge *Euphorbia lathyris*
carambola *Averrhoa carambola*
caraway *Carum carvi*
cardamon *Elettaria cardamomum*
cardinal-flower *Lobelia cardinalis*
cardoon *Cynara cardunculus*
carline thistle *Carlina*

carnanba palm *Copernica cerifera*
carnation *Dianthus caryophyllus*
carob *Ceratonia siliqua*
Carolina allspice *Calycanthus floridus*
carrion-flower *Smilax; Stapelia*
carrot *Daucus carota*
cashew nut *Anacardium occidentale*
cassava, bitter *Manihot esculenta*
cassava, sweet *Manihot dulcis*
castor bean; castor-oil plant *Ricinus communis*
catbrier *Smilax*
catchfly *Lychnis; Silene*
catchfly, Nottingham *Silene nutans*
catjang *Vigna sinensis cylindrica*
catmint; catnip *Nepeta cataria*
cat's claw *Doxantha unguis-cati*
cat's ear *Hypochoeris radicata*
cat-tail *Phleum; Typha*
cauliflower *Brassica oleracea botrytis*
cedar *Cedrus*
cedar, Atlas *Cedrus atlantica*
cedar, incense *Libocedrus*
cedar, Japanese *Cryptomeria japonica*
cedar of Lebanon *Cedrus libani*
cedar, northern white *Thuja occidentalis*
cedar, red *Juniperus virginiana*
cedar, stinking *Torreya taxifolia*
cedar, wattle *Acacia elata*
cedar, white *Chamaecyparis thyoides*
celandine, greater *Chelidonium majus*
celandine, lesser *Ranunculus ficaria*
celandine poppy *Stylophorum diphyllum*
celery *Apium graveolens dulce*
centaury *Centaurium*
centaury, American *Sabatia*
century plant *Agave*
cereus, night-blooming *Selenicereus grandiflorus*
ceriman *Monstera deliciosa*
chain-fern *Woodwardia*
chamomile *Anthemis nobilis*
chamomile, corn *Anthemis arvensis*
chamomile, false *Boltonia*
chamomile, yellow *Anthemis tinctoria*
chamomile, wild *Matricaria recutita*
chard, Swiss *Beta vulgaris cicla*
chaste tree *Vitex agnus-castus*
chayote *Sechium edule*
checker-berry *Gaultheria procumbens*
cherimoya *Annona cherimola*
Cherokee rose *Rosa laevigata*
cherry *Prunus arium* (sweet), *cerasus* (sour)
cherry, bird *Prunus padus*
cherry, bladder *Physalis*
cherry, ground *Physalis pruinosa*
cherry, Jerusalem *Solanum pseudo-capsicum*
cherry-laurel *Prunus laurocerasus*
cherry-pie *Heliotropium*
cherry, wild *Prunus avium*
chervil *Anthriscus*
chervil, bulbous *Chaerophyllum bulbosum*
chervil, salad *Anthriscus cerefolium*
chervil, wild *Cryptotaenia*
chestnut *Castanea*
chestnut, horse *Aesculus*
chestnut, Spanish *Castanea sativa*
chestnut-oak *Lithocarpus*
chick pea *Cicer arietinum*
chickweed *Stellaria*
chickweed, mouse-ear *Cerastium*
chicory *Cichorium intybus*
Chilean bellflower *Nolana*
Chilean boldo tree *Peumus boldus*
chili pepper *Capsicum annuum*
China aster *Callistephus chinensis*
chinaberry *Melia azedarach*
China-fir *Cunninghamia lanceolata*
China-grass *Boehmeria nivea*
Chinese gooseberry *Actinidia deliciosa*
Chinese hat *Holmskioldia sanguinea*
Chinese-lantern lily *Sandersonia aurantiaca*
Chinese moonseed *Sinomenium acutum*
Chinese sacred lily *Narcissus tazetta orientalis*
chinkerinchee *Ornithogalum thyrsoides*
chives *Allium schoenoprasum*
chives, Chinese *Allium tuberosum*
choctaw root *Apocynum cannabinum*
chokeberry *Aronia*
cholla *Opuntia*
Christmas-berry *Heteromeles; Schinus terebinthifolius*
Christmas-cactus *Schlumbergera buckleyi*
Christmas fern *Polystichum acrostichoides*

Christmas-rose *Helleborus niger*
Christ-thorn *Paliurus spina-christi*
chufa *Cyperus esculentus*
cicely, sweet *Myrrhis odorata*
cigar-plant *Cuphea ignea*
cinnamon fern *Osmunda cinnamomea*
cinnamon vine *Apios americana*
cinnamon, wild *Canella*
cinquefoil *Potentilla*
citron *Citrus medica*; *Citrullus lanatus citroides*
citronella *Collinsonia*
clary *Salvia sclarea*
cleavers *Galium aparine*
cliff-brake fern *Pellaea*
climbing cactus *Acanthocereus*
climbing fern *Lygodium*
climbing palas *Butea superba*
cloak fern *Notholaena*
clove tree *Syzygium aromaticum*
clover, alsike *Trifolium hybridum*
clover, bush *Lespedeza*
clover, hop *Trifolium agrarium*
clover, Mexican *Richardia scabra*
clover, sweet *Melilotus*
club of Hercules *Aralia spinosa*; *Xanthoxylum clava-herculis*
club-moss *Lycopodium*; *Selaginella*
cob-nut *Corylus avellana*
cockscomb *Celosia argentea cristata*
cock's-foot grass *Dactylus glomerata*
cockspur *Echinochloa crus-galli*
cockspur thorn *Crataegus crus-galli*
cocoa *Theobroma cacao*
coco-de-mer *Lodoicea maldavica*
coconut *Cocos nucifera*
coconut, double *Lodoicea maldavica*
coco-plum *Chrysobalanus icaco*
codlins-and-cream *Epilobium hirsutum*
coffee *Coffea*
coffee, wild *Psychotria*
cohosh *Actaea*
cohosh, black *Cimicifuga*
cohosh, blue *Caulophyllum thalictroides*
coltsfoot *Tussilago*
coltsfoot, alpine *Homogyne alpina*
columbine *Aquilegia*
Colville's glory *Colvillea racemosa*
comb fern *Schizaea*
comfrey *Symphytum*
comfrey, Russian *Symphytum uplandicum*
compass-plant *Silphium laciniatum*
cone-flower *Echinacea*; *Lepachys*; *Rudbeckia*
cone-plant *Conophytum*
Cootamunda wattle *Acacia baileyana*
copper leaf *Acalypha*
copper-pod *Peltophorum pterocarpum*
coral tree *Erythrina*
coral vine *Antigonum leptopos*
coriander *Coriandrum sativum*
cork oak *Quercus suber*
cork-tree *Phellodendron*
corn *Triticum* (Britain); *Zea mays* (U.S.A.)
corncockle *Agrostemma githago*
cornel *Cornus*
cornel, dwarf *Cornus suecica*, *C. canadensis*
cornelian cherry *Cornus mas*
cornflower *Centaurea cyanus*
corn-marigold *Chrysanthemum segetum*
corn mint *Mentha arvensis*
corn poppy *Papaver rhoeas*
corn-salad *Valerianella olitoria*
Cornish heath *Erica vagans*
costmary *Leucanthemum balsamita*
cottage pink *Dianthus caryophyllus*
cotton *Gossypium*
cotton, lavender *Santolina chamaecyparissus*
cottongrass *Eriophorum*
cotton-gum *Nyssa aquatica*
cotton tree *Bombax*
cotton-weed *Diotis candidissima*
cottonwood *Populus*
cowbane *Cicuta virosa*
cow-itch *Urera*
cow-parsley *Anthriscus sylvestris*
cow-parsnip *Heracleum*
cowpea *Vigna sinensis*
cowslip *Caltha palustris* (U.S.A.); *Mertensia* (Virginia); *Primula veris* (Britain)
cowslip, American *Dodecatheon*
cow tree *Brosimum*
crab-apple *Malus*
crab-apple, Siberian *Malus adstringens*; *Malus baccata*; *Malus prunifolia*
crack willow *Salix fragilis*

cranberry *Vaccinium oxycoccus*; *Vaccinum macrocarpon*
cranberry-bush *Viburnum trilobum*
cranesbill *Geranium*
crape-myrtle *Lagerstroemia*
cream cups *Platystemon californicus*
creeper, Rangoon *Quisqualis indica*
creeping-Charlie *Lysimachia nummularia*; *Pilea nummularifolia*
creeping-Jenny *Lysimachia nummularia*
creeping willow *Salix repens*
creosote bush *Larrea*
cress, American *Barbarea*
cress, bitter *Cardamine*
cress, garden *Lepidium sativum*
cress, Indian *Tropaeolum majus*
cress, rock *Arabis*
cress, spring *Cardamine*
cress, stone *Aethionema*
cress, upland *Barbarea verna*
cress, water *Rorippa*
cress, winter *Barbarea*
crested poppy *Argemone*
Cretan dittany *Amaracus dictamnus*
cricket-bat willow *Salix alba caerulea*
crocus *Crocus*
crocus, saffron *Crocus sativus*
crocus, Autumn *Colchicum*
cross-leaved heath *Erica tetralix*
cross-vine *Bignonia capreolata*
crosswort *Cruciata laevipes*
croton *Codiaeum variegatum*
crowberry *Empetrum nigrum*
crowfoot *Ranunculus*
crow garlic *Allium vineale*
crown-imperial *Fritillaria imperialis*
crown-vetch *Coronilla varia*
crystal wort *Riccia*
cuckoo-flower *Cardamine pratensis*
cuckoo-pint *Arum maculatum*
cucumber *Cucumis sativus*
cucumber, prickly *Echinocystis lobata*
cucumber, squirting *Ecballium elaterium*
cucuzzi *Lagenaria siceraria*
cudweed *Gnaphalium*
cup-flower *Nierembergia repens*
Cupid's dart *Catananche caerulea*
cup-and-saucer flower *Cobaea scandens*
curly-grass fern *Schizaea*
currant *Ribes*
currant, black *Ribes nigrum*
currant, buffalo *Ribes odoratum*
currant, flowering *Ribes sanguineum*
currant, golden *Ribes aureum*
currant, red *Ribes rubrum*
curry-leaf *Murraya paniculata*
custard-apple *Annona*
cypress *Cupressus*
cypress, bald *Taxodium distichum*
cypress, false *Chamaecyparis*
cypress, Italian *Cupressus sempervirens*
Cypress, Monterey *Cupressus macrocarpa*
cypress, pond *Taxodium ascendens*
cypress, summer *Bassia* (*Kochia*) *scoparia*
cypress, swamp *Taxodium*
cypress-pine *Callitris*

daffodil *Narcissus*
dahlia *Dahlia*
dahlia, climbing *Hidalgoa wercklei*
dahoon *Ilex cassine*
daikon *Raphanus sativus*
daisy *Bellis perennis*
daisy, African *Arctotis*; *Lonas*
daisy, Barberton *Gerbera jamesonii*
daisy, blue *Felicia amelloides*
daisy, bush *Olearia*
daisy, Michaelmas *Aster*
daisy, New Holland *Vittadenia triloba*
daisy, New Zealand *Celmisia*
daisy, ox-eye *Leucanthemum superbum*
daisy, Shasta *Chrysanthemum maximum*
daisy, Swan River *Brachycome iberidifolia*
daisy, Transvaal *Gerbera jamesonii*
daisy, white *Leucanthemum vulgare*
dame's violet *Hesperis matronalis*
dammar pine *Agathis*
dandelion *Taraxacum*
dandelion, dwarf *Krigia*
Darling River pea *Swainsona greyana*
dasheen *Colocasia esculenta*
date palm *Phoenix dactylifera*
date plum *Diospyros kaki*, *D. lotus*
date-yucca *Samuela*
dawn redwood *Metasequoia glyptostroboides*
day-flower *Commelina*
day-lily *Hemerocallis*

dead-nettle *Lamium*
dead-nettle, spotted *Lamium maculatum*
dead-nettle, white *Lamium album*
deadly nightshade *Atropa belladonna*
deerberry *Vaccinium stamineum*
deer-foot *Achlys*
deer-grass *Rhexia*
desert-candle *Eremurus*
desert rod *Enemostachys*
devil's bit *Chamaelirium luteum* (U.S.A.); *Succisa pratensis*
devil's paintbrush *Hieracium*
dewberry *Rubus caesius*
diamond-flower *Ionopsidium acaule*
dildo *Cereus*
dill *Anethum graveolens*
dish-cloth gourd *Luffa cylindrica*
distaff thistle *Carthamus tinctorius*
ditchmoss *Elodea*
dittany *Cunila origanoides*; *Dictamnus albus*
dittany, Cretan *Amaracus dictamnus*
divi-divi *Caesalpinia coriaria*
dock *Rumex*
dock, spinach *Rumex patientia*
dockmackie *Viburnum acerifolium*
dodder *Cuscuta*
dogbane *Apocynum*
dog rose *Rosa canina*
dog's-tooth violet *Erythronium*
dogwood *Cornus*
Dorset heath *Erica ciliaris*
Douglas fir *Pseudotsuga menziesii*
dove orchid *Peristeria*
dove tree *Davidia involucrata*
dragon arum *Arisaema*; *Dracunculus vulgaris*
dragon's mouth *Helicodiceros muscivorus*
dragonhead *Dracocephalum*
dragonhead, false *Physostegia*
dragon-root *Arisaema*
dropwort *Filipendula vulgaris*
dropwort, water *Oenanthe*
duckweed *Lemna*
dumb cane *Dieffenbachia seguine*
durian *Durio zibethinus*
durmast oak *Quercus petraea*
dusty miller *Centaurea*; *Artemisia stellerana*
Dutchman's breeches *Dicentra cucullaria*
Dutchman's pipe *Aristolochia durior*
dwale *Atropa belladonna*
dwarf alder *Fothergilla*
dwarf almond *Prunus tenella*
dwarf birch *Betula nana*
dwarf dandelion *Krigia*
dyer's greenweed *Genista tinctoria*

earthnut *Arachnis hypogaea*
ebony *Diospyros*
edelweiss *Leontopodium alpinum*
eelgrass *Zostera*; *Vallisneria*
eggplant *Solanum melongena*
eglantine *Rosa eglanteria*
elder *Sambucus*
elder, box- *Acer negundo*
elecampane *Inula helenium*
elephant's-ear *Colocasia*
elephant-ear fern *Elaphoglossum*
elm *Ulmus*
elm, American *Ulmus americana*
elm, Chinese *Ulmus parvifolia*
elm, Cornish *Ulmus angustifolia*
elm, Dutch *Ulmus hollandica*
elm, English *Ulmus procera*
elm, Huntingdon *Ulmus hollandica vegeta*
elm, Siberian *Ulmus pumila*
elm, slippery *Ulmus fulva*
elm, smooth-leaved *Ulmus carpinifolia*
elm, Spanish *Cordia*
elm, wych *Ulmus glabra*
empress of Brazil *Worsleya procera*
enchanter's nightshade *Circaea lutetiana*
endive *Cichorium endiva*
English oak *Quercus robur*
eryngo *Eryngium*
evening flower *Hesperantha*
evening primrose *Oenothera*
evening-star *Cooperia*
evergreen oak *Quercus ilex*
everlasting *Antennaria*; *Helichrysum*; *Helipterum*; *Xeranthemum*
everlasting, Australian *Helipterum*
everlasting, pearly *Anaphalis margaritacea*
everlasting, winged *Ammobium alatum*
everlasting pea *Lathyrus latifolius*
ever-ready onion *Allium cepa perutile*

fairy moss *Azolla*

fairybells *Disporum*
false acacia *Robinia pseudacacia*
false aralia *Dizygotheca*
false chamomile *Boltonia*
false cypress *Chamaecyparis*
false dragon-head *Physostegia*
false goatsbeard *Astilbe*
false gromwell *Onosmodium*
false heath *Fabiana imbricata*
false hellebore *Veratrum*
false indigo *Amorpha*
false mallow *Malvastrum*; *Sidalcea*
false mitrewort *Tiarella*
false plantain *Heliconia*
false saffron *Carthamus*
false Solomon's seal; false spikenard *Smilacina*
false spirea *Sorbaria*
false tamarisk *Myricaraia*
fancy anna *Delonix regia*
fanwort *Cabomba caroliniana*
farkleberry *Vaccinium arboreum*
feather-grass *Stipa*
feather hyacinth *Muscari comosum monstrosum*
fennel *Foeniculum vulgare*
fennel-flower *Nigella*
fenugreek *Trigonella foenumgraecum*
fern, adder's tongue *Ophioglossum*
fern, asparagus *Asparagus plumosus*
fern, bladder *Cystopteris*
fern, bracken or brake *Pteridium aquilinum*
fern, chain *Woodwardia*
fern, Christmas *Polystichum*
fern, cinnamon *Osmunda cinnamomea*
fern, cliffbrake *Pellaea*
fern, climbing *Lygodium*
fern, cloak *Notholaena*
fern, comb *Schizaea*
fern, curly-grass *Schizaea*
fern, elephant-ear *Elaphoglossum*
fern, flowering *Osmunda*
fern, gold *Pityrogramma*
fern, grape *Botrychium*
fern, hare's foot *Davallia*
fern, hart's tongue *Phyllitis scolopendrium*
fern, hayscented *Dennstaedtia*
fern, holly *Polystichum*
fern, interrupted *Osmunda*
fern, Japanese *Cyclophorus*
fern, lady *Athyrium filixfemina*
fern, lip *Cheilanthes*
fern, maidenhair *Adiantum*
fern, male *Dryopteris filix-mas*
fern, moonwort *Botrychium*
fern, ostrich *Matteucia struthiopteris*
fern, polypody *Polypodium*
fern, rattlesnake *Botrychium*
fern, rock brake *Cryptogramma*
fern, royal *Osmunda*
fern, rush *Schizaea*
fern, rusty-back *Ceterach officinarum*
fern, sensitive *Onoclea sensibilis*
fern, shield *Dryopteris*; *Polystichum*
fern, silver *Dryopteris*
fern, spleenwort *Asplenium*
fern, staghorn *Platycerium*
fern, swamp spleenwort *Athyrium*
fern, sweet *Comptonia*
fern, sword *Nephrolepis*
fern, tree *Cyathea*
fern, walking *Camptosorus rhizophyllus*
fern, wall *Polypodium*
fern, water *Salvinia*
fern, wood *Dryopteris*
fescue *Festuca*
feverbush *Lindera*
feverfew *Leucanthemum parthenium*
fiddle-wood *Citharexylum*
fig *Ficus*
fig-marigold *Mesembryanthemum*
fig-wort *Scrophularia*
filbert *Corylus maxima*
finger-grass *Digitaria*
finger-lime *Microcitrus*
finocchio *Foeniculum vulgare dulce*
fir *Abies*
fir, Douglas *Pseudotsuga menziesii*
firebush, Mexican *Bassia scoparia culta*
fire-thorn *Pyracantha*
fireweed *Epilobium* (*Chamenerium*) *angustifolium*
fish-grass *Cabomba caroliniana*
flag *Iris*
flag, sweet *Acorus calamus*
flamboyant tree *Delonix regia*
flame-lily *Pyrolirion*

flame-of-the-forest *Butea monosperma*
flamingo flower *Anthurium*
flannel-bush *Fremontodendron*
flannel-flower *Actinotus helianthi*
flax *Linum*
flax, New Zealand *Phormium*
flax-lily *Dianella*
fleabane *Erigeron*
fleur-de-lis *Iris*
floral firecracker *Brevoortia*
floss silk tree *Chorisia*
flower fern *Aneimia*
flowering fern *Osmunda*
flowering moss *Pyxidanthera*
flowering rush *Butomus umbellatus*
flowering spurge *Euphorbia corollata*
foamflower *Tiarella cordifolia*
fool's parsley *Aethusa cynapium*
forget-me-not *Myosotis*
forget-me-not, Antarctic *Myosotidium hortensia*
four-o'clock *Mirabilis jalapa*
foxglove *Digitalis*
foxglove, false *Gerardia*
fox grape *Vitis labrusca*
foxtail grass *Alopecurus*
foxtail lily *Eremurus*
frangipani *Plumeria*
fraxinella *Dictamnus albus*
French marigold *Tagetes patula*
French mulberry *Callicarpa*
fringe flower *Schizanthus*
fringe-tree *Chionanthus*
fringed orchis *Habenaria*
fritillary *Fritillaria*
frogbit *Hydrocharis morsus-ranae*
frost flower *Aster*
fruiting myrtle *Eugenia*
fumitory *Corydalis*; *Fumaria*
fumitory, climbing *Adlumia fungosa* (U.S.A.); *Corydalis claviculata* (Britain)
fuchsia *Fuchsia*
fuchsia, Australian *Correa*
funkia *Hosta*
furze *Ulex*

gale *Myrica gale*
galingale *Cyperus*
gallant soldier *Galinsoga*
gall-of-the-earth *Prenanthes*
garbanzo *Cicer arietinum*
garden cress *Lepidium sativum*
gardener's garters *Phalaris arundinacea* 'Picta'
gardenia *Gardenia*
garget *Phytolacca*
garland flower *Hedychium coronarium*
garlic *Allium sativum*
garlic, crow *Allium vineale*
garlic mustard *Alliaria petiolata*
gas plant *Dictamnus albus*
gay-wings *Polygala*
gean *Prunus avium*
geiger tree *Cordia sebestana*
genip *Melicocca bijuga*
genista, florist's *Cytisus canariensis*
gentian *Gentiana*
gentian, horse *Triosteum*
gentian, willow *Gentiana asclepiadea*
geranium, florist's *Pelargonium*
germander *Teucrium*
germander speedwell *Veronica*; *Chamaedrys*
ghostflower *Monotropa uniflora*
giant bamboo *Dendrocalamus*
giant bellflower *Ostrowskia magnifica*
giant fennel *Ferula*
giant scabious *Cephalaria*
giant summer hyacinth *Galtonia candicans*
giant water thyme *Lagarosiphon*
gill-over-the-ground *Glechoma hederacea*
ginger *Zingiber officinale*
ginger, wild *Asarum*
ginger-lily *Hedychium*
ginseng *Panax*
girasole *Helianthus tuberosus*
glasswort *Salicornia*
globe daisy *Globularia*
globe flower *Trollius*
globe-mallow *Sphaeralcea*
globe thistle *Echinops*
glory-bush *Tibouchina*
glory flower *Eccremocarpus scaber*
glory-lily *Gloriosa*
glory-of-the-snow *Chionodoxa*
glory-of-the-sun *Leucocoryne ixioides*
glory-pea *Clianthus*
gloxinia, florist's *Sinningia*

goat's-beard *Aruncus*; *Spiraea*; *Tragopogon*
goatsrue *Galega officinalis*
goat willow *Salix caprea*
gobo *Arctium lappa anagyroides*
gold fern *Pityrogramma*
gold thread *Coptis*
golden aster *Chrysopsis*
golden-bell *Forsythia*
golden chain *Laburnum*
golden club *Orontium*
golden drop *Onosma*
golden larch *Pseudolarix amabilis*
golden marguerite *Anthemis tinctoria*
goldenrod *Solidago*
goldenrod, plumed *Bigelovia*
golden saxifrage *Chrysosplenium*
golden seal *Hydrastis*
golden shower *Cassia fistula canadensis*
golden spider-lily *Lycoris aurea*
golden star *Chrysogonum virginianum*
golden wattle *Acacia pycnantha*
goldilocks *Ranunculus auricomus*; *Aster linosyris*
good-King-Henry *Chenopodium bonus-henricus*
gombo *Abelmoschus esculentus*
gooseberry *Ribes uva-crispa*
gooseberry, Barbados *Pereskia aculeata*
goosefoot *Chenopodium album*
gorse *Ulex*
gourd *Cucurbita*; *Lagenaria*
gourd, calabash *Lagenaria siceraria*
gourd, dish-cloth *Luffa aegytiaca*
gourd, ivy *Coccinea cordifolia*
gourd, snake *Trichosanthes anguina*
gourd, wax *Benincasa hispida*
goutweed *Aegopodium podagraria*
grape *Vitis vinifera*
grape, fox *Vitis labrusca*
grape, Oregon *Mahonia aquifolium*
grape, seaside *Coccoloba*
grape fern *Botrychium*
grapefruit *Citrus paradisi*
grape-hyacinth *Muscari*
grass, brome *Bromus*
grass, canary *Phalaris canariensis*
grass, Guinea *Panicum maximum*
grass, Kentucky *Poa pratensis*
grass, lemon *Cymbopogon flexuosus*, *C. citratus*
grass, Pampas *Cortaderia*
grass, pink *Calopogon*
grass, quaking *Briza*
grass-of-Parnassus *Parnassia palustris*
great burnet *Poterium*
Greek valerian *Polemonium*
green hellebore *Helleborus viridis*
green wattle *Acacia decurrens*
greenbrier *Smilax*
grey birch *Betula populifolia*
grey poplar *Populus canescens*
gromwell *Lithospermum officinale*
ground-cherry *Physalis pruinosa*
ground-ivy *Glechoma hederacea*
ground-laurel *Epigaea repens*
groundnut *Arachis hypogaea*
ground-oak *Chamaedrys*
ground pine *Lycopodium*
ground rattan *Rhapis*
groundsel *Senecio vulgaris*
groundsel-tree *Baccharis*
gru gru *Acrocomia aculeata*
guava *Psidium guavjava*
guava, strawberry *Psidium cattleianum*
guelder-rose *Viburnum*
Guernsey lily *Nerine sarniensis*
guinea flower *Hibbertia*
gul mohur *Delonix regia*
gum, black *Nyssa sylvatica*
gum myrtle *Angophora*
gum tree *Eucalyptus*
gumbo *Abelmoschus esculentus*
gumplant *Grindelia*

hackberry *Celtis*
hair grass *Aira*
handkerchief tree *Davidia involucrata*
harbinger-of-spring *Erigenia bulbosa*
hardhack *Spiraea tomentosa*
hardheads *Centaurea nigra*
harebell *Campanula rotundifolia*
hare's ear *Bupleurum*
hare's-foot fern *Davallia*
hare's-tail grass *Lagurus ovatus*
hart's-tongue fern *Phyllitis scolopendrium*
hatchet vetch *Securigera securidaca*
hawk's-beard *Crepis*

hawkbit *Leontodon*
hawkweed *Hieracium*
hawkweed, yellow *Tolpis*
hawthorn *Crataegus*
hay-scented fern *Dennstaedtia*
hazel *Corylus*
heal-all *Prunella vulgaris*
heartnut *Juglans ailantifolia cordiformis*
heart's-ease *Viola tricolor*
heath *Erica*; *Epacris* (Australia); *Styphelia* (Australia)
heath, Cornish *Erica vagans*
heath, cross-leaved *Erica tetralix*
heath, Dorset *Erica ciliaris*
heath, fake *Fabiana imbricata*
heath, Irish *Erica erigena*
heath, sea *Frankenia laevis*
heath, St. Dabeoc's *Daboëcia cantabrica*
heath, spike *Bruckenthalia spiculifolia*
heather *Calluna vulgaris*
heather, beach *Hudsonia*
heather, bell- *Erica cinerea*
heather, bog *Erica tetralix*
hedge-bindweed *Calystegia*; *Convolvulus*
hedgehog broom *Erinacea anthyllis*
hedge-hyssop *Gratiola*
hedge-mustard *Erysimum*
hedge-nettle *Stachys*
hedge-parsley *Torilis*
hedge wattle *Acacia armata*
heliotrope *Heliotropium*
heliotrope, garden *Valeriana*
heliotrope, winter *Petasites fragrans*
hellebore *Helleborus*
hellebore, false *Veratrum*
hellebore, green *Helleborus viridis*
hellebore, stinking *Helleborus foetidus*
helmet, policeman's *Impatiens glandulifera*
helmet flower *Coryanthes*
hemlock *Tsuga*
hemlock, poison *Conium maculatum*
hemlock, water *Cicuta virosa*
hemp *Cannabis sativa*
hemp-agrimony *Eupatorium cannabinum*
hemp, Indian- *Apocynum cannabinum*
hempweed, climbing *Mikania scandens*
hen-and-chickens houseleek *Sempervivum soboliferum*
henbane *Hyoscyamus niger*
henna *Lawsonia inermis*
herald's trumpet *Beaumontia*
herb-bennet *Geum urbanum*
herb-Christopher *Actaea spicata*
herb-grace *Ruta graveolens*
herb-Mercury *Mercurialis*
herb-Paris *Paris quadrifolia*
herb patience *Rumex patientia*
herb-Robert *Geranium robertianum*
herb-of-St. Barbara *Barbarea vulgaris*
Hercules' club *Aralia spinosa*; *Zanthoxylum clava-herculis*
heron's bill *Erodium*
herniary *Herniaria*
hickory *Carya*
hickory, shagbark *Carya ovata*
Himalaya-berry *Rubus procerus*
Himalayan poppy *Meconopsis*
hoarhound *Marrubium*
hoarhound, water *Lycopus*
hobblebush *Viburnum alnifolium*
hogweed *Heracleum sphondylium*
holly *Ilex*
holly, American mountain *Nemopanthus mucronatus*
hollyhock *Althaea*
holm oak *Quercus ilex*
holy grass *Hierochloë*
hondapara *Dillenia indica*
honesty *Lunaria annua*
honey-bell *Mahernia verticillata*
honey-berry *Melicocca bijuga*
honey-bush *Melianthus*
honey-locust *Gleditsia*
honeysuckle *Lonicera*
honeysuckle, Australian *Banksia*
honeysuckle, bush- *Diervilla*
honeysuckle, Cape *Tecomaria capensis*
honeysuckle, Japanese *Lonicera japonica*
honeywort *Cerinthe*
hop *Humulus*
hop-hornbeam *Ostrya carpinifolia*
horehound *Marrubium vulgare*
horehound, black *Ballota*
horehound, water *Lycopus*
hornbeam *Carpinus*
horned poppy *Glaucium*
horn-of-plenty *Fedia*
hornwort *Ceratophyllum demersum*

horse-balm *Collinsonia*
horse bean *Vicia faba*
horse-chestnut *Aesculus*
horse-chestnut, red *Aesculus rubra*
horse-chestnut, white *Aesculus hippocastanum*
horse-gentian *Triosteum*
horse-mint *Mentha aquatica*, *M. longifolia*
horse-radish *Armoracia rusticana*
horse-radish, Japanese *Eutrema wasabi*
horse-radish tree *Moringa*
horseshoe vetch *Hippocrepis comosa*
horsetail *Equisetum*
horse-weed *Collinsonia*
Hottentot fig *Carpobrotus edulis*
hound's tongue *Cynoglossum*
house-leek *Sempervivum*
huckleberry *Gaylussacia*
huckleberry, garden *Solanum melanocerasum*
hyacinth *Hyacinthus orientalis*
hyacinth, feather *Muscari comosum monstrosum*
hyacinth, grape *Muscari*
hyacinth, summer *Galtonia candicans*
hyacinth, tassel *Muscari comosum*
hyacinth, water *Eichhornia crassipes*
hyacinth, wild *Hyacinthoides*
hyssop *Hyssopus officinalis*
hyssop, giant *Agastache*

iceplant *Cryophytum crystallinum*; *Sedum spectabile*
immortelle *Helichrysum*; *Helipterum*; *Xeranthemum*
incense cedar *Libocedrus*
incense plant *Calomeria amaranthoides* (*Humea elegans*)
India hawthorn *Raphiolepis*
India lotus *Nymphaea*
India-rubber plant *Ficus elastica*
Indian almond *Terminalia catappa*
Indian cress *Tropaeolum*
Indian crocus *Pleione*
Indian cucumber-root *Medeola virginiana*
Indian fig *Opuntia*
Indian grass *Arundo*
Indian-hemp *Apocynum cannabinum*
Indian mulberry *Morinda*
Indian paint-brush *Castilleja*
Indian physic *Porteranthus trifoliatus*
Indian pink *Spigelia marilandica*
Indian pipe *Monotropa uniflora*
Indian poppy *Meconopsis*
Indian shot *Canna*
Indian strawberry *Duchesnea indica*
Indian tobacco *Lobelia inflata*
Indian turnip *Arisaema*
indigo *Indigofera*
indigo, false *Amorpha*
indigo, wild *Baptisia*
inkberry *Ilex glabra*; *Phytolacca americana*
innocence *Houstonia caerulea*
interrupted fern *Osmunda claytoniana*
Iris *Iris*
iris, English *Iris xiphioides*
iris, German *Iris germanica*
iris, mourning *Iris susiana*
iris, scarlet-seeded *Iris foetidissima*
iris, Spanish *Iris xiphium*
iris, stinking *Iris foetidissima*
iris, widow *Hermodactylus tuberosus*
Irish heath *Erica erigena*
ironbark *Eucalyptus*
iron tree *Metrosideros*
ironweed *Veronia*
ironwood *Carpinus*; *Ostrya virginiana*
ironwort *Sideritis*
Italian cypress *Cupressus sempervirens*
ivory-nut palm *Phytelephas macrocarpa*
ivy *Hedera*
ivy arum *Scindapsus*
ivy, Boston *Parthenocissus tricuspidata*
ivy gourd *Coccinia cordifolia*
ivy, ground *Glechoma hederacea*
ivy, poison *Rhus toxicodendron*

jaboticaba *Myrciaria cauliflora*
jackfruit *Artocarpus heterophyllus*
Jack-go-to-bed-at-noon *Tragopogon pratensis*
Jack-in-the-pulpit *Arisaema*
Jacobea *Senecio jacobaea*
Jacobean lily *Sprekelia formosissima*
Jacob's ladder *Polemonium*
Jacob's rod *Asphodeline lutea*
jade vine *Strongylodon macrobotrys*
jak *Artocarpus heterophyllus*

Jamaica pepper *Pimenta dioica*
Jamaica sorrel *Hibiscus sabdariffa*
Jamestown weed *Datura stramonium*
Japanese bellflower *Platycodon*
Japanese bitter orange *Poncirus trifoliata*
Japanese bunching onion *Allium fistulosum*
Japanese cedar *Cryptomeria japonica*
Japanese honeysuckle *Lonicera japonica*
Japanese horse-radish *Eutrema wasabi*
Japanese persimmon *Diospyros kaki*
Japanese quince *Chaenomeles*
Japanese raisin tree *Hovenia dulcis*
Japanese toad-lily *Tricyrtis*
Japanese yew *Cephalotaxus harringtoniana*
Japonica *Chaenomeles speciosa; Camellia japonica; Kerria japonica*
jasmine *Jasminum*
jasmine, bastard *Cestrum*
jasmine, Madagascar *Stephanotis floribunda*
Jerusalem artichoke *Helianthus tuberosus*
Jerusalem cherry *Solanum pseudocapsicum*
Jerusalem cowslip *Pulmonaria*
Jerusalem oak *Chenopodium botrys*
Jerusalem sage *Phlomis fruticosa*
Jerusalem thorn *Parkinsonia aculeata*; *Paliurus spinachristi*
jewel-weed *Impatiens*
Jew's mallow *Corehorus olitorios*
Job's tears *Coix lacryma-jobi*
Joe-Pye-weed *Eupatorium*
jonquil *Narcissus jonquilla*
Joshua tree *Yucca brevifolia*
joy-weed *Alternanthera*
Judas tree *Cercis siliquastrum*
June berry *Amelanchier*
juniper *Juniperus*
juniper-myrtle *Verticordia*
Jupiter's beard *Anthyllis*; *Centranthus ruber*
jute *Corchorus olitorius*

Kaffir lily *Clivia*; *Schizostylis*
kaki *Diospyros kaki*
kale *Brassica oleracea acephala*
kale, sea *Crambe maritima*
Kamila tree *Mallotus*
kangaroo apple *Solanum laciniatum*
kangaroo paw *Anigozanthus*
kangaroo vine *Cissus antarctica*
Kansas gay-feather *Liatris*
Kauri pine *Agathis*
kenaf *Hibiscus cannabinus*
kidney bean *Phaseolus vulgaris*
kidney vetch *Anthyllis vulneraria*
king-cup *Caltha palustris*
kiwifruit *Actinidia deliciosa*
knapweed *Centaurea*
knotweed *Polygonum*
kohlrabi *Brassica caulorapa*
kudzu-vine *Pueraria thunbergiana*
kumquat *Fortunella*
kumquat, Marumi *Fortunella japonica*
kumquat, Nagami *Fortunella margarita*

Labrador tea *Ledum groenlandicum*
laburnum *Laburnum*
laburnum, Indian *Cassia fistula*
lace-leaf *Aponogeton fenestralis*
ladies' fingers *Abelmoschus esculentus*; *Anthyllis vulneraria*
ladies' tresses *Spiranthes*
lad's love *Artemisia*
Lady Doorley *Ipomoea horsfalliae*
lady fern *Athyrium*
lady-of-the-night *Brunfelsia americana*
lady's finger *Abelmoschus esculentus*
lady's mantle *Alchemilla*
lady's slipper *Cypripedium*
lady's smock *Cardamine pratensis*
lady's thistle *Silybum marianum*
lady's thumb *Polygonum*
lamb-kill *Kalmia*
lamb's lettuce *Valerianella*
lamb's-quarters *Chenopodium*
lamb's tail grass *Alopecurus*
lantana, wild *Abronia*
larch *Larix*
larch, golden *Pseudolarix amabilis*
larkspur *Consolida*; *Delphinium*
lattice-leaf *Aponogeton fenestralis*
laurel *Laurus nobilis*
laurel, Alexandrian *Danae racemosa*
laurel, bay *Laurus nobilis*
laurel, cherry *Prunus laurocerasus*
laurel, mountain *Kalmia*
laurel, New Zealand *Corynocarpus*

laurel, Portugal *Prunus lusitanica*
laurel, sheep; laurel, swamp *Kalmia*
laurestinus *Viburnum tinus*
lavender *Lavandula*
lavender, sea *Limonium*
lavender-cotton *Santolina chamaecyparissus*
leadwort *Plumbago*
leather fern *Acrostichum danaeifolium*
leatherwood *Dirca palustris*; *Cyrilla racemiflora*
leek *Allium ampeloprasum porrum*
lemon *Citrus limon*
lemon balm *Melissa officinalis*
lemon verbena *Aloysia triphylla*
lentil *Lens culinaris*
leopard flower *Belamcanda chinensis*
leopard lily *Sansevieria*
leopard plant *Ligularia kaempferi aureomaculata*
leopard's bane *Doronicum*
lettuce *Lactuca sativa*
licorice *Glycyrrhiza glabra*
licorice, wild *Abrus precatorius*
life plant *Bryophyllum pinnatum*
lignum *Muehlenbeckia*
lignum vitae *Guiacum officinale*; *Guiacum sanctum*
lilac *Syringa*
lilly-pilly *Acmena smithii*
lily *Lilium*
lily, African *Agapanthus*
lily, Amazon *Eucharis grandiflora*
lily, blackberry *Belamcanda chinensis*
lily, blood *Haemanthus*; *Scadoxus*
lily, day *Hemerocallis*
lily, golden spider *Lycoris aurea*
lily, Guernsey *Nerine sarniensis*
lily, Jacobean *Sprekelia formosissima*
lily, Kaffir *Clivia*; *Schizostylis*
lily, leopard *Sansevieria*
lily, Madonna *Lilium candidum*
lily, Mariposa *Calochortus*
lily, Mediterranean *Pancratium maritimum*
lily, nankeen *Lilium testaceum*
lily, Peruvian *Alstroemeria*
lily, regal *Lilium regale*
lily, Scarborough *Vallota speciosa*
lily, St. Bernard's *Anthericum liliago*
lily, St. Bruno's *Paradisea liliastrum*
lily, spider *Hymenocallis*
lily, tiger *Lilium lancifolium* (*L. tigrinum*)
lily, zephyr *Zephyranthes*
lily-of-the-Nile *Agapanthus*
lily-of-the-valley *Convallaria*
lily-of-the-valley, wild *Maianthemum*
lily-thorn *Catesbaea spinosa*
lime *Citrus aurantifolia* (tropics); *Tilia* (Britain)
lime, finger *Microcitrus*
limeberry *Triphasia trifolia*
Lincolnshire asparagus *Chenopodium*
linden, or lime *Tilia*
ling, ling-heather *Calluna vulgaris*
lion's ear *Leonotis leonurus*
lion's foot *Leontopodium alpinum*
lion's-tail *Leonurus*
lip fern *Cheilanthes*
liquorice *Glycyrrhiza glabra*
litchi *Litchi chinensis*
little Mary *Bellium minutum*
liverleaf, liverwort *Hepatica*
lizard orchid *Himantoglossum hircinum* (Britain); *Burnettia cuneata* (Australia)
lizard's-tail *Saururus cernuus*
loblolly bay *Gordonia lasianthus*
loblolly magnolia *Magnolia grandiflora*
loblolly pine *Pinus taeda*
locust *Robinia*
locust, honey *Gleditsia*
loganberry *Rubus loganobaccus*
logwood *Haematoxylum campechianum*
Lombardy poplar *Populus nigra italica*
London plane *Platanus hybrida*
London pride *Saxifraga urbium*
loosestrife, purple *Lythrum salicaria*
loosestrife, swamp *Decodon verticillatus*
loosestrife, yellow *Lysimachia vulgaris*
lords-and-ladies *Arum maculatum*
lotus, blue *Nymphaea stellata*
lotus, sacred *Nelumbo nucifera*
lotus, white *Nymphaea lotus*
louse-wort *Pedicularis*
lovage *Levisticum officinale*; *Ligusticum scoticum*
love-apple *Lycopersicon esculentum*
love-grass *Eragrostis*

love-in-a-mist *Nigella damascena*
love-lies-bleeding *Amaranthus caudatus*
lucerne *Medicago sativa*
Lucombe oak *Quercus hispanica lucombeana*
lungwort *Pulmonaria; Mertensia*
lupin, lupine *Lupinus*

Macartney rose *Rosa bracteata*
Madagascar jasmine *Stephanotis floribunda*
Madagascar palm *Chrysalidocarpus lutescens*
Madagascar periwinkle *Catharanthus roseus*
madder *Rubia tinctoria*
Madeira vine *Boussingaultia gracilis pseudobasseloides*
madwort *Alyssum*
mahoe *Thespesia populnea*
mahogany tree *Swietenia mahogani*
maidenhair fern *Adiantum*
maidenhair tree *Gingko biloba*
maize *Zea mays*
Malabar nightshade *Basella alba*
male-fern *Aspidium; Dryopteris*
mallee *Eucalyptus*
mallee, white *Eucalyptus gracilis*
mallow *Malva*
mallow, false *Malvastrum*
mallow, globe *Sphaeralcea*
mallow, Jew's *Corchorus olitorius*
mallow, marsh *Althaea officinalis*
mallow, musk *Malva moschata*
mallow, rose, *Hibiscus*
mallow, tree *Lavatera arborea*
Maltese cross *Lychnis chalcedonica*
mammee-apple *Mammea americana*
manchineel *Hippomane mancinella*
mandrake *Mandragora* (Europe); *Podophyllum peltatum*(U.S.A.)
mango *Mangifera indica*
mangosteen *Garcinia mangostana*
manna grass *Glyceria*
maple *Acer*
maple, field *Acer campestre*
maple, flowering *Abutilon*
maple, hedge *Acer campestre*
maple, Japanese *Acer palmatum*
maple, Norway *Acer platanoides*
maple, sugar *Acer saccharum*
mare's tail *Ephedra; Hippuris vulgaris*
marguerite *Leucanthemum vulgare; Argyranthemum frutescens*
marguerite, golden *Anthemis tinctoria*
marigold *Calendula*
marigold, African *Tagetes erecta*
marigold, fig *Mesembryanthemum*
marigold, Cape *Dimorphotheca*
marigold, French *Tagetes patula*
marigold, marsh *Caltha palustris*
marigold, pot *Calendula officinalis*
Mariposa lily *Calochortus*
marjoram *Origanum vulgare*
marjoram, pot *Marjorana onites*
marjoram, sweet *Marjorana hortensis*
marjoram, wild *Origanum vulgare*
marking nut *Anacardium*
marlberry *Ardisia*
marram grass *Ammophila arenaria*
marrow, vegetable *Cucurbita pepo medullosa*
marsh felwort *Swertia*
marsh-marigold *Caltha palustris*
marsh rosemary *Limonium*
marsh samphire *Salicornia*
marvel-of-Peru *Mirabilis jalapa*
mask-flower *Alonsoa warscewiczii*
masterwort *Astrantia*
matrimony vine *Lycium barbarum*
may *Crataegus monogyna, C. laevigata*
May-apple *Podophyllum peltatum*
Mayflower *Epigaea repens*
mayweed *Anthemis; Matricaria; Tripleurospermum*
mayweed, scented *Matricaria discoidea*
mayweed, scentless *Tripleurospermum inodorum*
mayweed, stinking *Anthemis cotula*
meadow-beauty *Rhexia*
meadow foam *Limnanthes douglasii*
meadow foxtail *Alopecurus pratensis*
meadow rue *Thalictrum*
meadow saffron *Colchicum*
meadow sweet *Filipendula*
medick *Medicago*
medick, black *Medicago lupulina*
medick, spotted *Medicago arabica*
Mediterranean *Pancratium*

lily *maritimum*
medlar *Mespilus germanica*
melic grass *Melica*
melilot *Melilotus*
melon, bitter *Momordica charantia*
melon, water *Citrullus lanatus*
mercury *Chenopodium bonus-henricus; Mercurialis*
mercury, dog's *Mercurialis perennis*
mermaid weed *Proserpinaca*
mesquite *Prosopis*
Mexican bamboo *Polygonum cuspidatum*
Mexican buckeye *Ungnadia speciosa*
Mexican clover *Richardia scabra*
Mexican firebush *Kochia scoparia culta*
Mexican lily *Hippeastrum*
Mexican orange *Choisya ternata*
Mexican poppy *Argemone mexicana*
Michaelmas daisy *Aster*
mignonette *Reseda odorata*
mignonette-vine *Boussingaultia gracilis pseudobasseloides*
milfoil *Achillea millefolium*
milk thistle *Silybum marianum*
milk vetch *Astragalus*
milkweed *Asclepias; Gomphocarpus*
milkwort millet *Polygala; Eleusine; Panicum; Paspalum*
mimosa, florist's *Acacia*
mind-your-own-business *Soleirolia soleirolii*
mint, apple *Mentha villosa alopecuroides*
mint, Bowles *Mentha villosa alopecuroides*
mint-bush *Prostanthera*
mint, cat- *Nepeta cataria*
mint, corn *Mentha arvensis*
mint, horse *Mentha aquatica, M. longifolia*
mint, lemon *Monarda punctata, Monarda citriodora; Mentha piperata citrata*
mint, mountain *Pycnanthemum*
mint, pepper *Eucalyptus* (Australia); *Mentha piperita*
mint, spear- *Mentha spicata*
mint, water *Mentha aquatica*
missey-moosey *Sorbus americana*
mistletoe *Amyena*, etc. (Australia); *Phoradendron; Viscum* (Britain)
mitrewort *Mitella*
mitrewort, false *Tiarella*
moccasin flower *Cypripedium*
mock-orange *Philadelphus*
mock strawberry *Duchesnea indica*
mombin, purple *Spondias purpurea*
mombin, red *Spondias purpurea*
mombin, yellow *Spondias mombin*
monarch-of-the-East *Sauromatum venosum*
moneywort *Lysimachia nummularia*
monkey-bread tree *Adansonia digitata*
monkey-dinnerbell *Hura*
monkey-flower *Mimulus*
monkeynut *Arachis hypogaea*
monkey-pod tree *Samanea saman*
monkey-puzzle tree *Araucaria*
monkshood *Aconitum*
Monterey cypress *Cupressus macrocarpa*
moon-daisy *Leucanthemum vulgare*
moonflower *Ipomoea alba*
moonseed *Cocculus; Menispermum*
moonwort *Botrychium*
moonwort, blue *Soldanella*
moosewood *Dirca australe*
morning-glory *Ipomoea*
moss campion *Silene acaulis*
moth orchid *Phalaenopsis*
mother of thousands *Cymbalaria muralis; Saxifraga stolonifera*
mother-of-thyme *Thymus serpyllum*
motherwort *Leonurus*
Mount Atlas daisy *Anacyclus*
mountain ash *Sorbus aucuparia*
mountain avens *Dryas*
mountain-fringe *Adlumia fungosa*
mountain holly *Nemopanthus mucronatus*
mountain-laurel *Kalmia*
mountain mint *Pycnanthemum*
mountain spider-wort *Lloydia*
mountain spurge *Pachysandra*
mountain-tea *Gaultheria*
mourning bride *Scabiosa*
mouse-ear chickweed *Cerastium*
moutan *Paeonia suffruticosa*
mouthroot *Coptis*
mud plantain *Alisma; Heteranthera*
mudgee wattle *Acacia spectabilis*
mugga *Eucalyptus sideroxylon*
mugwort *Artemisia vulgaris*
mulberry, black *Morus nigra*

mulberry, Indian *Morinda*
mulberry, paper- *Broussonetia papyrifera*
mulberry, red *Morus rubra*
mulberry, white *Morus alba*
mullein *Verbascum*
mullein pink *Lychnis*
mung *Phaseolus aureus*
muscadine *Vitis rotundifolia*
musk hyacinth *Muscari*
musk orchis *Herminium monorchis*
muskmallow *Malva moschata*
muskmelon *Cucumis melo*
musk rose *Rosa moschata*
mustard, black *Brassica nigra*
mustard, brown *Brassica juncea*
mustard, garlic *Alliaria petiolata*
mustard, white *Brassica hirta; Sinapis alba*
myrobalan plum *Prunus cerasifera*
myrtle *Myrtus communis*
myrtle, running *Vinca*

nailwort *Draba; Paronychia*
naseberry *Manilkara zapota*
nasturtium, garden *Tropaeolum*
navelwort *Omphalodes; Umbilicus*
nectarine *Prunus persica nectarina*
nemu tree *Albizia*
netfern *Gleichenia*
nettle *Urtica*
nettle, tree *Dendrocnide*
New Holland daisy *Vittadenia triloba*
New Jersey tea *Ceanothus*
New Zealand bur *Acaena*
New Zealand daisy *Celmisia*
New Zealand edelweiss *Leucogenes*
New Zealand flax *Phormium*
New Zealand laurel *Corynocarpus*
New Zealand spinach *Tetragonia tetragonioides*
New Zealand wineberry *Aristotelia*
nicker-bean, grey *Caesalpinia bonducella*
nicker-bean, yellow *Caesalpina bonduc*
night-blooming cereus *Selenicereus grandiflorus*
nightshade *Solanum*
nightshade, deadly *Atropa belladonna*
nightshade, Malabar *Basella alba*
nightshade, woody *Solanum dulcamara*
nine bark *Physocarpus opulifolius*
none-so-pretty *Saxifraga urbium*
nopal *Opuntia megacantha*
Norfolk Island pine *Araucaria heterophylla*
northern white cedar *Thuja occidentalis*
nut, cob- *Corylus avellana*
nut, marking *Anacardium*

oak *Quercus*
oak, black *Quercus velutina*
oak, cork *Quercus suber*
oak, durmast *Quercus petraea (Q. sessiliflora)*
oak, English *Quercus robur* (*Q. pedunculata*)
oak, evergreen *Quercus ilex*
oak-fern *Thelypteris dryopteris*
oak, holm *Quercus ilex*
oak, Lucombe *Quercus hispanica lucombeana*
oak, poison *Toxicodendron*
oak, red *Quercus borealis*
oak, scarlet *Quercus coccinea*
oak, she- *Casuarina*
oak, Turkey *Quercus cerris*
oak, water *Quercus nigra*
oak, white *Quercus alba*
oak-of-Jerusalem *Chenopodium botrys*
oats *Avena*
obedient-plant *Physostegia*
okra *Abelmoschus esculentus*
old man *Artemisia abrotanum*
old-man cactus *Cephalocereus*
old man's beard *Clematis vitalba*
old woman *Artemisia stellerana*
oleander *Nerium oleander*
olive *Olea europaea*
olive, Russian *Elaeagnus*
olive, spurge *Cneorum*
olive, sweet *Osmanthus fragrans*
onion *Allium cepa*
onion, ever-ready *Allium cepa perutile*
onion, Japanese bunching *Allium fistulosum*
onion, potato *Allium cepa aggregatum*
onion, tree *Allium cepa proliferum*
onion, Welsh *Allium fistulosum*
opopanax *Acacia farnesiana*
open-arse *Mespilus germanica*

orach *Atriplex hortensis*
orange *Citrus*
orange, hardy *Poncirus trifoliata*
orange, mandarin *Citrus reticulata*
orange, Seville *Citrus aurantium*
orange, sour *Citrus aurantium*
orange, sweet *Citrus sinensis*
orange jessamine *Murraya*
orange root *Hydrastis canadensis*
orchid *Orchis* or members of *Orchidaceae* in general
Oregon grape *Mahonia aquifolium*
orpine *Sedum telephium*
osage orange *Maclura pomifera*
osoberry *Oemleria cerasiformis*
ostrich fern *Matteucia struthiopteris*
Oswego-tea *Monarda*
Otaheite apple *Spondias dulcis*
oxeye, yellow *Buphthalmum*
ox-eye daisy *Leucanthemum vulgare*
oxlip *Primula elatior*
oyster-plant *Tragopogon porrifolius*
oyster-plant, Spanish *Scolymus hispanicus*

paeony *Paeonia*
pagoda tree *Plumeria rubra*
paigle *Primula veris*
painted-cup *Castilleja*
palas, climbing *Butea superba*
palm, cabbage *Roystonea oleracea*
palm, date *Phoenix dactylifera*
palm, fishtail *Caryota*
palm, king *Archontophoenix*
palm, Madagascar *Chrysalidocarpus lutescens*
palm, oil *Elaeis guineensis*
palm, royal *Roystonea*
palm, pindo *Butia*
palm, queen *Arecastrum romanzoffianum*
palm, sago- *Cycas revoluta*
palm, toddy *Caryota urens*
palm, thatch *Thrinax*
palmetto *Sabal*
palmetto, saw *Serenoa repens*
pampas grass *Cortaderia*
Panama candle-tree *Parmentiera cereifera*
panamigo *Pilea involucrata*
pansy, garden *Viola wittrockiana*
papaya *Carica papaya*
paper bark *Melaleuca*
paper mulberry *Broussonetia papyrifera*
papyrus *Cyperus papyrus*
parsley *Petroselinum crispum*
parsley, fool's *Aethusa cynapium*
parsley, hedge- *Torilis*
parsnip *Pastinaca sativa*
parsnip, cow *Heracleum sphondylium*
partridgeberry *Mitchella repens*
partridge-pea *Cassia fasciculata*
pasque-flower *Pulsatilla*
passion-flower *Passiflora*
Paterson's curse *Echium plantagineum*
patience *Rumex patientia*
paw-paw *Asimina tribola; Carica papaya*
pea *Pisum sativum*
pea, asparagus *Psophocarpus tetragonolobus*
pea, bush *Pultenaea*
pea, chick *Cicer orietinum*
pea, Darling River *Swainsona greyana*
pea, everlasting *Lathyrus latifolius*
pea, partidge *Cassia faciculata*
pea, shamrock *Parochetus communis*
pea, swainson *Swainsona*
pea, sweet *Lathyrus odoratus*
pea, tree *Caragana*
pea, wedge *Gomphalobium*
peach *Prunus persica*
peanut *Arachis hypogaea*
pear *Pyrus communis*
pear, alligator *Persea americana*
pear, balsam *Momordica charantia*
pear, prickly- *Opuntia*
pear, wooden *Xylomelum pyriforme*
pearl-bush *Exochorda*
pearlwort *Sagina*
pearly everlasting *Anaphalis margaritacea*
pebble plants *Lithops*
pecan *Carya illinoensis*
pelican flower *Aristolochia grandiflora*
pennywort *Umbilicus rupestris*
peony *Paeonia*
pepino *Solanum muricatum*
pepper *Piper*
pepper-and-salt *Erigenia bulbosa*
pepper, bird *Capsicum annuum*
pepper, Tabasco *Capsicum frutescens*
pepperbush, sweet *Clethra*
peppergrass *Lepidium sativum*

pepperidge *Nyssa*
peppermint *Mentha piperita; Eucalyptus*
pepper-root *Dentaria*
pepper-tree *Schinus molle*
pepper-vine *Ampelopsis*
pepperwort *Lepidium sativum*
periwinkle *Vinca*
periwinkle, Madagascar *Catharanthus roseus*
persicaria *Persicaria*
persimmon *Diospyros*
Peruvian lily *Alstroemeria*
pheasant's-eye *Adonis annua*
physic-nut *Jatropha multifida*
pickaback plant *Tolmiea menziesii*
pickerel-weed *Pontederia cordata*
picotee *Dianthus caryophyllus*
pigeonberry *Duranta repens; Phytolacca americana*
pigeon pea *Cajanus cajan*
pignut *Carya; Conopodium majus; Omphalea triandra* (Jamaica)
pignut palm *Hyophorbe*
pimpernel, water *Samolus*
pincushion flower *Scabiosa*
pindo palm *Butia*
pine *Pinus*
pine, Austrian *Pinus nigra*
pine, bristlecone *Pinus aristata*
pine, jack *Pinus banksiana*
pine, loblolly *Pinus taeda*
pine, Scots *Pinus sylvestris*
pine, screw- *Pandanus*
pine, stone *Pinus cembra; Pinus pinea*
pineapple *Ananas comosus*
pineapple-weed *Matricaria discoidea*
pinguin *Bromelia pinguin*
pink *Dianthus*
pink, cottage *Dianthus caryophyllus*
pink, ground *Phlox*
pink, Indian *Spigelia marilandica*
pink, moss *Phlox*
pink, mullein *Lychnis*
pink, rose *Sabatia*
pink, sea *Armeria*
pink, swamp *Calopogon; Helonias*
pink-root *Spigelia marilandica*
pink siris *Albizia*
pinxterflower *Rhododendron nudiflorum*
pipsissewa *Chimaphila umbellata*
pistachio *Pistacia*
pitcherplant *Nepenthes; Sarracenia*
plane tree *Platanus*
plantain *Plantago*
plantain banana *Musa*
plantain, false *Heliconia*
plaintain, poor Robin's *Hieracium venosum*
plantain, rattlesnake *Goodyera*
plantain, water *Alisma*
plantain lily *Hosta*
pleurisy-root *Asclepias tuberosa*
plum *Prunus domestica*
plum, beach *Prunus maritima*
plum, hog- *Spondias mombin*
plum, myrobalan *Prunus cerasifera*
plum-yew *Cephalotaxus*
plumbago *Plumbago*
plume-grass *Erianthus*
plume poppy *Bocconia; Macleaya*
poison hemlock *Conium maculatum*
poison-ivy, -oak, -sumach *Rhus toxicodendron*
pokeberry; pokeweed *Phytolacca americana*
polyanthus *Primula polyantha*
polypody *Polypodium*
pomegranate *Punica granatum*
pond cypress *Taxodium ascendens*
pond-lily, yellow *Nuphar*
pondweed *Potamogeton*
pondweed, Cape *Aponogeton*
poor man's weatherglass *Anagallis arvensis*
poor Robin's plantain *Hieracium venosum*
poplar; popple *Populus*
poplar, balm-of-Gilead *Populus gileadensis*
poplar, balsam *Populus balsamifera*
poplar, black *Populus nigra*
poplar, grey *Populus canescens*
poplar, Lombardy *Populus nigra italica*
poplar, white *Populus alba*
poppy *Papaver*
poppy, blue *Meconopsis*
poppy, bush- *Dendromecon rigidum*
poppy, California *Eschscholzia*
poppy, celandine- *Stylophorum diphyllum*
poppy, corn *Papaver rhoeas*

poppy, crested *Argemone*
poppy, Himalayan *Meconopsis*
poppy, horned *Glaucium*
poppy, Iceland *Papaver nudicaule*
poppy, Indian *Meconopsis*
poppy, opium *Papaver somniferum*
poppy, oriental *Papaver orientale*
poppy, plume *Bocconia; Macleaya*
poppy, prickly *Argemone*
poppy, sea *Glaucium*
poppy, snow *Eomecon chionantha*
poppy, water *Hydrocleys nymphoides*
poppy, Welsh *Meconopsis cambrica*
poppy-mallow *Callirhoe*
portia-tree *Thespesia populnea*
Portugal laurel *Prunus lusitanica*
postman, running *Kennedia prostrata*
pot marigold *Calendula officinalis*
potato *Solanum tuberosom*
potato-bean *Apios americana*
potato, Irish *Solanum tuberosum*
potato onion *Allium cepa aggregatum*
potato, sweet *Ipomoea batatas*
powder-puff bush *Calliandra haematocephala*
prairie-button *Liatris*
prairie-clover *Petalostemon purpureum*
prairie-gentian *Eustoma*
prairie-lily *Cooperia; Mentzelia*
prickly ash *Zanthoxylum*
prickly cucumber *Echinocystis lobata*
prickly date-palm *Acanthophoenix*
prickly pear *Opuntia*
prickly-poppy *Argemone*
prickly thrift *Acantholimon*
primrose *Primula*
primrose, Cape- *Streptocarpus*
primrose, evening *Oenothera*
primrose willow *Ludwigia*
Prince Albert's yew *Saxegothaea conspicua*
prince's feather *Amaranthus hybridus hypochondriacus; Polygonum orientale*
prince's pine *Chimaphila*
privet *Ligustrum*
prophet-flower *Arnebia; Aipyanthus echioides*
puccoon *Lithospermum*
pulse *Lathyrus*
pummelo *Citrus grandis*
pumpkin *Cucurbita maxima*
purple loosestrife *Lythrum salicaria*
purple sage *Salvia officinalis purpurea*
purple wreath *Petrea*
purslane *Portulaca oleracea*
purslane, tree *Atriplex halimus*
pussy's toes *Antennaria*
puttyroot *Aplectrum hyemale*
pyrethrum, common *Chrysanthemum coccineum*
pyrethrum, Dalmatian *Chrysanthemum cinerariifolium*
pyrola, one-flowered *Moneses uniflora*
pyxie *Pyxidanthera barbulata*

Quaker ladies *Houstonia*
quaking grass *Briza*
Queen Anne's lace *Anthriscus sylvestris* (England); *Daucus carota* (U.S.A.)
queen-of-the-prairie *Filipendula rubra*
queen palm *Arecastrum romanzoffianum*
Queensland wattle *Acacia podalyriifolia*
Queen Victoria waterlily *Victoria amazonica*
quillwort *Isoetes*
quince *Cydonia oblonga*
quince, Japanese *Chaenomeles*

radish *Raphanus sativus*
radish, horse- *Armoracia rusticana*
radish, wild *Raphanus raphanistrum*
raffia *Raphia*
ragged Robin *Lychnis flos-cuculi*
ragweed *Ambrosia*
ragwort *Senecio jacobaea*
ragwort, African *Othonna*
ragwort, Oxford *Senecio squalidus*
rakkyo *Allium chinense*
rain-tree *Samanea saman*
ramie *Boehmeria nivea*
rampion *Campanula rapunculus*
rampion, horned *Phyteuma*
ramsons *Allium ursinum*
Rangoon creeper *Quisqualis indica*
raspberry *Rubus idaeus*
raspberry, strawberry *Rubus illecebrosus*
rat-tail cactus *Aporocactus flagelliformis*
rattan *Calamus*
rattan-vine *Berchemia*

rattle-box *Crotalaria*
rattlesnake-fern *Botrychium virginianum*
rattlesnake-master *Agave virginica; Eryngium yuccifolium; E.aquaticum*
rattlesnake-orchid *Pholidota*
rattlesnake-plantain *Goodyera*
rattlesnake-root *Chamaelirium luteum; Prenanthes*
rattlesnake-weed *Hieracium venosum*
redbud *Cercis canadensis*
red Cape tulip *Haemanthus*
red cedar *Juniperus*
red-hot poker *Kniphofia*
red mulberry *Morus rubra*
red oak *Quercus borealis*
redroot *Ceanothus*
red valerian *Centranthus ruber*
redwood *Sequoia sempervirens*
reed, common *Phragmites australis*
reed-mace *Typha*
rein orchis *Habenaria*
rest-harrow *Ononis*
resurrection plant *Anastatica hierochuntica*
rheumatism root *Jeffersonia diphylla*
rhododendron *Rhododendron*
rhodora *Rhododendron canadense*
rhubarb *Rheum*
ribbon-grass *Phalaris*
ribbon wood *Plagianthus*
rice *Oryza*
rice, Canadian wild *Zizania aquatica*
rice-flower *Pimelea*
rice paper tree *Tetrapanax*
rock-cress *Arabis*
rock-cress, purple *Aubrieta*
rocket, salad *Eruca sativa*
rocket, sweet *Hesperis matronalis*
rocket, yellow *Barbarea vulgaris*
rockfoil *Saxifraga*
rock jasmine *Androsace*
rock-rose *Cistus; Helianthemum*
rogation flower *Polygala*
rose *Rosa*
rose acacia *Robinia hispida*
rose, Austrian copper *Rosa foetida bicolor*
rose, burnet *Rosa spinosissima*
rose, cabbage *Rosa centifolia*
rose, Cherokee *Rosa laevigata*
rose, cinnamon *Rosa cinnamomea*
rose, confederate *Hibiscus mutabilis*
rosebay *Nerium oleander; Rhododendron maximum*
rosebay willowherb *Epilobium angustifolium*
rose, damask *Rosa damascena*
rose, dog *Rosa canina*
rose, guelder *Viburnum*
rose, Macartney *Rosa bracteata*
rose, musk *Rosa moschata*
rose, Scotch *Rosa spinosissima*
rose-campion *Lychnis coronaria*
roselle *Hibiscus sabdariffa*
rose-mallow *Hibiscus moscheutos*
rosemary *Rosmarinus officinalis*
rosemary, marsh- *Limonium*
rosemary, wild *Ledum*
rose-of-China *Hibiscus rosa-sinensis*
rose-of-Sharon *Hibiscus rosa-sinensis; Hypericum calycinum* (Britain)
rose-pink *Sabatia*
rose-root *Sedum rosea*
rosinweed *Silphium*
rouge plant *Rivina humilis*
rowan *Sorbus aucuparia*
royal fern *Osmunda regalis*
royal poinciana *Delonix regia*
royal waterlily *Victoria amazonica*
rubber plant *Ficus elastica*
rubber tree *Hevea*
rubber vine *Cryptostegia*
rue *Ruta graveolens*
rue-anemone *Anemonella thalictroides*
running postman *Kennedia prostrata*
rupturewort *Herniaria*
rush *Juncus*
rush fern *Schizaea*
rush, pool *Scirpus*
Russian comfrey *Symphytum uplandicum*
Russian olive *Elaeagnus*
Russian sage *Perovskia atriplicifolia*
Russian thistle *Salsola*
Russian vine *Fallopia baldschuanica*
Russian wormwood *Artemisia sacrorum*
rutabaga *Brassica napobrassica*
rye, wild *Elymus*
rye-grass *Lolium; Secale*

safflower *Carthamus tinctorius*

saffron *Crocus sativus*
saffron, false *Carthamus tinctorius*
saffron, meadow *Colchicum*
sage *Salvia*
sage, Bethlehem *Pulmonaria*
sage, Jerusalem *Phlomis fruticosa*
sago palm *Cycas revoluta*
saguaro *Carnegiea gigantea*
sainfoin, saintfoin, *Onobrychis viciifolia*
Saint Agnes' flower *Leucojum*
Saint Andrew's cross *Ascyrum*
Saint Barbara's herb *Barbarea*
Saint Barnaby's thistle *Centaurea solstitialis*
Saint Bernard's lily *Anthericum liliago*
Saint Bruno's lily *Paradisea liliastrum*
Saint Dabeoc's heath *Daboëcia cantabrica*
Saint George's herb *Valeriana*
Saint John's bread *Ceratonia siliqua*
Saint John's wort *Hypericum*
Saint Martin's flower *Alstroemeria*
Saint Mary's wood *Calophyllum*
Saint Patrick's cabbage *Saxifraga umbrosa*
Saint Peter's wort *Ascyrum;Hypericum; Primula veris*
salad chervil *Anthriscus*
sallow, common *Salix cinerea*
sallow, great *Salix caprea*
salsify *Tragopogon porrifolius*
salsify, black *Scorzonera hispanica*
salt-bush *Atriplex; Rhagodia* (Australia); *Enchylaena* (Australia)
salt tree *Halimodendron*
saltwort *Salsola*
samphire, marsh *Salicornia*
samphire, rock *Crithmum maritimum*
sandalwood *Santalum album*
sandalwood, red *Adenanthera pavonina*
sand-box tree *Hura crepitans*
sand lily *Leucocrinum montanum*
sand-myrtle *Leiophyllum buxifolium*
sand-verbena *Abronia*
sandwort *Arenaria*
sanicle, bear's-ear *Cortusa matthioli*
sapodilla *Manilkara zapota* (*Achras zapota*)
sapote *Pouteria sapota* (*Calocarpum sapota*)
sarsaparilla *Aralia*
sarsaparilla plant *Smilax*
sassafras *Sassafras albidum*
sassafras, southern *Atherosperma moschatum* (Australia)
satin-flower *Lunaria; Sisyrinchium*
satinwood tree *Murraya; Chloroxylon*
sausage-tree *Kigelia pinnata*
savin *Juniperus sabina*
savory *Satureja*
savory, summer *Satureja hortensis*
savory, winter *Satureja montana*
savoy *Brassica oleracea sabauda*
saw cabbage palm *Acoelorraphe*
saw-palmetto *Serenoa repens*
saw-wort *Serratula*
saxifrage *Saxifraga*
scabious *Scabiosa*
scallion *Allium cepa*
scammony *Convolvulus scammonia*
Scarborough-lily *Vallota speciosa*
scarlet oak *Quercus coccinea*
scarlet runner *Phaseolus coccineus*
sclarea *Salvia*
scolymus *Scolymus hispanicus*
Scotch rose *Rosa spinosissima*
Scotch thistle *Onopordum acanthium*
Scots pine *Pinus sylvestris*
screw pine *Pandanus*
scurfy pea *Psoralea*
scurvy grass *Cochlearia*
scythian lamb *Cibotium barometz*
sea-buckthorn *Hippophaë rhamnoides*
sea campion *Silene maritima*
sea daffodil *Pancratium maritimum*
sea-heath *Frankenia laevis*
sea-holly *Eryngium maritimum*
seakale *Crambe maritima*
sea-lavender *Limonium*
sea-pink *Armeria*
sea poppy *Glaucium*
sea-purslane *Atriplex*
sea-ragwort *Cineraria*
sea rocket *Cakile*
seaside grape *Coccoloba urifera*
sea-urchin cactus *Echinopsis*
sedge *Carex*
sedge, umbrella- *Cyperus*
seersucker plant *Geogenanthus*
self-heal *Prunella vulgaris*
senna, bladder *Colutea*
senna, wild *Cassia marilandica*

sensitive fern *Onoclea sensibilis*
sensitive-plant *Mimosa pudica*
service-berry *Amelanchier*
service-tree *Sorbus domestica*
sesame *Sesamum indicum*
setwall *Valeriana*
shadblow; shadbush *Amelanchier*
shaddock *Citrus grandis*
shagbark hickory *Carya ovata*
shallot *Allium cepa aggregatum*
shamrock *Oxalis acetosella* (England); *Trifolium dubium* (Ireland); *Trifolium repens* (Ireland, U.S.A.); see p. 316
sheep-laurel *Kalmia*
sheep's-bit scabious *Jasione*
shell-flower *Molucella*
she-oak *Casuarina equisetifolia*
shepherd's beard *Andropogon*
shepherd's club *Verbascum*
shepherd's purse *Capsella bursa-pastoris*
shield fern *Dryopteris; Polystichum*
shield-wort *Peltaria*
shingle-plant *Monstera delicosa*
shinleaf *Pyrola*
shoe-black *Hibiscus rosa-sinensis*
shooting star *Dodecatheon*
shower, golden *Cassia fistula*
shrimp plant *Drejerella guttata (Justicia brandegeana)*
side-saddle flower *Sarracenia*
silk-cotton tree *Bombax ceiba (Salmalia malabarica)*
silk-cotton tree, yellow *Cochlospermum religiosum*
silk-grass *Yucca filamentosa*
silk-tassel bush *Garrya elliptica*
silk-tree *Albizia*
silk-vine *Periploca graeca*
silver bell tree *Halesia*
silver-berry *Elaeagnus commutata*
silver birch *Betula pendula* etc.
silver fern *Pityrogramma*
silver morning-glory *Argyreia speciosa*
silver-rod *Solidago; Asphodeline*
silver tree *Leucadendron*
silverweed *Argyreia speciosa* (tropics); *Potentilla anserina*
simpler's joy *Verbena*
sisal *Agave sisalana*
skirret, skirwort *Sium sisarum*
skullcap *Scutellaria*
skunk-cabbage *Symplocarpus foetidus*
slipper-flower *Pedilanthus*
slipper-wort *Calceolaria*
slippery elm *Ulmus fulva*
sloe-tree *Prunus spinosa*
smallage *Apium graveolens*
smartweed *Fallopia hydropiper*
smilax, florist's *Asparagus asparagoides*
smoke bush *Comesperma*
smoke-tree *Cotinus coggygria*
snail-flower *Phaseolus caracalla*
snail-seed *Cocculus carolinus*
snakegourd *Trichosanthes anguina*
snake plant *Sansevieria*
snakeroot *Aristolochia; Cimicifuga; Eupatorium; Liatris*
snake's head fritillary *Fritillaria meleagris*
snake's-head iris *Hermodactylus tuberosus*
snake's mouth orchid *Pogonia*
snake's-tongue *Ophioglossum*
snapdragon *Antirrhinum*
snapweed *Impatiens*
sneezeweed *Achillea ptarmica*
sneezewort *Achillea ptarmica*
snowball tree *Viburnum macrocephalum sterile: Viburnum plicatum plicatum*
snowbell *Styrax*
snowberry *Chiococca; Symphoricarpos*
snowberry, creeping *Chiogenes hispidula*
snowdrop *Galanthus*
snowdrop-tree *Halesia*
snowflake *Leucojum*
snow-glory *Chionodoxa*
snow-in-summer *Cerastium tomentosum*
snow-on-the mountain *Euphorbia marginata*
snow poppy *Eomecon chionantha*
snow wreath *Neviusia alabamensis*
snowy mespilus *Amelanchier*
soap-bark tree *Quillaja saponaria*
soap-berry *Sapindus saponaaria; Shepherdia canadensis*
soap-bush *Noltea africana*
soapwort *Saponaria officinalis*
Solomon's seal *Polygonatum*
Solomon's seal, false *Smilacina*
sorghum *Sorghum*

sorrel, garden *Rumex rugosus*
sorrel, Jamaica *Hibiscus sabdariffa*
sorrel, sheep's *Rumex acetosella*
sorrel, wood *Oxalis acetosella*
sorrel-tree *Oxydendrum arboreum*
sour gum *Nyssa*
soursop *Annona muricata*
sour-wood *Oxydendrum arboreum*
southern beech *Nothofagus*
southern cane *Arundinaria*
southernwood *Artemisia abrotanum*
sowbread *Cyclamen*
sowthistle *Sonchus*
soybean *Glycine max*
Spanish bayonet *Yucca*
Spanish bluebell *Hyacinthoide hispanica*
Spanish broom *Spartium junceum*
Spanish elm *Cordia*
Spanish iris *Iris xiphioides*
Spanish moss *Tillandsia usneoides*
Spanish oyster-plant *Scolymus hispanicus*
spatter-dock *Nuphar*
spearflower *Ardisia*
spear-grass *Stipa*
spear lily *Doryanthes*
spearmint *Mentha spicata*
spearwort, great *Ranunculus lingua*
spearwort, lesser *Ranunculus flammula*
speedwell *Veronica*
speedwell, Buxbaum's *Veronica persica*
speedwell, Derwent *Veronica derwentiana*
speedwell, germander *Veroncia chamaedrys*
spice-bush *Lindera benzoin*
spider-flower *Cleome*
spider lily *Hymenocallis*
spider lily, golden *Lycoris aurea*
spider plant *Chlorophytum comosum*
spiderwort *Tradescantia*
spike heath *Bruckenthalia spiculifolia*
spike rush *Eleocharis*
spikemoss *Selaginella*
spikenard, American *Aralia racemosa*
spikenard, false *Smilacina*
spinach *Spinacia oleracea*
spinach beet *Beta vulgaris cicla*
spinach, New Zealand *Tetragonia tetragonioides*
spindle-tree *Euonymus*
spine-areca *Acanthophoenix*
spiral-flag *Costus*
spire lily *Galtonia candicans*
spirea *Spiraea; Filipendula*
spirea, blue *Caryopteris*
spirea, false *Sorbaria*
spleenwort *Asplenium*
spleenwort, swamp *Athyrium*
spoonwood *Kalmia latifolia*
spring-beauty *Claytonia*
spruce *Picea*
spruce, black *Picea mariana*
spruce, Norway *Picea abies*
spruce, red *Picea rubens*
spruce, Sitka *Picea sitchensis*
spruce, white *Picea glauca*
spurge *Euphorbia*
spurge, caper *Euphorbia lathyris*
spurge, flowering *Euphorbia corollata*
spurge, wood *Euphorbia amygdaloides*
spurge-laurel *Daphne laureola*
spurge-olive *Cneorum*
squash *Cucurbita*
squaw-berry *Mitchella repens*
squaw-weed *Senecio*
squill *Scilla; Urginea*
squill, striped *Pushkinia scilloides*
squirrel-corn *Dicentra canadensis*
squirrel-tail grass *Hordeum*
squirting cucumber *Ecballium elaterium*
stagbush *Viburnum prunifolium*
stagger-bush *Lyonia mariana*
staghorn fern *Platycerium*
staghorn sumach *Rhus typhina* (*R. hirta*)
star, golden *Chrysogonum virginianum*
star-apple *Chrysophyllum cainito*
star-flower *Trientalis borealis*
star-flower, spring (*Tristagma* (or *Ipheion*) *uniflorum*
star-grass *Aletris; Hypoxis*
star-of-Bethlehem *Ornithogalum*
star thistle *Centaurea calcitrapa*
star wattle *Acacia verticillata*
starwort *Aster; Stellaria*
statice *Limonium*
steeple-bush *Spiraea*
stinging nettle *Urtica*
stinkweed *Pluchea*
stitchwort *Stellaria*

stock *Matthiola*
stock, Brompton *Matthiola incana maritima*
stock, night-scented *Matthiola longipetala bicornis*
stock, Virginia *Malcomia maritima*
Stokes' aster *Stokesia laevis*
stone cress *Aethionema*
stonecrop *Sedum*
stonecrop, biting *Sedum acre*
stone pine *Pinus cembra; Pinus pinea*
stone-root *Collinsonia*
storax-tree *Styrax*
stork's-bill *Pelargonium*
straw flower *Helichrysum*
strawberry *Fragaria*
strawberry, barren *Waldsteinia*
strawberry, Indian or mock *Duchesnea indica*
strawberry tree *Arbutus*
striped squill *Puschkinia scilloides*
stud flower *Helonias bullata*
succory *Cichorium intybus*
sugar-apple *Annona squamosa*
sugarberry *Celtis laevigata*
sugarcane *Saccharum officinarum*
sugar maple *Acer saccharum*
sumac *Rhus*
sumac, poison *Toxicodendron*
sumac, staghorn *Rhus typhina*
summer-cypress *Bassia scoparia culta*
summer savory *Satureja hortensis*
sundew *Drosera*
sundrops *Oenothera*
sunflower *Helianthus*
sunray *Helipterum*
sunrose *Helianthemum*
sunshine wattle *Acacia botrycephala*
supple-Jack *Paullinia*
Swainson pea *Swainsona*
swamp bay *Magnolia virginiana*
swamp cypress *Taxodium*
swamp loosestrife *Decodon verticillatus*
swamp pink *Arethusa bulbosa; Calopogon; Helonias bullata*
swamp wattle *Acacia elongata*
swampweed *Selliera radicans*
Swan River daisy *Brachycome iberidifolia*
swede turnip *Brassica napobrassica*
sweet Alice *Pimpinella anisum*
sweet alyssum *Lobularia maritima*
sweet basil *Ocimum basilicum*
sweet bay *Laurus nobilis; Magnolia virginiana* (U.S.A.)
sweet-box *Sarcococca*
sweet briar *Rosa eglanteria*
sweet broom *Scoparia dulcis*
sweet cicely *Myrrhis odorata* (Britain); *Osmorhiza* (U.S.A.)
sweet clover *Melilotus*
sweet corn *Zea mays*
sweet fern *Comptonia peregrina*
sweet flag *Acorus calamus*
sweet gale *Myrica gale*
sweet gum *Liquidambar styraciflua*
sweet marjoram *Majorana hortensis*
sweet olive *Osmanthus*
sweet pea *Lathyrus odoratus*
sweet pepper-bush *Clethra alnifolia*
sweet potato *Ipomoea batatas*
sweet rocket *Hesperis matronalis*
sweetsop *Annona squamosa*
sweet sultan *Centaurea moschata*
sweet vernal grass *Anthoxanthum odoratum*
sweet William *Dianthus barbatus*
sweet wormwood *Artemisia annua*
sword fern *Nephrolepis*
sword-lily *Gladiolus*
sycamore *Acer; Platanus*
sycamore-maple *Acer pseudoplatanus*
Sydney golden wattle *Acacia longifolia*
tacamahac *Populus balsamifera*
tamarack *Larix laricina*
tamarind *Tamarindus indica*
tamarisk *Tamarix*
tamarisk, false *Myricaria*
tanbank *Lithocarpus*
tangerine *Citrus reticulata*
tansy *Tanacetum vulgare*
tapegrass *Vallisneria*
tapioca *Manihot*
tarata *Pittosporum eugenioides*
tare *Vicia*
taro-root *Colocasia esculenta*
tarragon *Artemisia dracunculus*
tarweed *Madia*
tassel flower *Emilia sagittata*

tawny-berry *Ilex krugiana*
tea *Camellia sinensis*
tea-berry *Gaultheria procumbens*
tea-tree *Leptospermum* (Australia); *Lycium barbarum* (Britain)
teak tree *Tectona grandis*
tear-thumb *Polygonum arifolium; Polygonum sagittatum*
teasel *Dipsacus fullonum*
teonsite *Euchlaena mexicana*
thatch palm *Thrinax*
thimbleberry *Rubus parviflorus*
thistle *Carduus; Cirsium; Cnicus*
thistle, blessed *Cnicus benedictus; Silybum marianum*
thistle, cotton *Onopordum acanthium*
thistle, creeping *Cirsium arvense*
thistle, distaff *Carthamus*
thistle, globe *Echinops*
thistle,golden *Scolymus*
thistle, hedgehog *Echinocactus*
thistle, plumed *Cirsium*
thistle, plumeless *Carduus*
thistle, Russian *Salsola*
thistle, saffron *Carthamus*
thistle, Scotch *Onopordum acanthium*
thistle, South African *Berkheya*
thistle, sow *Sonchus*
thistle, star *Centaurea calcitrapa*
thorn, Jerusalem *Paliurus spinachristi; Parkinsonia aculeata*
thorn, kangaroo *Acacia armata*
thorn, lily- *Catesbaea spinosa*
thorn-apple *Datura*
thoroughwort *Eupatorium*
thunder-plant *Sempervivum*
thrift *Armeria*
throatwort *Trachelium caeruleum*
thyme *Thymus*
thyme, common *Thymus vulgaris*
thyme, mother-of- *Thymus drucei; Thymus serpyllum*
tickseed *Bidens; Coreopsis*
tick-trefoil *Desmodium*
tidy-tips *Layia elegans*
tiger-flower *Tigridia*
tiger-lily *Lilium lancifolium* (*L. tigrinum*)
tiger's chaps *Faucaria*
timothy *Phleum pratense*
toadflax *Linaria*
toadflax, ivy-leaved *Cymbalaria muralis*
toad-lily *Tricyrtis*
tobacco *Nicotiana*
toddy palm *Caryota urens*
tomatillo *Physalis philadelphica*
tomato *Lycopersicon esculentum*
tomato, cherry *Lycopersicon esculentum cerasiforme*
tomato, currant *Lycopersicon pimpinellifolium*
tomato, tree- *Cyphomandra betacea*
tooth-ache-tree *Zanthoxylum americanum*
toothwort *Dentaria*
torch-lily *Kniphofia*
touch-me-not *Impatiens noli-tangere*
toyon *Heteromeles arbutifolia*
trailing arbutus *Epigaea repens*
trailing cactus *Acanthocereus*
Transvaal daisy *Gerbera*
traveller's joy *Clematis*
traveller's tree *Ravenala madagascariensis*
tree-aster *Olearia*
tree-fern *Cyathea* etc.
tree-mallow *Lavatera arborea*
tree-of-heaven *Ailanthus altissima*
tree onion *Allium cepa proliferum*
tree-poppy *Dendromecon rigidum*
tree-tomato *Cyphomandra betacea*
trefoil *Trifolium*
trefoil, bird's foot *Lotus corniculatus*
trout-lily *Erythronium*
true-love *Paris quadrifolia*
trumpet creeper *Bignonia; Campsis*
trumpet flower *Beaumontia; Clytostoma; Datura*
trumpet, herald's *Beaumontia*
trumpet tree *Cecropia peltata*
tuberose *Polianthes tuberosa*
tulip *Tulipa*
tulip, lady *Tulipa clusiana*
tulip tree *Liriodendron tulipifera*
tupelo *Nyssa*
turkey's beard *Xerophyllum*
turmeric *Curcuma*
turnip *Brassica rapa*
turnip, Indian *Arisaema*
turnip, swede *Brassica napobrassica*
turtlehead *Chelone*

turtle-vine *Bauhinia cumanensis*
tutsan *Hypericum androsaemum*
twayblade *Liparis; Listera*
twinflower *Linnaea borealis*
twin-leaf *Jeffersonia diphylla* (U.S.A.); *Zygophyllum* (Australia)
twist-arum *Helicodiceros muscivorus*
twisted stalk *Streptopus*

udo *Aralia cordata*
umbrella fern *Gleichenia*
umbrella palm *Hedyscepe canterburyana*
umbrella pine *Sciadopitys verticillata*
umbrella plant *Cyperus alternifolius; Darmera peltata*
umbrella tree *Magnolia tripetala; Melia azedarach; Thespesia populnea*
unicorn plant *Proboscidea louisianica*
upas tree *Antiaris toxicaria*

valerian *Valeriana*
valerian, African *Fedia*
valerian, red *Centranthus ruber*
vanilla *Vanilla planifolia*
vanilla, Carolina *Trilisa*
vanilla-grass *Hierochloe odorata*
vanilla-leaf *Achlys*
varnish tree *Aleurites moluccana*
vegetable-oyster *Tragopogon porrifolius*
vegetable-sponge *Luffa aegyptiaca*
velvet-bean *Stizolobium*
Venus flytrap *Dionaea muscipula*
Venus' hair *Adiantum capillus-veneris*
Venus' looking-glass *Specularia*
Venus' navelwort *Omphalodes*
verbena, lemon *Aloysia triphylla*
verbena, sand *Abronia*
vervain *Verbena officinalis*
vetch *Vicia*
vetch, crown *Coronilla varia*
vetch, kidney *Anthyllis*
vetch, milk *Astragalus*
vine *Vitis*
vine, kangaroo *Cissus antarctica*
vine, kudzu *Pueraria thunbergiana*
vine, mignonette *Boussingaultia gracilis*
vine, rattan- *Berchemia*
vine, Russian *Fallopia baldschuanica*
vine, wonga *Pandorea pandorana*
violet *Viola*
violet, African *Saintpaulia*
violet cress *Ionopsidium acaule*
viper's bugloss *Echium*
Virginia cowslip *Mertensia virginica*
Virginia creeper *Ampelopsis; Parthenocissus*
Virginia-stock *Malcomia*
Virginian poke *Phytolacca americana*
virgin's bower *Clematis*
viscaria *Lychnis*

wahoo *Euonymus atropurpurea*
wake-robin *Trillium*
wallcress *Arabis*
wall fern *Polypodium*
wallflower *Cheiranthus cheiri*
wall-pennywort *Umbilicus rupestris*
wall-pepper *Sedum acre*
wall-rue *Asplenium ruta-muraria*
walnut *Juglans*
walnut, black *Juglans nigra*
walnut, English *Juglans regia*
walnut, white *Juglans cinerea*
wandering-Jew *Tradescantia; Zebrina*
wandflower *Dierama; Sparaxis*
wand plant *Galax urceolata*
warratah *Telopea speciosissima*
Washington plant *Cabomba caroliniana*
water-arum *Calla palustris*
water carpet *Chrysosplenium*
water-chestnut *Trapa natans*
water convolvulus *Ipomoea aquatica*
water-cress *Nasturtium; Rorippa*
water-elm *Planera aquatica*
water-flag *Iris pseudacorus*
water hemlock *Cicuta virosa*
water hoarhound, horehound *Lycopus*
water-hyacinth *Eichhornia crassipes*
water-leaf *Hydrophyllum*
water-lettuce *Pistia stratiotes*
water-lily *Nuphar; Nymphaea*
water-lily, yellow *Nuphar*
watermelon *Citrullus lanatus*
water-milfoil *Myriophyllum*
water-moss *Fontinalis*
water oak *Quercus nigra*
water-pennywort *Hydrocotyle*
water-pimpernel *Samolus*
water plantain *Alisma*

water poppy *Hydrocleys nymphoides*
water-shield *Brasenia schreberi*
water-soldier *Stratiotes aloides*
water starwort *Callitriche*
water thyme *Anacharis*
water trumpet *Cryptocoryne*
water violet *Hottonia*
waterweed *Anacharis canadensis*
waterwillow *Justicia americana*
wattle *Acacia*
wattle, black *Acacia decurrens mollis*
wattle, blue *Acacia cyanophylla*
wattle, cedar *Acacia elata*
wattle, Cootamunda *Acacia baileyana*
wattle, golden *Acacia pycnantha*
wattle, green *Acacia decurrens*
wattle, hedge *Acacia armata*
wattle, mudgee *Acacia spectabilis*
wattle, Queensland *Acacia podalyriifolia*
wattle, star *Acacia verticillata*
wattle, sunshine *Acacia botrycephala*
wattle, swamp *Acacia elongata*
wattle, Sydney golden *Acacia longifolia*
wattle, winged *Acacia alata*
wax-flower *Hoya carnosa; Stephanotis floribunda*
wax gourd *Benincasa hispida*
wax-myrtle *Myrica cerifera*
wax-palm *Ceroxylon*
wax-tree *Rhus succedanea*
waxweed *Cuphea*
waxwork *Celastrus scandens*
wayfaring tree *Viburnum*
weavers' broom *Spartium junceum*
wedding bush *Ricinocarpos*
wedge-pea *Gompholobium*
weed, pineapple *Matricaria discoidea*
weeping willow *Salix babylonica*
weld *Reseda luteola*
wellingtonia *Sequoiadendron giganteum*
Welsh onion *Allium fistulosum*
Welsh poppy *Meconopsis cambrica*
West Indian kale *Colocasia*
wheat *Triticum*
whin *Genista; Ulex*
white-alder *Clethra*
white-beam *Sorbus aria*
white byrony *Bryonia dioica*
white campion *Silene alba*
white cedar, northern *Thuja occidentalis*
white dead-nettle *Lamium album*
white mulberry *Morus alba*
white oak *Quercus alba*
white poplar *Populus alba*
white willow *Salix alba*
whitlow-grass *Draba*
whitlow-wort *Paronychia*
whorl-flower *Morina*
willow *Salix*
willow, basket *Salix viminalis*
willow, bay *Salix pentandra*
willow, black *Salix nigra*
willow, crack *Salix fragilis*
willow, creeping *Salix repens*
willow, cricket-bat *Salix alba caerulea*
willow, desert *Chilopsis linearis*
willow, goat *Salix caprea*
willow moss *Fontinalis*
willow, weeping *Salix babylonica*
willow, white *Salix alba*
willow-herb *Epilobium*
widow tears *Tradescantia*
wild cinnamon *Canella*
wild rice *Zizania aquatica*
wind-flower *Anemone*
wing-nut *Pterocarya*
winged everlasting *Ammobium alatum*
winged wattle *Acacia alata*
winter aconite *Eranthis*
winterberry *Ilex verticillata*
winter daffodil *Sternbergia lutea*
wintergreen *Gaultheria procumbens; Pyrola*
winter heliotrope *Petasites fragrans*
wintersweet *Chimonanthus praecox; Acokanthera spectabilis* (tropics)
wire-plant *Muehlenbeckia complexa*
witch-grass *Agropyron repens*
witch-hazel *Hamamelis*
witch-hobble *Viburnum alnifolium*
withe-rod *Viburnum cassinoides*
woad, dyer's *Isatis tinctoria*
woad, waxen *Genista*
woad, wild *Reseda luteola*
wolfberry *Symphoricarpus*
wolfchop *Faucaria*
wolf's bane *Aconitum lycoctonum*
wolf's-milk *Euphorbia*

woman's-tongue-tree *Albizia lebbek*
wonga vine *Pandorea pandorana*
wood-apple *Feronia limonia*
wood-betony *Pedicularis*
woodbine *Lonicera; Ampelopsis; Parthenocissus*
wood fern *Dryopteris*
wood-rose, wooden rose *Merremia tuberosa*
woodruff *Asperula odorata* (*Galium odoratum*)
wood-sorrel *Oxalis acetosella*
wood spurge *Euphorbia amygdaloides*
woody nightshade *Solanum dulcamara*
wormwood *Artemisia*
wormwood, common *Artemisia absinthium*
wormwood, Russian *Artemisia sacrorum*
wormwood, sweet *Artemisia annua*

yam *Dioscorea*
yam bean *Pachyrhizus erosus*
yarrow *Achillea millefolium*
yellow birch *Betula lutea*
yellow ox-eye *Buphthalmum*
yellow-wood, Kentucky *Cladastris lutea*
yellowwort *Blackstonia perfoliata*
yerna buena *Micromeria*
yew *Taxus*
yew, plum *Cephalotaxus*
Yorkshire fog *Holcus lanatus*
Youth-and-old-age *Zinnia elegans*
youth-on-age *Tolmiea menziesii*
yulan *Magnolia denudata*

zephyr-lily *Zephyranthes*